6th International Disaster Mitigation Specialty Conference 2018

Held as Part of the Canadian Society for Civil Engineering Annual Conference 2018

Fredericton, Canada
13-16 June 2018

Editors:

Alan Lloyd
Ioan Nistor

ISBN: 978-1-5108-8973-6

Printed from e-media with permission by:

Curran Associates, Inc.
57 Morehouse Lane
Red Hook, NY 12571

CURRAN ASSOCIATES INC. proceedings .com

Some format issues inherent in the e-media version may also appear in this print version.

Printed by Curran Associates, Inc. (2019)

For permission requests, please contact Canadian Society for Civil Engineering at the address below.

Canadian Society for Civil Engineering
521-300 rue St-Sacrement
Montréal, QC
H2Y 1x4

Phone: (514) 933-2634
Fax: (514) 933-3504

info@csce.ca

Additional copies of this publication are available from:

Curran Associates, Inc.
57 Morehouse Lane
Red Hook, NY 12571 USA
Phone: 845-758-0400
Fax: 845-758-2633
Email: curran@proceedings.com
Web: www.proceedings.com

TABLE OF CONTENTS

Building Tomorrow's Society
Bâtir la Société de Demain

Fredericton, Canada
June 13 to 16, 2018 / *13 au 16 juin, 2018*

AN EXPERIMENTAL AND NUMERICAL INVESTIGATION OF THE CYCLIC PERFORMANCE OF A SMART SELF-CENTERING BRACING SYSTEM

Anas, Salem Issa[1], A.B.M. Rafiqul Haque[2]; and M. Shahria Alam[3]
[1] The University of British Columbia (Okanagan Campus), Canada
[2] Parsons Corporation, Vancouver, BC, Canada
[3] anasalem@mail.ubc.ca

Abstract: Buildings with traditional structural systems experience large residual deformation after an earthquake, and often lose their serviceability and need to be demolished incurring huge economic losses. In order to resolve this issue, various smart structural systems have been developed by many researchers. One such system is the novel Piston Based Self-Centering bracing (PBSC) system. This study investigates the cyclic performance of this bracing system experimentally to predict its load-deformation response during seismic events. This newly developed bracing system employs Nickel Titanium (NiTinol) based Shape Memory Alloy (SMA) bars inside a sleeve-piston assembly for its self-centering mechanism. During cyclic tension-compression loading, the bars are pulled from opposite directions in order to avoid compressive loading on them. The energy dissipation is achieved through nonlinear load deformation hysteresis. Furthermore, the PBSC bracing system is designed to be fully buckling restrained. The system exhibits flag shaped force deformation hysteresis. The experimental results are compared with previously generated numerical results, where very good agreement is achieved.

1 INTRODUCTION

A recent study conducted by the Insurance Bureau of Canada estimated the overall loss after a 9.0-magnitude earthquake in British Columbia at almost $75 billion and a $61 billion loss after a 7.1-magnitude earthquake in the Quebec City-Montreal-Ottawa corridor (IBC, 2013), which clearly reflects the vulnerability of Canadian civil infrastructure. To avoid such scenarios in Canada, it is imperative to take immediate measures. Since seismic load, in the form of ground shaking, generates one of the most devastating forces that our infrastructure can experience, designing structures against these large forces are often uneconomic. In various building and infrastructures, different structural elements and systems resist and dissipate earthquake-induced energy by means of deformations. Once permanent deformations take place, a structure becomes difficult to fix. After a major earthquake, these structures may have to be demolished and re-built acquiring huge economic losses. For example, in the Maule (Chile) Earthquake in 2010, the economic losses were estimated to be $30 billion (loss of infrastructure alone was $20.9 billion) which is equivalent to 17% of the GDP of Chile (Elnashai et al. 2010). In the Christchurch (New Zealand) Earthquake in 2011, about $20 billion economic losses (equivalent to 13% of New Zealand's GDP) was estimated. The destruction was enormous, including demolition of around 70% of downtown buildings, loss of more than 50% of heritage structures, closure of the major business district for over 18 months, and outmigration of thousands of residents (Elwood 2013). Such seismically damaged infrastructure become a major economic obligation.

Engineers have so far come up with many different concepts to evade or resist seismic forces. The most notables include base isolation system, ductile structural connection, and bracing systems. The base isolation system allows a structure to get partially isolated from the foundation by using elements of very low stiffness in between the structure and the foundation. They can reduce the damage experienced from the earthquake by a big margin. Unfortunately, this option is very expensive and difficult to repair once damaged. So, the cost is a major concern for this system. The ductile beam-column connection is the most widely practiced method for designing against earthquake. This method is performing well in reducing fatality all over the world. The only problem with this system is that the design allows nonlinear/plastic deformation in the beam plastic hinge region of the beam-column joints/structural connections. To reduce such structural damages, one option is to resist the earthquake forces directly (A. Haque, 2017).

Unfortunately, it is very expensive to build a structure to resist earthquake deformation in the elastic range of response. To solve this problem, self-centering devices can be used in a structure to resist the seismic load by axial action. This reduces deformation demand for the structural components and reduces their damage by a big margin. Such a self-centering device can be used in the form of bracing, and restraining device (e.g. in bridges against unseating, in buildings at beam-column joint) against earthquake movements. While considering a bracing system in a building there are various kinds available where Concentrically Braced Frames (CBFs) are considered as one of the most widely used bracing systems (Roeder et al., 2012). Unfortunately, the traditional tension compression bracing system cannot perform well under earthquake loads. Seismic load induces cyclic tension-compression load in the braces, which can cause them to buckle if buckling load exceeds at any point of time during the seismic event. After the braces buckle, the deformation of the frame increases significantly and causes the beam-column joints to go into nonlinear range. It sometimes causes irreversible damage to the structure. Furthermore, a CBF structure is much stiffer compared to moment resisting frame counterpart. This extra stiffness attracts much more seismic load as well as floor acceleration. Excessive acceleration can cause damage to non-structural components and can also cause weakly connected non-structural components to fall on the occupants causing serious injury or fatality (A. Haque, 2017).

In order to solve these issues, many bracing systems have been developed by researchers in the past few decades such as Buckling Restrained Bracing (BRBs) (Takeuchi et al., 2004), CastConnex Scorpion Yielding Brace (Christopoulos et al., 2008), Memory Alloys for New Seismic Isolation Devices (MANSIDE) project braces (Dolce et al., 2000), and Self Centering Energy Dissipation Device (SCED) (Tremblay & Christopoulos, 2012). The BRB and the Scorpion Yielding Brace resist seismic force by going into nonlinear range. They exhibit fat hysteresis loops which contribute to a higher amount of damping and thus can reduce the velocity and the acceleration of a structural system. Unfortunately, they do not have the self-centering property. Therefore, if there is permanent deformation, it is very difficult to push the structure back to its original position.

The other two options, the SCED and the MANSIDE braces offer re-centering capability; however, their construction is complicated and were not adopted by the construction industry. Similar to the latter two options, several researchers recently conducted experimental as well as numerical studies testing and validating newly developed self-centering bracing systems including but not limited to (Choi et al., 2017; C.-X. Qiu & Zhu, 2017; C. Qiu & Zhu, 2016; Speicher et al., 2017; Wu & Phillips, 2017; Xie et al., 2016; L.-H. Xu et al., 2016; L. H. Xu et al., 2016). These include the use of SMA wires and rods, friction dampers, Fiber Reinforced Polymer (FRP) rods and other techniques. All the adopted methods were targeting enhanced seismic performance of buildings in terms of maximizing energy dissipation and minimizing residual drifts. All of these systems showed good aspects in terms of self-centering, efficient energy dissipation and flag shape hysteresis response. However, issues related to complexity and availability of some material are still the main drawbacks. To solve the above-mentioned problems, this research is focused on developing a novel Piston Based Self-Centring (PBSC) bracing system. This study investigates the cyclic performance of this bracing system experimentally to predict its load-deformation response during seismic events. This newly developed bracing system employs Nickel Titanium (Nitinol) based shape memory alloy (SMA) bars inside a sleeve-piston assembly for its self-centering mechanism. The energy dissipation is achieved through nonlinear load deformation hysteresis. Furthermore, the PBSC bracing system is designed to be fully buckling restrained. The system exhibits flag shaped force deformation hysteresis. The PBSC braced frames are designed and their seismic performance is evaluated in terms of inter-story drift responses. The developed system surpasses other systems in terms of constructability and self-centring ability.

2 DESCRIPTION AND DETAILS OF THE SYSTEM

The idea for the PBSC device is anticipated to work mainly in a Chevron/V configuration bracing in buildings as shown in Figure 1. This system can be employed for both new/existing steel or concrete structures. Other applications include a single configuration as bridge restrainer, or parallel to the beam and attached to the beam and bracing which works like a shear panel device.

The proposed PBSC system is employed using a device commonly seen in mechanical systems, which is a cylinder-piston assembly. As shown in Figure 1, the tensile and compressive strength of a brace should be almost equal. The brace's design strength should be kept below the buckling and yield capacity of the shaft. In order to limit compression load in an individual brace, the shaft and tie arrangement shall be constructed as a piston system where the ties and part of the shaft shall be held inside a larger metal sleeve (Circular, square, rectangular or any other geometric cross section). For fabrication purposes, design drawings including the details of the PBSC arrangement is given in Figure 2. In this system, separate ties are used to connect front plate to piston plate and back plate to piston plate. Both ends of the sleeve shall have thick metal caps to provide support for the front ties (when the whole system is under tension) and the back ties (while the system is under compressive load) (A. Haque, 2017).

Two sets of ties (front and back) are connected at the shaft end plate and the other ends of the ties are connected to the cylinder/sleeve end plates using movable joints. The movable joints are constructed in such a way that they only allow the ties to go out. Inward movement of the ties will be allowed up to the ties end points. Locks/couplers placed at the tie ends are introduced to prevent the ties from fully entering the cylinder/sleeve. This way brace tensile and compressive loads will go through the front and back ties alternately during a cyclic loading. The joint between the ties and the plates also allows rotational movement without any moment generation (hinge joint). This will ensure bar straightness in the event of any kind of plate bending. The piston plate is made slightly smaller in size than the sleeve's inner dimension so that plate bending/rotation during loading does not affect the sleeve. The ties are designed for a load lower than the buckling and yield strength of the shaft and the cylinder/sleeve. When the system is under compression and the load reaches the yield load of the ties, the ties will yield and deform significantly thus lowering the axial stiffness of the system. This will in turn limit the axial force of the system and keep it below the buckling capacity of the shaft. In this study, the ties are made of super elastic shape memory alloy (SMA) (i.e. NiTinol) bars. The Super elastic SMA bars are known for their unique property of regaining original/undeformed shape upon load removal. The use of super elastic SMA bars will ensure the full self centering capability of the brace. The dimension of the different components of the systems are shown in Figure 2.

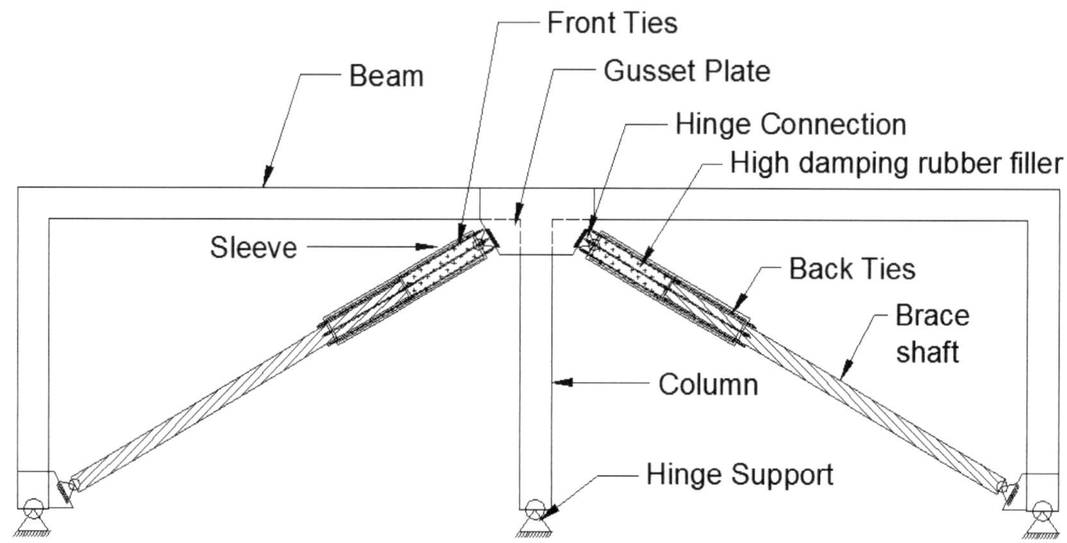

Figure 1: Basic Components of a Piston Based Self-Centring brace

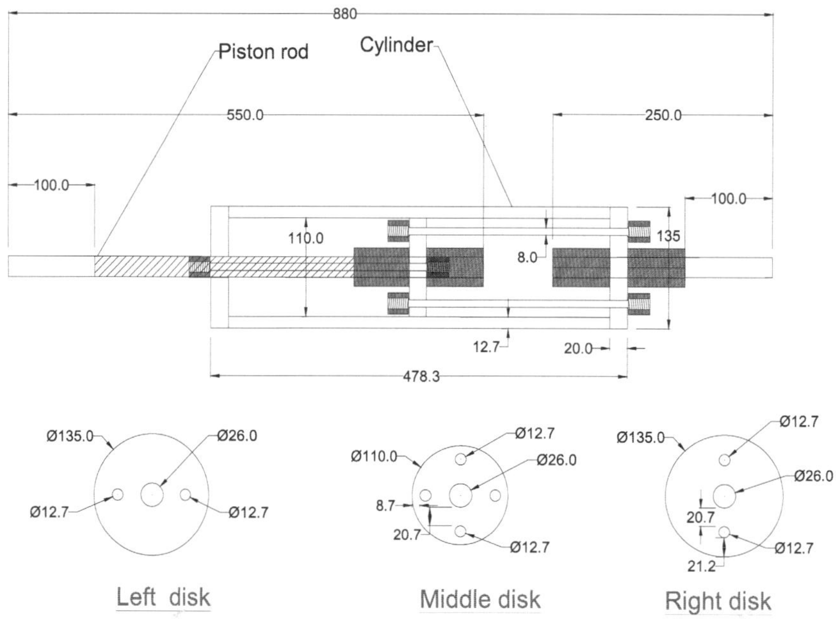

Notes:
1- All dimensions are in mm

Figure 2: PBSC brace components

2.1 SMA Bars Machining

Two SMA rods with the diameter of 12.7mm and length of 600mm were used in the PBSC brace specimen. Each rod is first cut in half to have 4 SMA rods of 300mm long each. Each of the 300mm rod was then machined down to 8mm diameter except for the two ends of the rods. Each end of each rod with length of 25mm was threaded with fine thread. The machined ends are done to install the lock nuts where they will be the support means when loading the SMA rods in tension. The steps involved in the abovementioned machining process along with one of the machined bars are presented in Figure 3. It is worth mentioning that high temperature can change the microstructure of the SMA and negatively affect its superelastic response. Therefore, a cooling oil was applied to the SMA rods throughout the entire machining process in order to keep the temperature low enough so that it retains its self-centering ability.

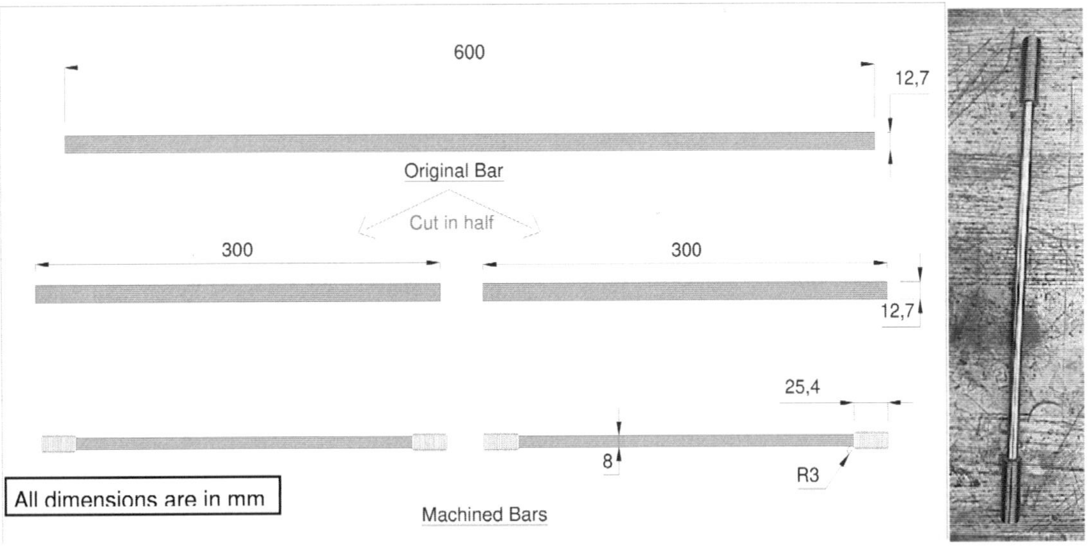

Figure 3: SMA bar machining process

2.2 Specimen Details

The fabrication process of the PBSC brace specimen is carried out by firstly generating 3D models including the different components. Figure 4 shows the internal view/arrangement of the elements. Alternatively, the structural details of the cylinder, piston and internal components once the brace is fully constructed are shown in Figure 5. As shown in the latter figure, the lock nuts are installed and tightened enough to prevent the SMA rods from sliding or moving in any direction. However, this tightening was applied manually so that no significant tension force is applied to the SMA. Noteworthy, the outer body cylinder of the specimen, as well as, the supporting discs are fabricated using 12.7mm thick steel cylinder and plates, respectively. It is critical to keep the supporting components of the system strong enough and in the elastic range as the SMA rods reach their maximum load carrying capacity. Weak components, especially the outer encasing cylinder, can experience some permanent deformations which defies the entire concept of self-centring device.

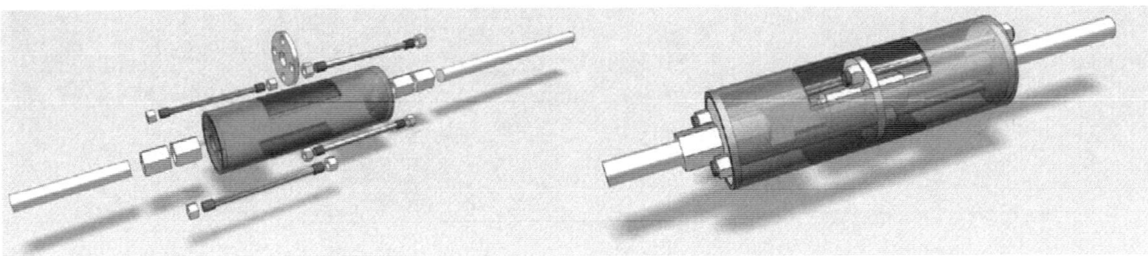

Figure 4: Internal view of the constructed brace

Figure 5: PBSC brace specimen

3 EXPERIMENTAL INVESTIGATION

3.1 Test Setup

The objective of the experimental investigation was to investigate the cyclic behavior of bracing members in concentrically braced frames by means of cyclic axial tests. Quasi-static tests were carried out at the High Head Laboratory for Structural Engineering of the University of British Columbia Okanagan (UBCO). The specimen was tested using the MTS universal testing machine with a capacity of 500 kN. Taking into consideration the machine dimensional limits and characteristics, together with the ease of specimen handling, the experimental set-up described in Figure 6 was adopted.

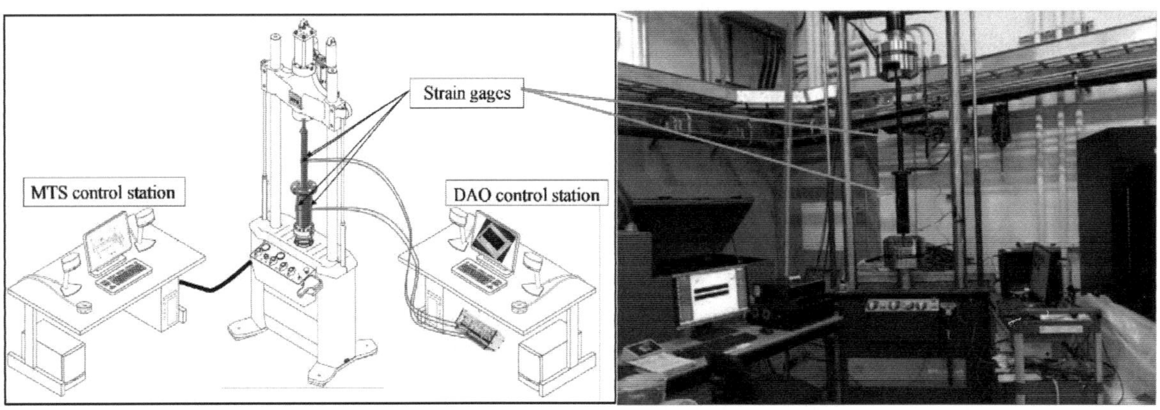

Figure 6: Test setup

MTS control system and the data acquisition system were both connected to the specimen to measure the different test parameters. The MTS machine is equipped with a load cell to measure the axial force as well as vertical movement transducers to measure the movement of the MTS head. Additionally, four SMA strain gauges in total were attached to the four SMA rods to get more insights into the stress-strain behaviour of the SMA bars, as shown in Figure 6.

3.2 Loading Protocol

Figure 7 represents the cyclic loading time history used for the PBSC brace specimen. This time history with a maximum value of 15mm was associated with the moving head of the universal testing machine. A maximum value of 15mm is considered so that the maximum strain in the SMA rods does not exceed 6%. This type of Nickel-Titanium base SMA rods experience some residual deformation after exceeding this threshold. The following loading protocol was employed by A. R. Haque and Alam (2017) to numerically validate the PBSC brace; the acquired numerical simulation results are presented and compared in the following section.

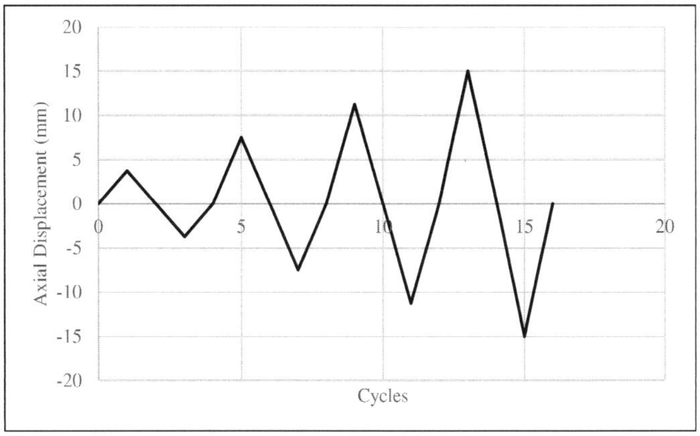

Figure 7: Quasi-static deformation loading protocol

4 TEST RESULTS

Employing the above test setup and the loading protocol, the cyclic test was conducted. Stable and symmetric self-centring hysteresis loops were obtained as shown in Figure 8(a). The PBSC brace showed negligible permanent deformation throughout the entire testing procedure. As mentioned previously, the permanent deformation can be avoided by applying deformations that do not exceed the SMA superelastic threshold. Therefore, a design engineer should consider such criteria when designing this bracing system. The SMA bar lengths should be selected based on the expected maximum deformation that the brace will experiences during the design seismic event. Figure 8(b) depicts a stress-strain plot obtained for one of the SMA rod specimen. Marginal residual deformation was observed in the test result. However, it does not have significant impact on the overall performance of the brace element. Additionally, the test results are compared with the numerical ones obtained by A. R. Haque and Alam (2017). As shown in Figure 9, very good agreement between the experimental and numerical results is achieved. The energy dissipation calculated from the experimental results is 10% higher than those obtained from the numerical simulation. This can be attributed, as shown in the figure to the maximum deformation obtained from both results which were higher in the case of the experiment compared to the numerical simulation.

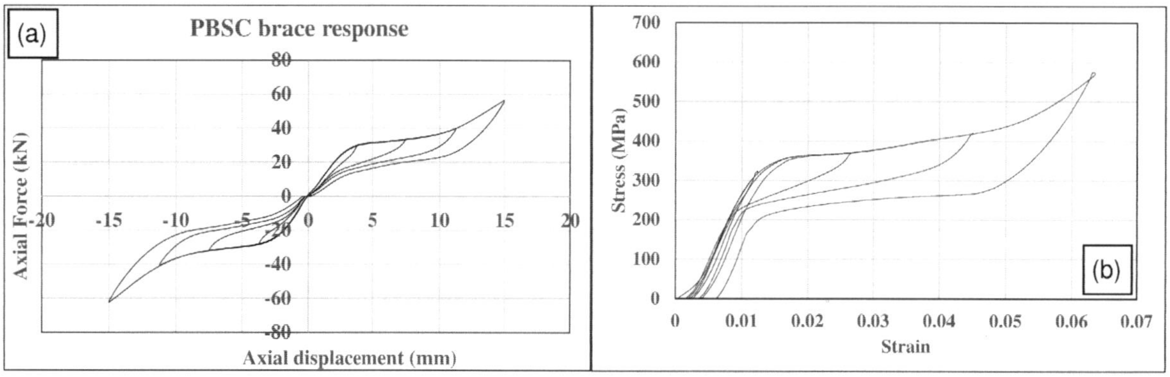

Figure 8: (a) Hysteretic response of the PBSC bracing system, (b) Stress-Strain response of a single 8 mm diameter SMA bar

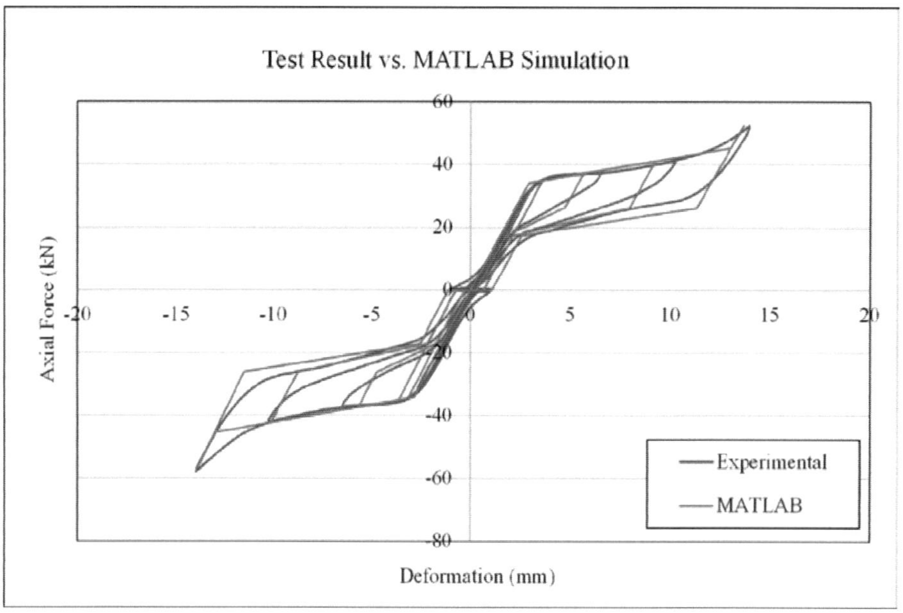

Figure 9: Test results vs. MATLAB simulation for the PBSC brace specimen

5 CONCLUSIONS

This paper has presented and discussed the seismic performance evaluation of a Piston Based Self-Centering brace for the civil engineering structures. Such bracing will not only be a cost-effective and efficient technique for new buildings but also for retrofitting older deficient structures. Design drawings were generated, and the test specimen was fabricated and tested under quasi-static loading protocol using the MTS machine with increasing loading rates. Experimental results were compared with previously developed numerical model for the brace. Very good agreement in the seismic response of the proposed system for the experimental and numerical results was achieved. Based on the results obtained from the experimental test and the numerical simulation, it was concluded that the force-displacement hysteresis loops of the PBSC brace are repeatable, stable, and identical in tension and compression. The proposed system overcomes the other available self-centering systems in its simplicity and constructability.

Acknowledgements

The financial contribution of Natural Sciences and Engineering Research Council of Canada (NSERC) through Discovery Grant was critical to conduct this study and is gratefully acknowledged.

References

Ingold, T.S. and Miller, K.S. 1983. Drained Axisymmetric Loading of Reinforced Clay. *Journal of Geotechnical Engineering*, ASCE, **109**(2): 883-898.

Choi, E., Youn, H., Park, K., & Jeon, J.-S. (2017). Vibration tests of precompressed rubber springs and a flag-shaped smart damper. *Engineering Structures, 132*, 372-382.

Christopoulos, C., Tremblay, R., Kim, H.-J., & Lacerte, M. (2008). Self-centering energy dissipative bracing system for the seismic resistance of structures: development and validation. *Journal of Structural Engineering, 134*(1), 96-107.

Dolce, M., Cardone, D., & Marnetto, R. (2000). Implementation and testing of passive control devices based on shape memory alloys. *Earthquake Engineering & Structural Dynamics, 29*(7), 945-968.

Haque, A. (2017). *Seismic performance evaluation and design of a novel piston based self-centering bracing system.* University of British Columbia.

Haque, A. R., & Alam, M. S. (2017). *Hysteretic Behaviour of a Piston Based Self-centering (PBSC) Bracing System Made of Superelastic SMA Bars–A Feasibility Study.* Paper presented at the Structures.

IBC. (2013). Study of impact and the insurance and economic cost of a major earthquake in British Columbia and Ontario/Québec. *Boston, MA, 345.*

Qiu, C.-X., & Zhu, S. (2017). Performance-based seismic design of self-centering steel frames with SMA-based braces. *Engineering Structures, 130*, 67-82.

Qiu, C., & Zhu, S. (2016). Shake table test and numerical study of self-centering steel frame with SMA braces. *Earthquake Engineering & Structural Dynamics.*

Roeder, C. W., Lumpkin, E. J., & Lehman, D. E. (2012). Seismic performance assessment of concentrically braced steel frames. *Earthquake spectra, 28*(2), 709-727.

Speicher, M. S., DesRoches, R., & Leon, R. T. (2017). Investigation of an articulated quadrilateral bracing system utilizing shape memory alloys. *Journal of Constructional Steel Research, 130*, 65-78.

Takeuchi, T., Nakamura, H., Kimura, I., Hasegawa, H., Saeki, E., & Watanabe, A. (2004). Buckling restrained braces and damping steel structures: Google Patents.

Tremblay, R., & Christopoulos, C. (2012). Self-centering energy dissipative brace apparatus with tensioning elements: Google Patents.

Wu, J., & Phillips, B. M. (2017). Passive self-centering hysteretic damping brace based on the elastic buckling mode jump mechanism of a capped column. *Engineering Structures, 134*, 276-288.

Xie, Q., Zhou, Z., Huang, J., Zhu, D., & Meng, S. (2016). Finite-Element Analysis of Dual-Tube Self-Centering Buckling-Restrained Braces with Composite Tendons. *Journal of Composites for Construction*, 04016112.

Xu, L.-H., Fan, X.-W., & Li, Z.-X. (2016). Development and experimental verification of a pre-pressed spring self-centering energy dissipation brace. *Engineering Structures, 127*, 49-61.

Xu, L. H., Fan, X. W., & Li, Z. X. (2016). Cyclic behavior and failure mechanism of self-centering energy dissipation braces with pre-pressed combination disc springs. *Earthquake Engineering & Structural Dynamics.*

Building Tomorrow's Society
Bâtir la Société de Demain

Fredericton, Canada
June 13 to 16, 2018 / *13 au 16 juin, 2018*

ADAPTING PAVEMENT INFRASTRUCTURE TO FLOOD RISK UNDER CLIMATE CHANGE — A REVIEW OF ADAPTATION STRATEGIES

Lu, Donghui[1,2], Tighe, Susan[1], and Xie, Wei-Chau[1]
[1]University of Waterloo, Canada
[2] corresponding author donghui.lu@uwaterloo.ca

Abstract: Flooding is one of the most common natural hazards in Canada. Climate change is expected to increase flood risk caused by frequent and intense precipitation, and rise of sea level. The implications of climate change influence the planning, design, construction, and management of pavement infrastructure due to flooding risk. In order to live better with the consequence of global warming, appropriate actions should be taken to prevent or minimize pavement damage, and to ensure a satisfactory pavement performance. Climate change adaptation helps to reduce pavement vulnerability to extreme weather events, and thus to offset the effects of global warming. This paper reviews adaptation strategies for managing pavement flooding risk from various aspects. Research literature, current inventories, policies, and adaptation framework and plans from Canada and other developed nations are reviewed. An adaption framework and technical adaptation strategies are developed, which serve as a reference for researchers, pavement managers, and other stakeholders to make better adaptation decisions and to increase pavement system resiliency.

1 INTRODUCTION

Climate adaptation refers to adjustment in natural or human systems to climate change in order to moderate harm. The Intergovernmental Panel on Climate Change (IPCC) defines adaptation as "*the process of adjustment to actual or expected climate and its effects*". "*In human systems, adaptation seeks to moderate or avoid harm or exploit beneficial opportunities*", while mitigation is "*an anthropogenic intervention to reduce the sources or enhance the sinks of greenhouse gases*" (IPCC, 2014). Climate mitigation is to take actions to permanently eliminate or reduce the long-term risk of climate change. Hence, tackling the causes of climate change is a proactive strategy, while adaptation is to plan to live with the consequences of climate change. They are complementary strategies for managing the risks of climate change. As the impact of climate change is already acting on pavement infrastructure and mitigation is not an immediate measure, for the next several decades, the ability of living with the consequences is critical. This paper will review current practices of the adaptation strategies.

2 IMPACT OF FLOODING ON PAVEMENT INFRASTRUCTURE IN THE CHANGING CLIMATE

2.1 Flooding from Extreme Precipitation

As predicted by IPCC, it is very likely that in the mid-latitude areas, there will be more intense and frequent precipitation events. Most parts of Canada are within the mid-latitude region in the northern hemisphere. This increases the potential of flooding events in the changing climate. Figure 1 shows floodwater on pavement structure and flood characteristics. When floodwater stands on pavement surface, pavements can be subjected to accelerated deterioration (Willway et al., 2008). In addition, the increase in groundwater level affects the subgrade. When subgrade moisture levels become too high, the structural integrity of pavements could be affected. Acceleration in the degradation of pavement materials increases instability of ground. More frequent extreme downpours cause more landslides disrupting roadways. The debris removal from intense storms will likely become even greater in the future (Peterson *et al.*, 2008). In general, the increased risk of moisture damages leads to greater needs in pavement maintenance, preservation, and rehabilitation. Furthermore, more precipitations or flooding could increase weather-related crashes and traffic disruptions (road closures) leading to safety issues.

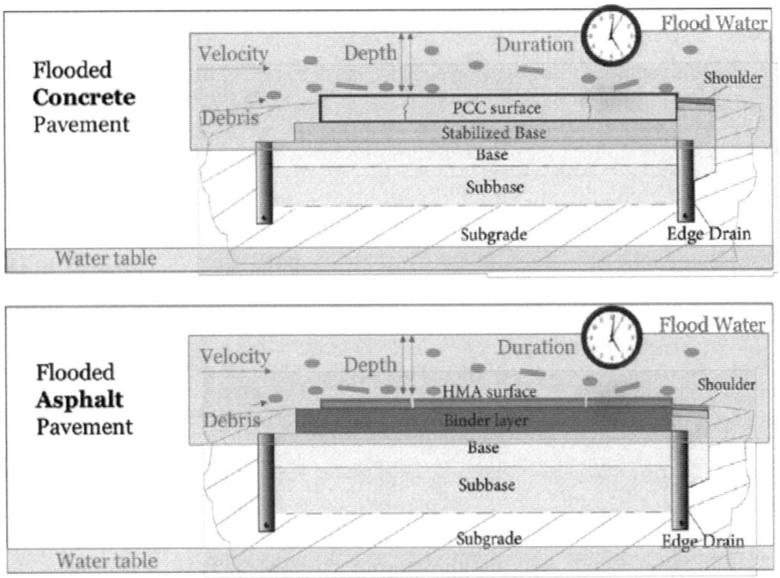

Figure 1: Floodwater on pavement structure and flood characteristics

2.2 Flooding from Sea-level Rise

Pavement infrastructure in coastal communities is vulnerable from the rise of sea level caused by climate change. Low lying pavement infrastructure in coastal areas is especially at risk from flooding. The encroachment of saltwater could result in accelerated pavement degradation. The consequence of coastal flooding could be road closures and increased life cycle cost. Hurricane events are estimated to increase with more intense precipitations. In this case, pavement infrastructure would be subjected to increased flood inundation period. The weakening of pavement from floodwater saturation will lead to accelerated deterioration. Debris removal could increase operation cost and user cost. If groundwater rises with the rise of sea level, groundwater will intersect the subgrade and unbounded layers weakening the pavement structure.

3 CURRENT ADAPTATION POLICIES AND STRATEGIES

The IPCC "*Special Report on Managing the Risks of Extreme Events and Disasters to Advance Climate Change Adaptation*" gives options on reducing exposure and infrastructure vulnerability, and improving climate resilience (IPCC, 2012). The report also recommends that an iterative process, including monitoring, evaluation, learning, research, and innovation can reduce the risk from climate extremes (IPCC, 2012). Wall and Meyer (2013) presented a review of climate change adaptation strategies that focused on the transportation sector. Some of the important frameworks from different countries are listed in Table 1.

Table 1: Adaptation frameworks used Canada and other countries

Framework	Country	Agency/Organization
Climate Risks And Adaptation Practices For The Canadian Transportation Sector 2016	Canada	Government of Canada (Palko and Lemmen, 2017)
Infrastructure Vulnerability Assessment and Adaptation to a Changing Climate	Canada	Public Infrastructure Engineering Vulnerability Committee (PIEVC) (Engineers Canada, 2015)
Federal Adaptation Policy Framework	Canada	Government of Canada (2011)
Adapting to Climate Change: A Risk Based Guide for Local Governments	Canada	Natural Resources Canada (Black *et al.*, 2010)
Adapting to Climate Change: A Risk Based Guide for Ontario Municipalities	Canada	Ontario Ministry of Municipal Affairs and Housing (Bruce *et al.*, 2006)
Risk Management for Roads in a Changing Climate: A Guidebook to the RIMAROCC Method	European Union	ERA-NET ROAD (Bles *et al.*, 2010)
Climate Change Effects on the Land Transport Network Volume Two: Approach to Risk Management	New Zealand	NZ Transport Agency (Gardiner *et al.* 2009)
Climate Change Adaptation Strategy and Framework	United Kingdom	Highways Agency (2009)
Climate Change, Extreme Weather Events, and the Highway System Practitioner's Guide and Research Report	United States	National Cooperative Highway Research Program Report 750 (NCHRP, 2014)
Climate Change Risks for Coastal Buildings and Infrastructure: A Supplement to the First Pass National Assessment	Australia	Department of Climate Change and Energy Efficiency (2011)

3.1 Canada

Federal Adaptation Policy Framework was developed in 2011. It guides domestic actions by the Government of Canada to address adaptation to the impacts of climate variability and change. The framework sets out a vision of adaptation, objectives, roles of the federal government, and provides criteria for setting priorities for action. The document notes that the "*costs associated with future climate-related failures in infrastructure could potentially be avoided by changing current infrastructure design protocols to become more resilient to predicted future changes in climate*" (Government of Canada, 2011). An adaptation Platform from Natural Resource Canada was launched in March 2012 bringing together key groups from government, industry, and professional organizations in Canada to collaborate on climate change adaptation priorities. The related investments to be made are to help Canadians adapt to climate change, including protecting the health of Canadians, assessing key vulnerabilities in Northern/Inuit populations, improving predictions of climate changes, and disseminating adaptation tools for regional adaptation. Figure 2 shows the engineering protocol of climate change vulnerability assessment from Public Infrastructure Engineering Vulnerability Committee (PIEVC).

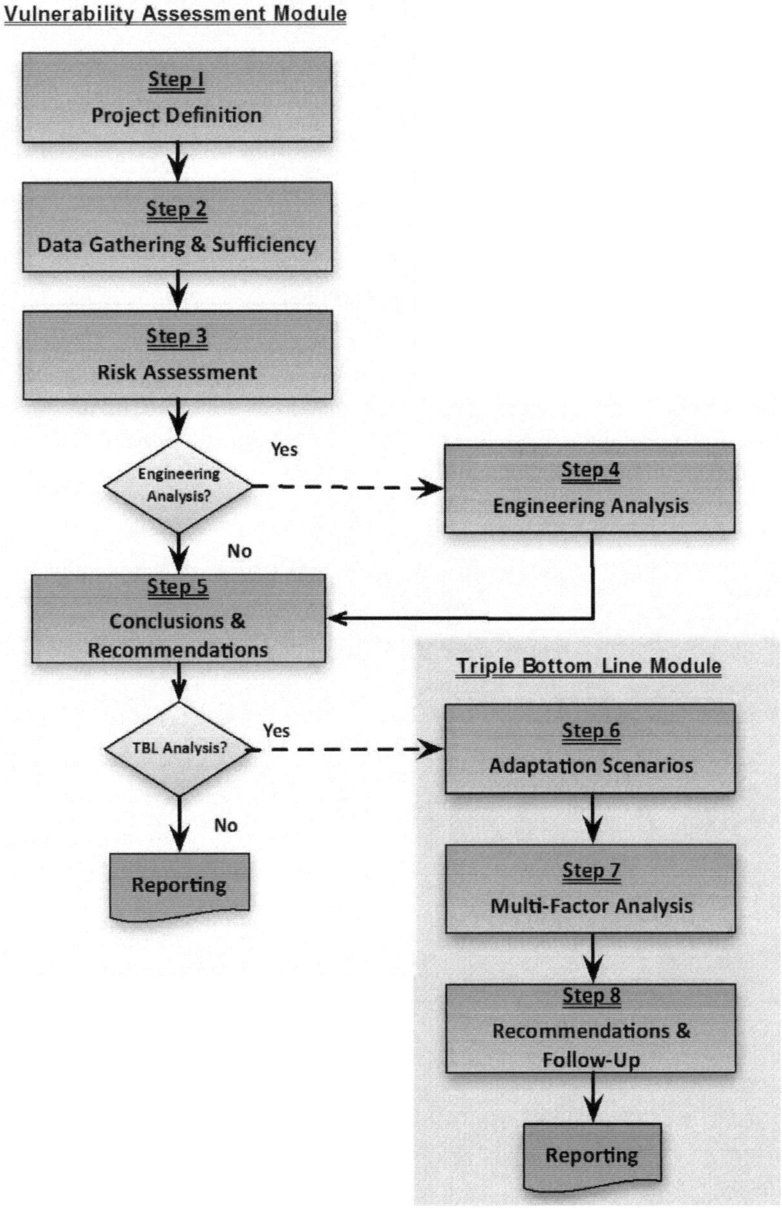

Figure 2: PIEVC Engineering Protocol (Engineers Canada 2015)

3.2 United States

The Council on Environmental Quality (CEQ) issued "Instructions for Implementing Climate Change Adaptation Planning in Accordance with Executive Order 13514" in 2011. The Order requires US federal agencies to evaluate climate change risks and manage the risks. A climate change adaptation policy was adopted by the U.S. DOT in 2012. The policy also encourages state, regional, and local transportation agencies to consider climate change impacts in the decision-making process. In 2014, "Climate Change, Extreme Weather Events, and the Highway System: Practitioner's Guide and Research Report" was released, which comprehensively addressed the strategic issues facing transportation. There are 8 steps in the US transportation adaptation framework (NCHRP, 2014): (1) Identify key goals and performance measures; (2) Define policies on assets, asset types, or locations that will receive adaptation consideration; (3) Identify climate changes and effects on local environmental conditions; (4) Identify the vulnerabilities of asset(s) to changing environmental conditions; (5) Conduct risk appraisal of asset(s) given vulnerabilities; (6) Identify adaptation options for high-risk assets and assess feasibility, cost effectiveness, and defensibility of options; (7) Coordinate agency functions for adaptation program implementation (and optionally identify agency/public risk tolerance and set trigger thresholds; (8) Conduct site analysis or modify design standards, operating strategies, maintenance strategies, construction practices, etc.

3.3 United Kingdom

The 2008 Climate Change Act in United Kingdom created a framework for building the ability to adapt to climate change. The United Kingdom has also developed the Climate Change Risk Assessment (CCRA) including analysis of climate hazard occurrence, the potential consequences, and the prioritizations.

England developed an adaptation framework named Highway Agency's Climate Change Adaptation Strategy. The framework (Figure 3) includes seven steps for increasing the highway resiliency including methods for prioritizing risk and implementation. These steps are summarized as: (1) Define objectives and decision-making criteria; (2) Identify climate trends that affect the highways agency; (3) Identify highways agency vulnerabilities; (4) Risk appraisal; (5) Options analysis to address vulnerabilities; (6) Develop and implement adaptation action plans; (7) Adaptation program review.

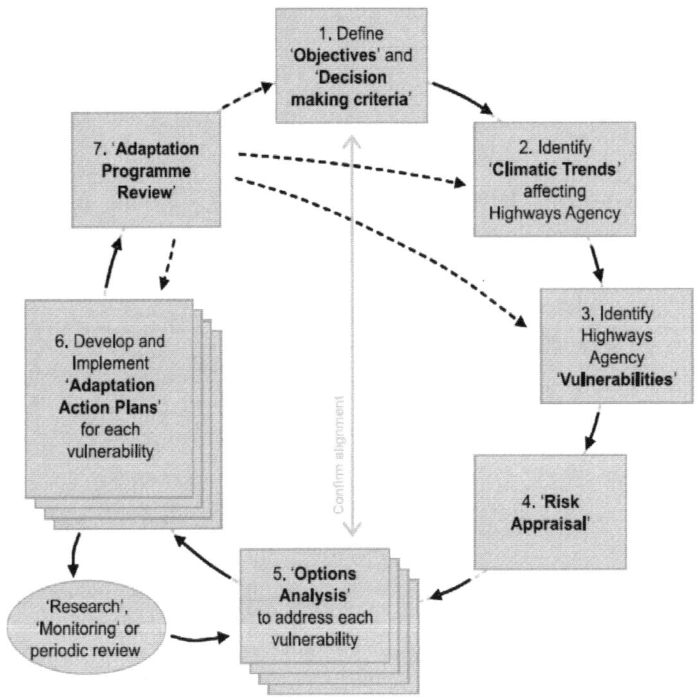

Figure 3: Highway agency adaptation framework model (Highways Agency, 2009)

3.4 Australia

In Australia, climate change adaption is managed under the Department of Climate Change and Energy Efficiency. The policy related to adaptation aims to adapt to the climate change that cannot be avoided. The key goal is to ensure that the infrastructure can provide continued and uninterrupted functioning of these assets, which are critical to support the economy (Commonwealth of Australia, 2010). The Australian Government released a *National Climate Resilience and Adaptation Strategy* in 2015, which outlines how Australia is managing its climate risks for the benefit of the community, economy and environment, now and into the future (Commonwealth of Australia, 2015).

3.5 Pavement Flood Adaptation Framework in the Changing Climate

In general, most of the adaptation frameworks include: (1) define objectives; (2) identify climate stressors; (3) identify vulnerabilities; (4) risk assessment; (5) adaptation alternative appraisal; (6) develop implementation plans; (7) monitoring. Based on the review of frameworks, a new framework is established in this paper. Pavement flood risk management framework includes risk assessment and risk adaptation, which is illustrated in Figure 4. From the aspect of asset management, setting goals and policies is the first step in the management procedure. Flood hazard analysis and pavement vulnerability analysis are the integral components of risk assessment. The potential impact of hazard on pavement asset, including pavement asset value loss and future deterioration trend, can then be identified. Identification of "needs" is the next step which is the basis for risk adaptation. Prioritization is for optimization of resource allocation in the short-term and long-term. Budgeting, project design, and implementation are required for achieving adaptation goals. Monitoring and evaluation involve collecting feedbacks and verifying good practices for pavement flooding risk management. These steps ensure that vulnerability can be reduced, and resiliency can be built.

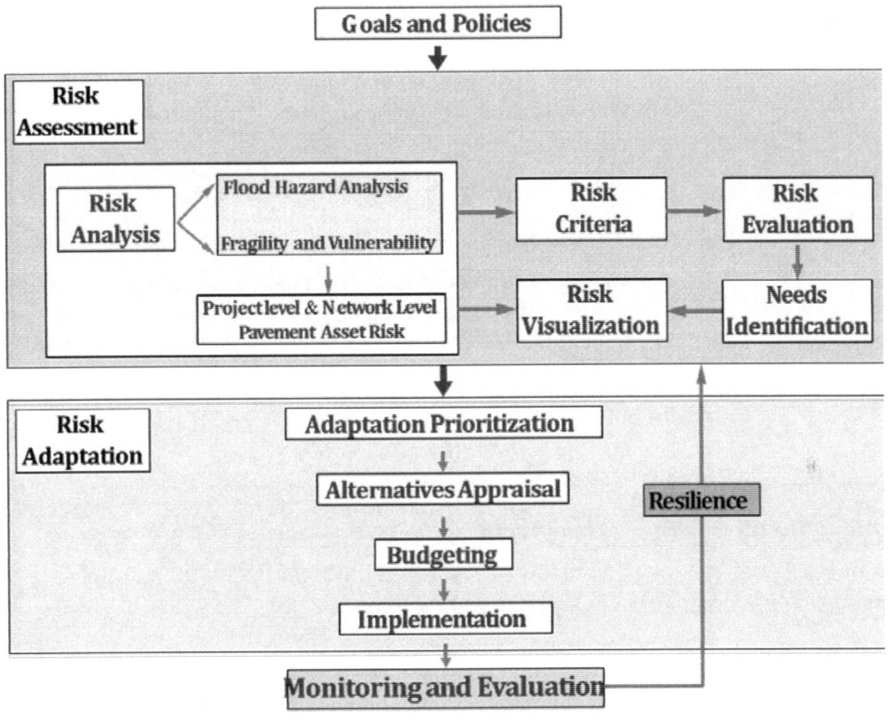

Figure 4: Adaptation Framework

4 PAVEMENT FLOODING ADAPTATION

Flooding events can result in inundation of pavement infrastructure, leading to reduction of pavement performance. The flooding magnitude can be reduced through different ways (Silva and Costa, 2016):

Infiltrate: measures to 'infiltrate' excessive water includes trenches, basins or permeable pavements to reduce the inundation of pavement structure.

Convey: 'convey' is related to the process of transporting rainwater through channels. These channels may vary in size and nature. The sizes of these channels can be upgraded according to extent of climate change and risk.

Store: the storage capacity to collect rainwater before distributing to storm water runoff. This capacity can compensate the capacity of convey system, especially when there is heavy downpour of rain in a short time.

Avoid: measures to prevent water from contacting pavement structure. Increased elevation of pavement can potentially prevent the ingress of water from getting into pavement structures. Relocation is a more expensive way to avoid flood hazards. In addition, upgrade or maintain great culverts so that flood water can pass through under pavement can alleviate the issues.

Considering pavement structure design, the adaptation in terms of design and construction improvement in the changing climate is summarized in Table 2.

Table 2: Extreme precipitation adaptation and pavement design (NCHRP, 2014)

Climate Change Impact	Affected Components and Strategies
More Extreme Rainfall Events	• Increased need for surface friction (Maintain positive cross slope to facilitate flow of water from surface; Increase resistance to rutting; Reduce splashing/spray through porous surface mixtures; Increased need for functioning sub drainage) • Need to improve visibility and pavement marking demarcation • High levels of precipitation may threaten embankment stability • Develop a better understanding of how submergence affects pavement layer structural capacity and strategies to address it
Higher Average Annual Precipitation	• Reduce moisture susceptibility of unbound base/subgrade materials through stabilization • Ensure resistance to moisture susceptibility of asphalt mixes • Investigate better construction processes

4.1 Pavement Flooding Adaptation Measures

Figure 5 shows measures to reduce flood risk with strategies emphasizing flood adaptation (on the left) and strategies emphasizing pavement design and construction (on the right).

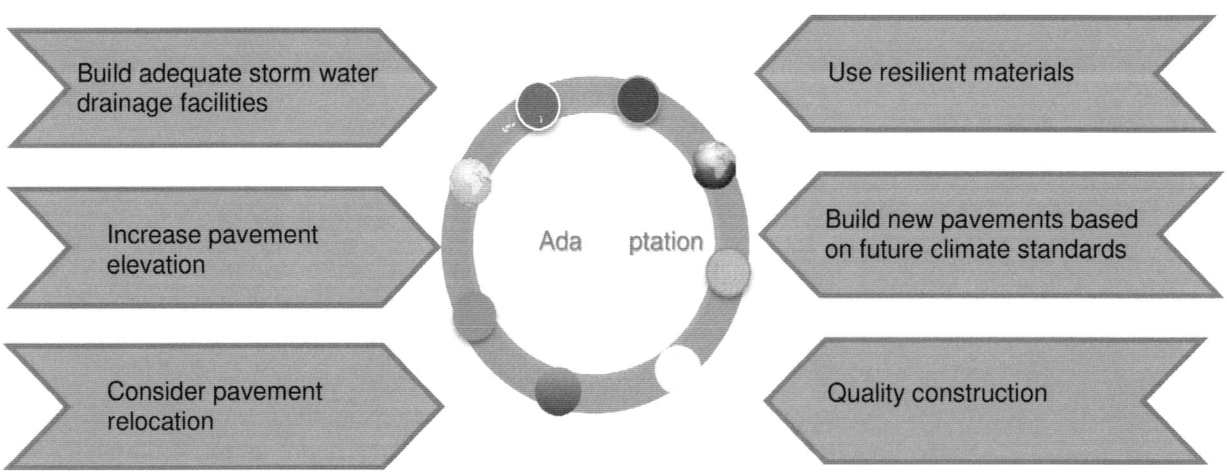

Figure 5 Summary of pavement flooding adaptation measures

For different climate regions and budget reality, different strategies should be utilized. When evaluating adaptation options, cost-effectiveness adaptation designs should be the preferred program.

5 SUMMARIES AND CONCLUSIONS

This paper reviews adaptation strategies for managing pavement flooding risk from various aspects. Research literature, current inventories, policies, and adaptation framework and plans from Canada and other developed nations are reviewed. Pavement flooding risk adaptation framework and adaptation measures are reviewed and summarized. This study can provide a reference for researchers, pavement managers, and other stakeholders to make better adaptation decisions and increase pavement system resiliency for flood hazard in the changing climate.

Acknowledgements

The research was supported by the Natural Sciences and Engineering Research Council of Canada (NSERC), Norman W. McLeod Chair Partners, and Centre for Pavement and Transportation Technology (CPATT) Partners.

References

Australian Goveronment. 2011. Climate Change Risks to Coastal Buildings and Infrastructure: A supplement to the first pass national assessment. http://www.climatechange.gov.au/publications/coastline/climate-change-risks-to-coastal.aspx.

Bles, T., Yves E., Jean-Jacques F., Stefan F., Bo L., Marjolein M., Michel R., and Frode S. 2010. Risk Management for Roads in a Changing Climate: A Guidebook to the RIMAROCC Method. http://www.climatechange.gov.au/publications/coastline/climate-change-risks-to-coastal.asp

Bruce, J. P., Egener I.D.M., and Noble D. 2006. Adapting to Climate Change A Risk-Based Guide for Ontario Municipalities." *Natural Resources Canada*. http://www.coastalchange.ca/download_files/external_reports/Bruce_(2006)_AdaptingtoClimateChange_ARisk-basedGuideforONMunicipalities.pdf.

Commonwealth of Australia. 2010. Adapting to Climate Change in Australia: An Australian Government Position Paper. 2010. http://www.environment.gov.au/climate-change/adaptation/strategy

Commonwealth of Australia. 2015. National Climate Resilience and Adaptation Strategy. http://www.environment.gov.au/climate-change/adaptation/strategy

Engineers Canada. 2015. PIEVC Engineering Protocol For Infrastructure Vulnerability Assessment and Adaptation to a Changing Climate - Principles and Guidelines. https://pievc.ca/sites/default/files/pievc-protocol-principles-guidelines-june-2016-part_1-e.pdf.

Gardiner, L., Firestone D., Osborne A., Kouvelis B., Clark A., and Tait A. 2009. Climate Change Effects on the Land Transport Network Volume Two : Approach to Risk Management. Atmospheric Research. NZ Transport Agency Research Report 378, Newzland.

Government of Canada. 2011. Federal Adaptation Policy Framework. Environment Canada, Gatineau, QC.

Highways Agency. 2009. Climate Change Adaptation Strategy and Framework.Department for Transport, UK. http://assets.highways.gov.uk/about-us/climate-change/CCAF_Strategy_and_Vol_1_Rev_B_Nov.pdf

IPCC. 2012. Managing the Risks of Extreme Events and Disasters to Advance Climate Change Adaptation. https://doi.org/10.1017/CBO9781139177245.

IPCC. 2014. Climate Change 2014: Synthesis Report. Contribution of Working Groups I, II and III to the Fifth Assessment Report of the Intergovernmental Panel on Climate Change. https://doi.org/10.1017/CBO9781107415324.004.

Meyer, M. D. 2012. Risk-Based Adaptation Frameworks For Climate Change Planning In The Transportation Sector: A Synthesis Of Practic. https://doi.org/10.17226/22462.

NCHRP. 2014. Climate Change, Extreme Weather Events, and the Highway System: Practitioner's Guide and Research Report. National Cooperative Highway Research Program. Transportation Research Board. Washington DC, US. http://www.national-academies.org/trb/bookstore

Palko, K. G., and Donald S. L. 2017. Climate Risks & Adaptation Practices For the Canadian Transportation Sector 2016. Ottawa, ON: Government of Canada. http://www.nrcan.gc.ca/sites/www.nrcan.gc.ca/files/earthsciences/pdf/assess/2016/ClimatRisk-E-ACCESSIBLE.pdf.

Peterson, T. C., Marjorie M., Tamara G. H., Andrew H. H., and Wehner M.I. F. 2008. Climate Variability and Change with Implications for Transportation. Transportation Research Board, Washington, USA. http://onlinepubs.trb.org/onlinepubs/sr/sr290_variability.pdf%5Cnhttp://files/597/Climate Variability and Transportation.pdf.

Silva, M. M., and João P. C. 2016. Flood Adaptation Measures Applicable in the Design of Urban Public Spaces: Proposal for a Conceptual Framework. *Water (Switzerland).* **8** (7). https://doi.org/10.3390/w8070284.

Willway, T., L. Baldachin, S. Reeves, M. Harding, M. McHale, and M. Nunn. 2008. The Effects of Climate Change on Highway Pavements and How to Minimise Them: Technical report. Impacts of Climate on Highway Maintenance PPR 184, U.K.: Department of Transportation

Leadership in Sustainable Infrastructure

Leadership en Infrastructures Durables

Vancouver, Canada

June 13 to 16, 2018 / *13 au 16 juin, 2018*

BLAST LOADS ON STRUCTURES

Elassaly Mohamed[1,4], Salem Mohamed[2,5], Mohsen Alaa.[3]
[1]Prof. Of Structure, Faculty of Engineering, Fayoum University
[2]Lecturer of Reinforced concrete, Housing and Building National Research Center
[3]Graduate student, Structural Engineering Dept., Fayoum University, Fayoum, Egypt
[4] mohamed.elassaly@gmail.com
[5] mm.aladawy@gmail.com

Abstract: Over the last decades, using of explosives by terrorist groups around the world that target high occupancy and public buildings, has become a growing problem in the world. Explosive devices have become smaller in size and more powerful than some years ago, leading to structural failure or massive damage in buildings. It also could result in extensive life loss or serious injuries. Terrorists usually use vehicle bombs in order to increase the number of injuries and fatalities and cause extensive damages to properties. The present research presents a thoroughly documentation of the most significant terrorism events in Egypt that happened in government officials, police, tourists and religious buildings. In addition, the research examines the pressure time history that results from the explosions using the computer program Vector-Blast which is based on the blast wave characteristics of TNT scale. This would help in predicting future potential damages that could happen with different explosives types and quantities, for RC structures having various parameters.

1 INTRODUCTION

Most of the damaged structures by bombs or impact loading are not designed to resist blast loads. Many countries, all over the world, have experienced increase in terrorism events; thus, there is a great need to better understanding of the effects of explosives on structures. These effects include shock wave physics and pressure, besides thermochemistry of explosives. In order to understand a structure's resistance to explosives, pressure-time history must be predicted accurately at various points on the structure. When the atmosphere surrounding the explosion is pushed back, an external blast wave will be created due to a massive energy coming outside from the center of the explosion. The front of the wave has a pressure greater than the region behind it; then, it immediately begins to decay as the shock propagates outward. Explosives create an incident blast wave, characterized by instantaneous rise from atmospheric pressure to a peak overpressure. As the shock front expands, pressure decays back to ambient pressure, leading to a negative pressure phase, that occurs usually in longer duration than the positive phase as shown in Figure 1. The negative phase is usually less important in the design process than the positive phase. When the incident pressure wave, on a structure, is not parallel to the direction of the wave's travel path, it is reflected producing what is known as reflected pressure. The reflected pressure is always greater than the incident pressure, at the same distance from the explosion. The reflected pressure varies with the angle of incidence of the shock wave and the incident pressure, as shown in Figure 2. When the shock wave is perpendicular to the exposed surface, the point of impact will experience the maximum reflected pressure. When the reflecting surface is parallel to the blast wave, minimum reflected pressure will occur.

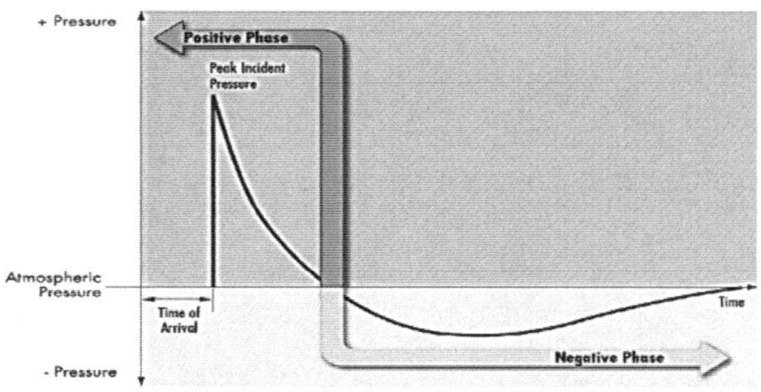

Figure 1: Blast pressure-time history (FEMA-426 2003)

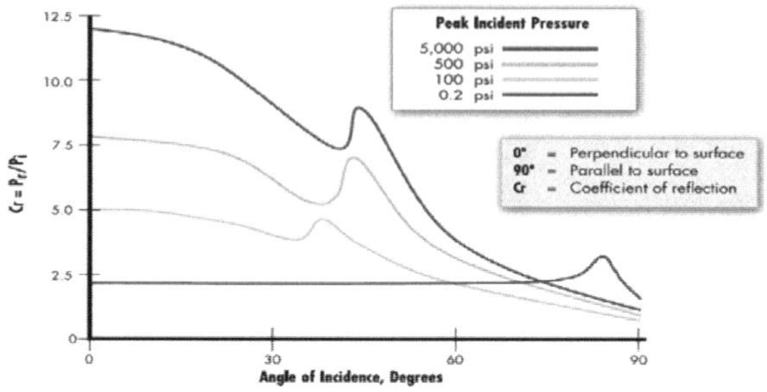

Figure 2: Reflected pressure coefficient vs. angle of incidence (FEMA-426 2003)

Explosions are classified into two major categories (TM 5-1300 1990): External and Internal. External explosions are outside blasts, in an open environment, while internal explosions occur inside a covered container or building. An external blast wave can be classified in to air burst, free air burst and surface burst, depending on whether the point of detonation of the explosive is above, at or below the ground surface (Figure 3).

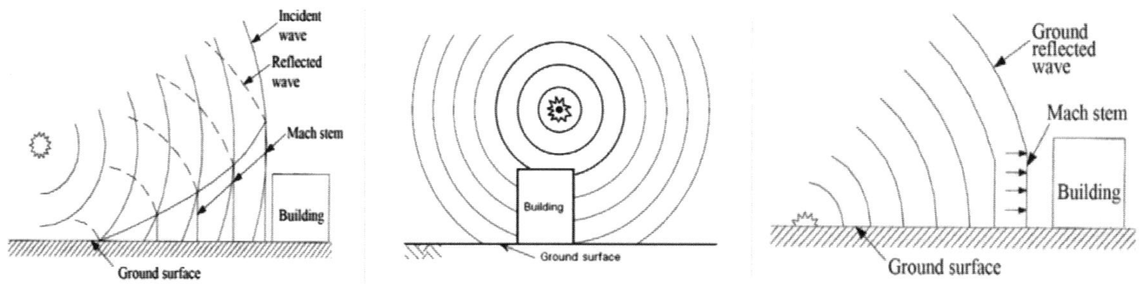

(a) Air burst with ground reflections

(b) Free-air burst explosion

(c) Surface burst

Figure 3: Classifications of external blast load (Jayasooriya 2010)

All blast parameters depend on the quantity of energy released by the explosion (or charge weight) and distance from the origin of the explosion to the building. This distance is called stand-off distance, as shown in Figure 4. The threat of the explosion will rapidly decrease over the stand-off distance. Scaled distance defined by cube root method can be calculated from the following equation:

[2] Scaled distance, (z) in m/kg^1/3 $= \dfrac{R}{W^{1/3}}$

[1] $R = \sqrt{(x_b - x_{AP})^2 + (Y_b - y_{AP})^2 + (z_b - z_{AP})^2}$

Where R is ray path distance (Figure 5), W is the mass of TNT charge equivalent

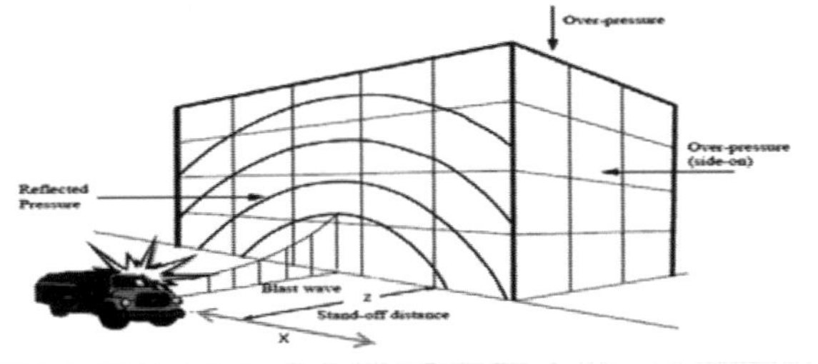

Figure 4: Stand-off distance (Moon 2009)

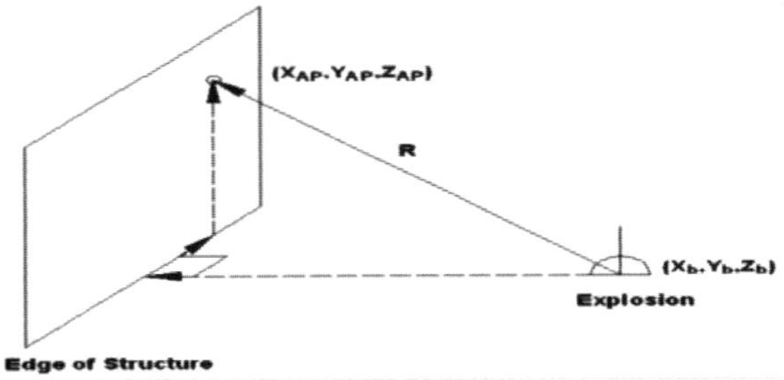

Figure 5: Simple ray path calculation (Miller 2004)

When a point on the wave front hits a corner, it diffracts around it. The process of diffraction causes energy to be sent into all directions. The pressure and impulse loading on the structure are greatly reduced as part of the energy from the incident wave ray is transferred to the structure.

Dharaneepathy et al. 1995, tested models of cylindrical structures, having heights of 100, 200, and 300 m and with 5m diameters. The charge weight was 125 kg of TNT, at distances varying from 30 to 110 m, using numerical simulations. The results indicated that there exists a critical ground zero distance at which the blast response reaches the maximum values; this distance should be used as the design datum distance. Ripley et al. 2004, examined the effects of wave reflection and diffraction angle on a structure. They used a charge mass of 50g of C4 having dimensions (2.5x2.5x5.0cm), located at three separate stand-off distances. Good numerical values of the pressure-time histories were recorded. The effect of diffraction angle was investigated. The results were considered acceptable, since the estimated pressure and impulse results were different by almost 19% and 15%, respectively from the experimental ones. For all charge locations, numerical results were found to be less than experimental values but were considered acceptable. They concluded that the Chinook code is able to capture the effects of diffraction, blast channeling and complex wave reflection, accurately. Kakogiannis et al. 2010, studied the blast wave numerically by two types of finite element methods: Eulerian multi-material modeling and pure Lagrangian using CONWEP (Hyde, 1992). They compared it with the experimental ones. In the first type of simulations, pressure waves were calculated by multi-material Eulerian formulation. For the Lagrangian finite element models, the load was applied as an equivalent triangular pulse. The Eulerian models provided results closer to the experimental ones. In general comparison with experimental, results showed that the combination of both versions CONWEP implementation and Eulerian multi material modeling were considered efficient design tools. Pranata and Madutujuh 2012, designed a blast resistant single door to bear 0.91 bar blast pressure and 44 ms blast duration. Dynamic elastic finite element method was carried out with 900 nodes using computer software ADINA. The dynamic time history analysis modeled blast load as Impact load for the given duration. The numerical analysis was done in order to know the behavior under blast load and estimate the safety margin of the door. Shallan et al. 2014, investigated the effects of blast loads on three buildings using the finite element program AUTODYN. They studied two story buildings with three different aspect ratios 0.5, 1.0 and 1.5; explosive load was equivalent to 1000 kg of TNT, placed at 2.0m height from the ground. They concluded that the reflected overpressure temperature of different points in the building increase with decrease of the standoff distance of the blast load from the building. The arrival time of reflected over pressure and temperature increase with increasing standoff distance. The blast load, at distance, equals 1.5m from the building, made a total failure in the column in the face of the blast load. Netherton and Stewart 2016, compared deterministic and probabilistic methods in blast loading. They noticed that deterministic blast loading methods did not fully account for society's usual acceptance (or rejection) of the risks associated with damage, safety and injury, as a result of an explosive blast-load. The authors concluded from the prediction of blast-loads using probabilistic models, three important forms of risk-advice: Risk Mitigation Advice, the Probability of Safety Hazards and the cost–benefit analysis of risk mitigation proposals.

In the present research work, some of the most important explosions in Egypt and other countries have been documented. An analysis program Vector-Blast (Miller 2004) is applied on those buildings. Vector-Blast is an analytical tool designed to calculate pressure-time histories at specified points on the front, rear faces and on the sides of the structure. Blast load characteristics and dissipation with time, are calculated.

Pressure time histories are calculated for different TNT charges and standoff distances, in order to predict peak reflected pressure for any building. The outcomes of this study should be used in the design of blast resistance building, with graphical interfaces.

2 EXPLOSIVES AND TNT EQUIVALENT

Explosives are different from one to another by their explosion characteristics such as detonation rate, effectiveness and amount of energy released. TNT is considered the datum for explosions "Explosion Bench Mark". Explosives are often expressed in terms of standard TNT equivalent mass (R.E.) in the process of prediction, as presented in Table 1.

Table 1: Different equivalent TNT common explosives (https://en.wikipedia.org/wiki/TNT_equivalent)

No.	Explosive Type	Density (lbs./ft^3)	Detonation Vel.(ft/s)	R.E.
1	HMTD (hexamine peroxide)	54.8	14829.2	0.74
2	ANFO (94% AN + 6% fuel oil)	57.3	17289.8	0.74
3	Nitromethane (NM)	70.4	20865.9	1.10
4	ANNMAL (66% AN + 25% NM + 5% Al + 3% C + 1% TETA)	72.3	17585.1	0.87
5	Dynamite, Nobel's (75% NG + 23% diatomite)	92.2	23621.8	1.25
6	Amatol (50% TNT + 50% AN)	93.4	20636.2	0.91
7	Semtex 1A (76% PETN + 6% RDX)	96.6	25163.7	1.35
8	Amatol (80% TNT + 20% AN)	96.6	21554.9	1.10
9	Composition C-4 (91% RDX)	99.1	26377.6	1.34
10	Trinitrotoluene (TNT)	99.7	22637.5	1.00
11	Trinitrobenzene (TNB)	99.7	23949.8	1.20
12	Tetrytol (70% tetryl + 30% TNT)	99.7	24179.5	1.20
13	Gelignite (92% NG + 7% nitrocellulose)	99.7	26148.0	1.60
14	Composition C-3 (78% RDX)	99.7	25032.5	1.33
15	Composition A-5 (98% RDX + 2% stearic acid)	102.8	27788.4	1.60
16	Pentolite (56% PETN + 44% TNT)	103.4	24671.6	1.33
17	Tritonal (80% TNT + 20% aluminium)*	105.9	21817.3	1.05
18	Composition B (63% RDX + 36% TNT + 1% wax)	107.2	25721.5	1.33
19	Ammonium nitrate (AN + <0.5% H2O)	107.2	8858.2	0.42
20	Hexogen (RDX)	110.9	28543.0	1.60
21	Torpex (aka HBX, 41% RDX + 40% TNT + 18% Al + 1% wax)	112.1	24409.2	1.30
22	PBXW-11 (96% HMX, 1% HyTemp, 3% DOA)	112.8	28608.6	1.60
23	Hexanitrobenzene (HNB)	122.7	30839.5	1.85
24	Mercury(II) fulminate	275.4	13943.4	0.51

3 EXAMINED CASE STUDIES

A sample of ten buildings are chosen as case studies for the application of program Vector-Blast. Those buildings were affected by bombs in different forms of destruction. The weights of bombs in those events ranged from 176 to 20062 lbs. of TNT equivalent. Table 2 presents a summary of properties of those buildings, date of events and dimensions of buildings. In addition, characteristics of explosives are presented, including ways or methods of implementing the explosives, equivalent TNT charge and stand-off distances. Finally, Table 2 presents brief descriptions of the resulting destructions for those buildings as well as the pertained number of fatalities and injuries. The associated damages that occurred to some of those buildings, are also depicted in Figures (6 to 12).

4 ANALYSIS RESULTS

The computer program Vector–Blast (Miller 2004) is applied on all the above case study events. Vector-Blast estimates the reflected, incident and negative side-on peak pressure, for the different case studies having varying scaled-off distances and charge weight. Table 3 presents a summary of sample study outcomes of case study 2, for hypothetical ascending stand-off and scaled distances. Case study 2 resembles EL-Khobar towers, Saudi Arabia (1996) having the largest charge weight among investigated cases. Figures [13 and 14] depict the variation of reflected and incident peak pressure with scaled distance, respectively, for case study 2.

Implementing Vector-Blast for other investigated case studies, results in similar behaviour and outcomes for the reflected, incident and negative side-on pressure. Table 4 presents a summary of outcomes of all investigated case studies, arranged ascending according to scaled distances of each event. Figure 15 depicts the reflected pressure for all investigated case studies, combined in a single graph.

Table 2: Summary of the examined case studies

No	Event	Date	Building (Dim, ft.)	Method	TNT(lbs.)	Standoff distance (ft.)	Structural damage	Other related damages	Fatalities and Injuries	Ref.
1	Alfred P. Murrah Federal Building Oklahoma, U. S.	19-4-1995	W= 239 D= 88.6 H= 131	Truck Bombing	3968	Z= 16.4 X= 141 Y= 0	Shear cracking of the floor slabs Two columns were buckled at third floor. Girder rotate inward. One third of the building damaged.	A hole (D=7.87, W=98.4 ft.). 324 buildings within a 4-block radius damaged 86 cars were burnt The glass was shattered in 258 nearby buildings. $652 million of damage.	168 Fatalities, 680 Injuries	https://en.wikipedia.org/wiki/Oklahoma_City_bombing
2	Al Khobar towers Saudi Arabia	25-6-1996	W= 148 D= 39.4 H= 78.7	Truck bomb	20062	Z = 72 X= 65.6 Y = 0	Security fence windows and concrete building Windows were shattered up to (1.6 km) away	A hole (W=85, D=36 ft.) Felt in state of Bahrain; 32 km away Destruction of: Six high rise buildings Several military vehicles	19 Fatalities, 498 Injuries	https://en.wikipedia.org/wiki/Khobar_Towers_bombing
3	Hilton Taba South Sinai Egypt	7-10-004	W= 243 D= 59 H= 118	Truck bomb	794	Z=2.79 X = 29.5 Y = 0	Ten floors of the hotel collapsed following the blast.		34 Fatalities, 171 Injuries	https://en.wikipedia.org/wiki/2004_Sinai_bombings
4	Islamabad Marriott Hotel Pakistan	20-9-2008	W= 410 D= 82 H= 98.4	Truck bomb	~ 2866	Z = 98.4 X = 220 Y = 0	A gas pipe blown up Fire engulfed The reception area	Crater (W=65.6, D=19.7 ft.) Windows in nearby buildings hundreds of yards away.	54 Fatalities, 266 Injuries	https://en.wikipedia.org/wiki/Islamabad_Marriott_Hotel_bombing
5	Saints, St. Mark & Pope. Peter I Alex. Egypt.	1-1-2011	W=197 D=115 H=32.8	Vehicle bomb	176	Z = 32.8 X = 82 Y = 0	Glass broke but remained in the window frame Doors stayed in frames but will not be reusable.	Two cars exploded next to the original car. Windows and doors broke in the opposite mosque.	21 Fatalities, 43 Injuries	https://www.christianheadlines.com/news/trauma-grips-survivors-of-church-blast-in-alexandria-egypt-
6	South Sinai Security Directorate Egypt	7.10.2013	W=95 D=95 H=62	Car bomb	441	X= 82 Z = 46 Y = 0	Glazing broke and Shattered away. Doors were broken.	A 3.28 ft. depth crater Damages in: Façade of the General Sec. Directorate of Education. 3 schools damaged. 6 cars were burnt.	3 Fatalities, 63 Injuries	https://www.ahram.org.eg/NewsPrint/251330.aspx

7	Mansoura Security Directorate Egypt	24.12.2013	W=164 D=144 H=52.5	Van bomb	3307	Z = 13 X = 98.4 Y = 0	Collapse of: Side facade First floor completely Second and third floors, partially.	A crater (D=9.84, W=16.4 ft.) Partial collapse in Mansoura City Council, National Theater, United Bank and police dept. Destroying police cars	16 Fatalities, 150 Injuries	https://en.wikipedia.org/wiki/December 2013 Mansoura bombing
8	Cairo Security Directorate Egypt	24.1.2014	W=262 D=82 H=72	Pick-up bomb	3858	Z = 49 X= 105 Y= 0	Destruction of: Seven floor building Some RC roofs Chairs, offices, computers and air conditioner Facade, windows and doors	A hole (D=11.5, W=32.8 ft.) Destruction of: Islamic museum's poss. Museum's decorations, facade The Book House building Cairo Appeal Prison Criminal Investigation D. 50 police cars	4 Fatalities, 76 Injuries	http://www.albawabhnews.com/354381
9	Police dept. EL-ARISH	12-4-2015	W=98.4 D=82 H=32.8	Pick-up B.	1102	Z=32.8 Y=0 X=26.3	Destruction of: The glass facades Offices in three floors. Office furniture	A hole (D=39.4, W=6.56 ft.) Destruction of façades of adjacent buildings. 7 vehicles were burnt.	7 Fatalities, 44 Injuries	https://www.mobtada.com/details/317956
10	Italian Council, Cairo Egypt	11-7-2015	W=279 D=32.8 H:32.8	Car Bomb	992	Z=19.7 Y=0 X=197	Destruction some of: Consulate offices Glass and concrete Doors and fence Entrance and surrounding wall	Some damages in: The October 6th bridge. Visa office & Water pipe Journalists' Syndicate B. The Higher Judicial Dept. Egyptian Museum Façade	1 Fatality, 44 Injuries	https://www.cbsnews.com/news/italian-consulate-bombed-in-

Figure 6: Murrah Federal Building in Oklahoma City, 1995 (before and after attack)

Figure 7: Al Khobar tower, Saudi Arabia, 1996

Figure 8: Hilton Taba in south Sinai, Egypt, 2004

(a) Mansoura security directorate

(b) The National Theater

Figure 9: Destruction of Mansoura Security Directorate and National Theater in Egypt, 2013

(a) Cairo security directorate

(c) Islamic Museum

Figure 10: Destruction of Cairo Security Directorate and Islamic Museum in Egypt, 2014

Figure 11: Third Police Department of EL-ARISH in Egypt, 2015

Figure 12: Italian council in Egypt, 2015

Table 3: Reflected, incident, dynamic, negative side-on peak pressure for hypothetical ascending stand-off and scaled distances of case study 2 (Al Khobar towers, Saudi Arabia, 1996)

Charge Weight (lbs.)	Standoff Distance (ft.)	Scaled Distance (ft./lb^1/3)	Reflected Peak PR (psi)	Incident PR (psi)	Dynamic PR (psi)	Negative side on PR (psi)
20062	8.20	0.302063609	53511.7	4604	19224.8	14.935
20062	9.84	0.362476331	43006.5	3819.5	15331.8	14.935
20062	13.12	0.483301775	29157	2737.8	10154.6	14.9205
20062	16.40	0.604127218	20700.8	2059.2	7081.55	14.8915
20062	19.69	0.724952662	15203.1	1610.5	5149.15	14.8335
20062	22.97	0.845778106	11667.5	1297.9	3858.45	14.7755
20062	26.25	0.966603549	9264.4	1069.7	2952.24	14.6885
20062	29.53	1.087428993	7536.78	896.67	2314.21	14.6015
20062	32.81	1.208254437	6247.62	761.64	1868.51	14.4855
20062	39.37	1.449905324	4487.74	567.27	1272.64	14.1955
20062	49.21	1.812381655	2680.48	390.6	772.343	13.485
20062	59.06	2.174857986	1830.07	279.46	487.751	12.818
20062	72.18	2.658159761	1083.93	184.22	273.209	11.2955
20062	82.02	3.020636092	757.176	138.14	182.338	9.831
20062	98.43	3.62476331	434.217	89.871	98.948	7.1775
20062	114.83	4.228890529	267.656	62.176	55.9845	4.872
20062	131.23	4.833017747	176.668	45.458	33.379	3.7845
20062	147.64	5.437144965	125.904	34.786	20.9525	3.19
20062	164.04	6.041272184	92.771	27.579	13.7895	2.7695
20062	196.85	7.249526621	53.911	18.72	6.7425	2.2185
20062	229.66	8.457781057	36.337	13.659	3.683	1.885
20062	262.47	9.666035494	26.1	10.382	2.204	1.711
20062	328.08	12.08254437	16.0515	6.8585	1.0295	1.276
20062	393.70	14.49905324	11.136	4.959	0.5655	1.044
20062	459.32	16.91556211	8.2795	3.8425	0.348	0
20062	524.93	19.33207099	6.5685	3.132	0.232	0
20062	590.55	21.74857986	5.423	2.639	0.1595	0
20062	656.17	24.16508874	4.5965	2.2765	0	0
20062	787.40	28.99810648	3.5235	1.7545	0	0
20062	984.25	36.2476331	2.61	1.305	0	0
20062	1148.29	42.28890529	2.175	1.0875	0	0
20062	1354.99	49.900908024	1.8705	0.928	0	0

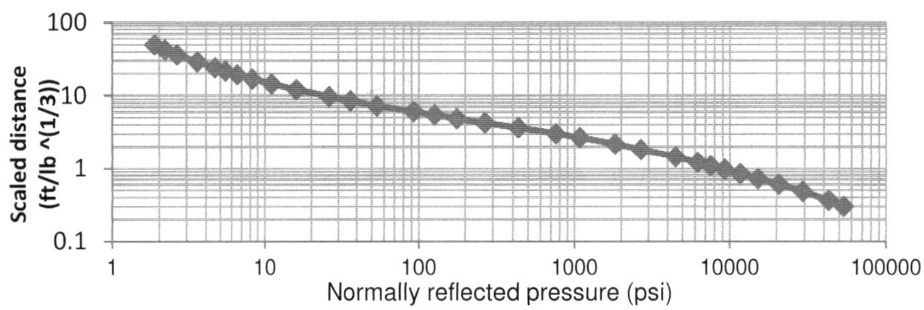

Figure 13: Variation of reflected peak pressure with hypothetical ascending scaled distance

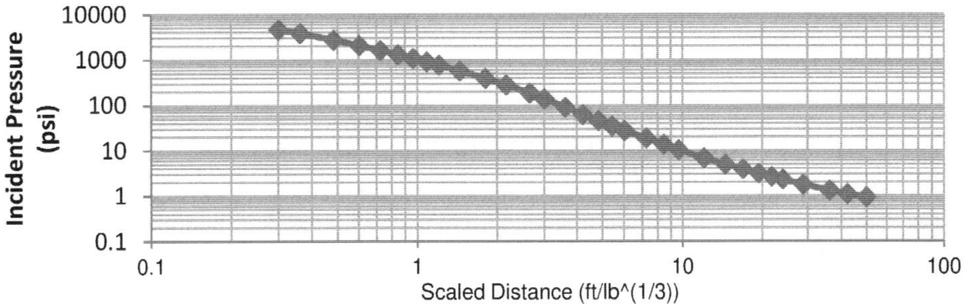

Figure 14: Variation of incident peak pressure with hypothetical ascending scaled distance

Table 4: Summary of outcomes, arranged ascending according to scaled distances of case study event

Case study No	Charge weight (lbs.)	Standoff Distance (ft.)	Scaled Distance (ft./lb^1/3)	Reflected Peak PR (psi)
3	794	2.79	0.3014	53644.9
7	3307	13.12	0.8815	10865.981
1	3968	16.40	1.0369	8196.89
10	992	19.69	1.9751	2265.886
2	20062	72.18	2.6582	1083.93
8	3858	49.21	3.1399	673.206
9	1102	32.81	3.1782	648.933
5	176	32.81	5.8543	102.6165
4	2866	98.43	6.9339	60.958
6	441	82.02	10.7837	20.4885

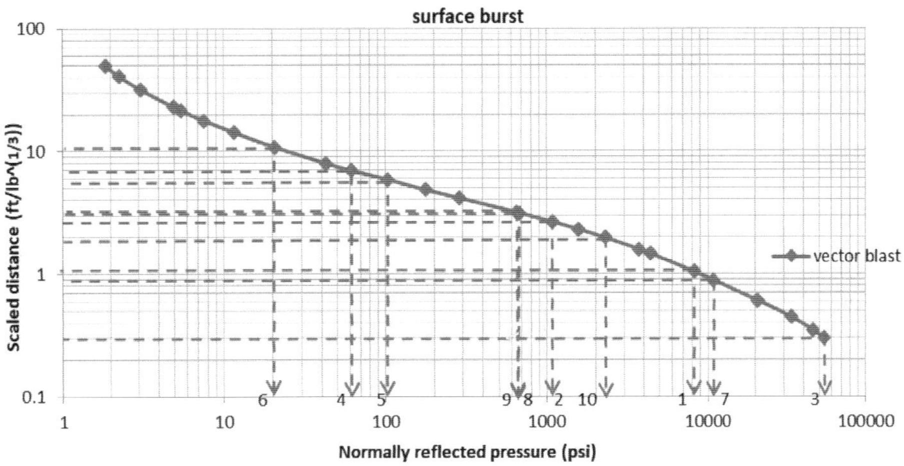

Figure 15: Scaled distance vs. normally reflected peak pressure

5 BLAST EFFECTS PREDICTIONS

The estimated capacity of TNT equivalent, pertained to the different ways, methods and vehicles used to deliver the weapon to the site of interest, is presented in Table 5. The expected level of damage in a building is estimated in Table 6, according to the incident pressure coefficient, Cr (FEMA-426 2003); Cr equals the

DM41-9

ratio of reflected to incident pressure. Accordingly, the level of damage of case study buildings, is predicted for the calculated pressure levels. Tables 7 summaries the predicted level of damage of case study building, which are conforming to the observed damage, during actual events.

Table 5: Capacity of TNT correlated with the way of explosive (Homeland Security Digital Library), https://www.hsdl.org/?abstract&did=4506

Diagram	Way of explosive	Capacity of TNT equivalent (lbs.)
	Pipe bomb	5
	Suicide bomber	20
	Brief case	50
	car	500
	suv	1000
	Small moving van Delivery truck	4000
	Water truck Large moving van	10000
	Semi-trailer	60000

Finally, in order to evaluate the range of likely charge weight that would lead to a specific damage mode or level, a correlation is established between the charge weight, mode of damage, and standoff distance for different cases. Table 8 and Figure 16 present this correlation, in which modes of damages are classified in to: Total destruction, column buckling, concrete wall failure, minor damage and finally glass breakage mode. Therefore, the expected damage mode in a building, could be estimated for a specific charge weight, way of explosive and standoff distance.

Table 6: Correlation of damage to Reflected Pressure Coefficient, Cr (FEMA-426 2003)

Damage	Reflected Pressure Coefficient, Cr=Pr/Pi
Typical window glass breakage	0.15 -0.22
Minor damage to some buildings	0.5 – 1.1
Panels of sheet metal buckled	1.1 – 1.8
Failure of concrete block walls	1.8 – 2.9
Collapse of wood framed buildings	Over 5
Serious damage to steel framed buildings	4 - 7
Severe damage to reinforced concrete framed buildings	6 - 9
Probable total destruction of most buildings	10 - 12

Table 7: Classification of level of damage of case study buildings according to reflected and incident pressures

Case study building	Level of Damage	Reflected peak Pressure, Pr (psi)	Incident Peak Pressure, Pi (psi)	Calculated Pressure coefficient, Cr_i	Typical range, Cr
Hilton Taba	Failure	53644.9	4613.639	11.63	10-12
Mansoura Directorate		10865.98	1334.29	8.14	
Murrah building	Severe	8196.89	963.67	8.5	6-9
Italian Council		2265.89	335.327	6.76	
Al-Khobar tower	Serious	1083.93	184.2225	5.88	
Cairo Directorate	to	673.206	126.266	5.33	>5
El-Arish police department	Severe	648.933	122.7425	5.29	
The Saints, St. Mark &	Serious	102.6165	29.5365	3.47	4-7

DM41-10

Peter I					
Islamabad Hotel	Partial	60.958	20.5465	2.97	1.8-2.9
South Sinai Directorate	Failure	20.489	8.4535	2.42	

Table 8: Blast damages rate correlated to charge weight

Charge weight	Threshold, total destruction	Threshold, column buckling	Concrete wall failure	Minor damage	Glass break
5	0.5130	1.7100	5.1299	10.2599	18.8097
20	0.8143	2.7144	8.1433	16.2865	29.8586
50	1.1052	3.6840	11.0521	22.1042	40.5243
100	1.3925	4.6416	13.9248	27.8495	51.0575
500	2.3811	7.9370	23.8110	47.6220	87.3071
1000	3.0000	10.0000	30.0000	60.0000	110.0000
4000	4.7622	15.8740	47.6220	95.2441	174.6141
10000	6.4633	21.5443	64.6330	129.2661	236.9878
60000	11.7446	39.1487	117.4460	234.8921	430.6354

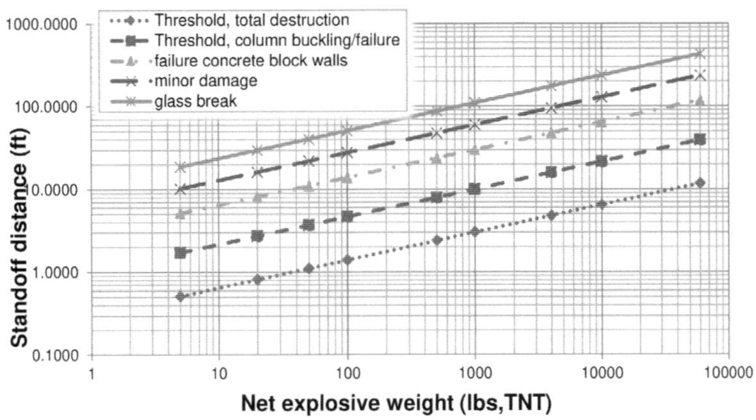

Figure 16: Blast range effects on reinforced concrete structures

6 CONCLUSIONS

The present study presents a documentation for the damage occurred to some events that occurred in Egypt lately, due to blast loads, resulting from different explosion scenarios. The damage level that occurred to 10 case study buildings, is investigated using Vector-blast analysis program. Blast loads varied between (176~20062 lbs.) of TNT charges and were located at different standoff distances (2.79~82 ft.). The observed damage of those case study buildings, conforms to that obtained by Vector-blast analysis. The expected blast damage as well as the level of protection needed for a reinforced concrete structure, can be well predicted for a specific charge weight, way of implementing explosive charge and standoff distance. The study presents tables and figures that would help in predicting the damage level for various charge weights and standoff distances.

References

Dharaneepathy, M.V., Keshava Rao, M.N.and Santhakumar, A.R. 1995. Critical Distance for Blast Resistant Design. *International Journal of Computers and Structures*, **54**(4), pp. 587-595.

FEMA 426, 2003. Reference Manual to Mitigate Potential Terrorist Attacks against Buildings, Risk Management Series, *US department of Homeland Security*, Washington, USA.
Hyde, D.W. 1992. CONWEP, Conventional Weapons Effects Program. U.S. Army Engineer Waterways Experiment Station, Vicksburg, M.S.
Jayasooriya, R. 2010. Vulnerability and Damage analysis of reinforced concrete framed buildings subjected to near field blast events. Ph.D. thesis, *Faculty of Built Environment and Engineering*, Queensland University of Technology, Brisbane, Australia.
Kakogiannis, D., Van Hemelrijck, D., Wastiels, J. Palanivelu, S., Van Paepegem, W., Vantomme, J. Kotzakolios, A. and V. Kostopoulos. 2010. Assessment of Pressure Waves Generated by Explosive Loading. *Tech Science Press*, CMES, **65**(1), pp.75-92.
Miller, P. 2004.Towards the modeling of blast loads on structures. M.A. Sc. thesis, *Department of Civil Engineering*, University of Toronto, , Canada.
Moon, N.N. 2009. Prediction of blast loading and its impact on buildings. M.T thesis, *National Institute of Technology*, Department of Civil Engineering, Rourkela-769008.
Netherton, M.D. and Stewart, M.G. 2016. Risk-based blast-load modeling: Techniques, models and benefits. *International Journal of Protective Structures 2016*, **7**(3), pp. 430–451.
Pranata, Y.A. and Madutujuh, N. 2012. Dynamic time history analysis of blast resistant door using blast load modeled as impact load. *Civil Engineering Forum.* **XXI/1** - January 2012.
Ripley, R.C, von Rosen, B., Ritzel, D.V., and Whitehouse, D.R. 2004. Small-Scale Modeling of Explosive Blasts in Urban Scenarios. *21st International Symposium on Ballistics*, Adelaide, Australia.
Shallan, O., Eraky, A., Sakr, T., and Emad, S. 2014. Response of Building Structures to Blast Effects. International *Journal of Engineering and Innovative Technology (IJEIT),* **4**(2), ISSN: 2277-3754.
TM 5-1300, Department of the Army. 1990. Structures to resist the effects of accidental explosions.
https://en.wikipedia.org/wiki/TNT_equivalent
https://en.wikipedia.org/wiki/Oklahoma_City_bombing
https://en.wikipedia.org/wiki/Khobar_Towers_bombing
https://en.wikipedia.org/wiki/2004_Sinai_bombings
https://en.wikipedia.org/wiki/Islamabad_Marriott_Hotel_bombing
https://www.christianheadlines.com/news/trauma-grips-survivors-of-church-blast-in-alexandria-egypt-11644990.html
https://www.ahram.org.eg/NewsPrint/251330.aspx
https://en.wikipedia.org/wiki/December_2013_Mansoura_bombing
http://www.albawabhnews.com/354381
https://www.mobtada.com/details/317956
https://www.cbsnews.com/news/italian-consulate-bombed-in-cairo-egypt/
https://www.hsdl.org/?abstract&did=4506

Building Tomorrow's Society
Bâtir la Société de Demain

Fredericton, Canada
June 13 to 16, 2018 / *13 au 16 juin, 2018*

BLAST PEROFRMANCE OF CROSS-LAMINATED TIMBER PANELS WITH REALISTIC BOUNDARY CONDITIONS

Côté, Dominic[1], Viau, Christian[1,3], Lacroix, Daniel[2] and Doudak, Ghasan[1]
[1] University of Ottawa, Canada
[2] Carleton University, Canada
[3] cviau037@uottawa.ca

Abstract: This paper presents a recent study which investigated the effects of realistic boundary conditions on the behaviour of cross-laminated timber (CLT) panels when subjected to out-of-plane blast loads. Experimental testing was conducted at the University of Ottawa Blast Research Laboratory, where shock waves were generated and applied through the use of a shock tube test apparatus. Connection configurations with seismic detailing were considered, in order to evaluate whether existing structures have adequate capacities to resist a given blast load with desired damage in the primary structural elements. Typical connections used in construction to resist gravity and lateral loads, as well as connections designed specifically to resist a given blast load, were also investigated. The results indicate that the detailing of the connections appears to significantly affect the behaviour of the CLT panel. Bearing type connections performed better than those composed of only wood screws, which tended to fail in a brittle manner. It was observed that under-designed steel angles provided significant energy dissipation through steel yielding and wood crushing, resulting in reduced panel damage. The study also concluded that using simplified tools such as single degree-of-freedom (SDOF) modelling together with current available material models for CLT may not be sufficient to adequately describe the behaviour and approximate the damage.

1 INTRODUCTION

Over the past decades, the effect of blast loads on structural members has steadily been gaining research interest due to high-profile deliberate attacks and accidental explosions, such as the September 11 attack in New York and the Oklahoma City Bombing. Coincidentally, the recent interest in developing and utilizing massive wood panels in construction and the advancement in technology to produce high-performance connections has pushed the boundaries for what can be achieved in timber construction. Despite the meticulous effort to produce complete design requirements for CLT panels under gravity, seismic and wind loads, little information on how to design such members for blast loading currently exists.

Recent research efforts have established key dynamic characteristics of CLT including strength increase factor (SIF) and dynamic increase factor (DIF), both of which are key parameters used in design and research to develop suitable material predictive models (Poulin et al. 2017). Although this study has significantly contributed to understanding the behaviour of the material, it was conducted on CLT panels with idealized simply supported boundary conditions. Connections between the wall and the floor systems above and below are critical to fully describe the overall behaviour of CLT structures when subjected to blast loads. The end conditions used in the construction of cross-laminated timber structures may affect the failure mode of the elements, where the boundary connection elements could fail prematurely, and thereby prevent the wall panel from reaching its intended maximum flexural capacity. Alternatively, a failure in the connections, when adequately detailed, may yield ductile failure and limit the damage to the CLT wall panel. Both design approaches are investigated in the current study and outcomes from the experimental investigation is reported.

This paper reports on the effects of realistic boundary conditions on the behaviour of cross-laminated timber wall panels when subjected to simulated out-of-plane blast loads. Both typical connection details as well as connections that are expected to perform well under blast loading are evaluated experimentally. Analytical modelling methodologies for blast design are discussed and their appropriateness for timber systems with connections is evaluated.

2 EXPERIMENTAL PROGRAM

Dynamic testing was undertaken using a shock tube, which is a testing apparatus capable of generating pressure waves to simulate far-field detonations created by high explosives. An example of such test setup is shown in Figure 1 (a). In order to simulate floor diaphragms as part of creating the boundary conditions for the panels, pieces of CLT specimens were cut and clamped at the panel ends to the shock tube, as shown in Figure 1 (b). These end blocks were then attached to the wall panels using the connection detail being investigated, simulating the connection at the ends of a CLT wall and the in-plane stiffness of the floors above and below.

A total of eleven dynamic tests were performed on five-layer Nordic X-Lam Cross-Laminated Timber panels with varying connection detailing. The specimens comprised of spruce-pine-fir (SPF) plies measuring 35 mm by 80 mm with 1950Fb machine stress rated lumber used in layers oriented in the major direction and No. 3/Stud lumber in layers oriented in the minor direction. Heco-Topix screws were used to connect the panel ends to the end blocks directly or through two different types of steel angles. Three types of screws were used, including 300 mm long screws, with 6 mm and 8 mm in diameter; 120 mm long screws, with 8 mm and 10 mm diameter; and 350 mm long double-threaded screws, with 8.5 mm diameter. Two steel angle types were used; one consisted of ML24Z steel angles from Simpson Strong-Tie®; and the other of 6.6 mm thick manufactured steel angle. For some of the configurations, the connections were designed based on the capacity of the CLT panels. For the purpose of this research project, the designed connections were selected by considering a 20 % increase on the dynamic reaction obtained from tests on CLT panels with simply supported boundary conditions (Poulin et al. 2017).

Table 1 presents a summary of the specimens which were tested dynamically, including the type of connection tested, specimen identification (ID), and construction details of the top and bottom connections. The nomenclature of each specimen conveys type of connection (i.e. EG – End-Grain; DT – Double-Threaded; DA – Ductile Angle, and; SA – Stiff Angle), and a numerical indicator representing the order of which the specimen was tested within its group (e.g. DT3 was the third specimen tested with double-threaded screw connections).

(a)

(b)

Figure 1: (a) Experimental test setup; (b) Boundary condition

Table 1: Outline of Experimental Program

Connection Group	Specimen ID	Bottom Connection	Top Connection
End-grain screws	EG1	4 – ML24Z angle	25 – 6x300 Heco-Topix Zn Gelb A2L
	EG2	2 – ML24Z angle	12 – 6x300 Heco-Topix Zn Gelb A2L w/ reinforcement
	EG3	27 – 8x300 Heco-Topix Zn Gelb A2L w/ reinforcement	27 – 8x300 Heco-Topix Zn Gelb A2L w/ reinforcement
Double-threaded screws at 60°	DT1	2 – 8.5x350 ZnA2K+DS-Beschichtet	2 – 8.5x350 ZnA2K+DS-Beschichtet
	DT2	8 – 8.5x350 ZnA2K+DS-Beschichtet	8 – 8.5x350 ZnA2K+DS-Beschichtet
Ductile angles	DA1 – DA4	2 – ML24Z angle	2 – ML24Z angle
Stiff angles	SA1 – SA2	Manufactured angle	Manufactured angle

3 EXPERIMENTAL RESULTS

3.1 Screw Connections

End-grain screw connections are typically used in construction to resist gravity loads. The purpose of this set of tests was to investigate whether typical connections have any inherent blast resisting capabilities and ductility. The connection capacity of specimens EG1 and EG2 was calculated using the blast standard requirements, which requires overdesigning the connections by a factor of 1.2. The connection of specimens EG3 was intentionally overdesigned using an additional factor of 1.25, for a total overdesign ratio of 1.5.

Significant damage was observed in the connection, where tension perpendicular to grain failure occurred at the row of screws closest to the tension face. Little-to-no damage occurred in the angled connections nor the CLT panel. In an attempt to strengthen the end-grain connection joints, reinforcing screws were installed normal to the panel face. This reinforcement provided a slight increase in the connection capacity by shifting the location of splitting towards the compression face. Figure 2 (a) and (b) show the failure modes for the unreinforced and reinforced end-grain connections, respectively.

Double-threaded screws, which have a discontinuity at the middle of the screw thread, were installed at 60° from horizontal. The number of screws in the first specimen tested under this connection group (DT1) was selected based on seismic design. It was ensured that the point of thread discontinuity was located at the interface between the panel and end block. No damage to the CLT panel was observed in any of the tests, however, the screws at the top and bottom connections all failed in combination of shear and withdrawal, as shown in Figure 2 (c).

The same double-threaded screws were used in a similar test specimen (DT2), which contained connections designed based on the Canadian blast design standard (CSA 2012). This was an attempt to increase the connection capacity beyond what was deemed adequate for the purpose of seismic design of the panel. Whereas fracturing of the screws was observed in the test with only two screws per joint (DT1), the additional screws in DT2 caused rolling shear and tension perpendicular to grain failure, as shown in Figure 2 (d).

(a) (b) (c) (d)

Figure 2: Screw connection failure modes: (a) EG1; (b) EG2; (c) DT1; (d) DT2

3.2 Angle Connections

The connections tested in this group consisted of bearing type connectors in order to reduce the perpendicular to grain stresses and failure. Two types of angles were tested: a relatively thin prefabricated steel angle, and an in-house-designed heavy angle.

Four tests were replicated using the ductile angles. In these tests, the connection at both the top and bottom floors consisted of two ML24Z angle brackets. The capacity of this connection was sufficient for seismic detailing, however, was under-designed based on the dynamic out-of-plane capacity considerations. The aim of this test was to try to achieve ductile damage in the connection without causing complete failure in the panel or connections. The thin angles (D1 to D4) resulted in a ductile behaviour, where bending in the angles helped absorb the energy of the blast, with very little to no visible damage in the CLT panel. All four angle brackets deformed significantly and in a similar manner, as shown in Figure 3 (a).

In order to investigate the possibility of attaining flexural failure in the CLT panels, a 6.6 mm thick steel angle bracket was manufactured and tested. No permanent yielding in the angles was observed, but screw deformation and crushing of wood allowed the ends of the panel to rotate and displace slightly. Signs of rolling shear were apparent near the end supports, which is consistent with the level of loading present in specimens S1 and S2. Unlike the specimens with ductile angles, the two specimens tested with the stiff angle caused flexural failure in the panel (Figure 3 (b)).

(a) (b)

Figure 3: (a) Crushing of wood and yielding of ductile angle; (b) Flexural failure of CLT panel

4 DISCUSSION

4.1 Effect of Connections

The goal of the experimental program presented in this paper was to qualitatively determine the effects of realistic boundary connections on the response of CLT panels to blast loads. While this paper presents the case where realistic connections are present, identical five-ply CLT panels were tested under idealized pin-ended conditions in order to establish their behaviour (Poulin et al. 2017). The results from that study serve as a reference to the measurements made in the current study since the material in both studies consists of the same CLT size, grade, and was obtained from the same manufacturer. The ultimate failure observed for the CLT panels with pin-ended support was consistently characterized by flexure failure. The ultimate failure occurred at an average reflected pressure and impulse of 54.8 kPa and 596.2 kPa-ms, respectively (Poulin et al. 2017). Using these experiments as an average baseline for comparison with the current study's results, the effects of realistic boundary conditions on CLT panels subjected to blast loads is evaluated in the following section.

The end-grain connections used in specimens EG1 to EG3 were observed to cause premature failure in the system at the connections, where the CLT panels were not able to reach their complete flexural capacity. Although it is well-known that wood is weak in tension perpendicular to grain and that splitting failure should be avoided even when the structural element is subjected to static loading, failure in the connections was not expected because the connection was designed to resist 20% more loads than the expected dynamic reaction. The observed behaviour is of great concern due to the fact that it may lead to collapse of the structure and loss of life. Although based on very limited data, such connection detailing is strongly cautioned against for blast design considerations.

In an attempt to improve the behaviour observed in the specimen with end-grain connections (EG1), screws were installed from the tension face to reinforce the connection joints, a practice that has shown promising results in terms of improving connection capacities and behaviour (Trautz and Koj 2009). Additionally, the number of screws in the end grain was reduced to emphasize the positive effect of the reinforcing screws. Without the reinforcing screws, this connection would be under-designed according to the Canadian blast design standard (CSA 2012). Although the connection was under-designed, the failure shifted towards the compression face, engaging more wood material and thereby increasing the splitting capacity. Despite this, the brittle failure obtained in this test emphasizes the fact that end grain screw connections are not suitable for blast design. This is further underlined in the test conducted on specimen EG3, where the number of screws was increased to resist 50% more load compared to the unreinforced specimen. Despite such effort, splitting failure was still observed. Based on these findings, it can be concluded that providing reinforcing screws only slightly enhances the behaviour of the CLT panels. It is therefore not advisable to employ such retrofit technique to existing CLT wall systems for platform type construction. In general, it can be noted that end grain connections are vulnerable to out of plane dynamic pressure waves, and should be avoided in new construction where a risk of blast is present.

The use of double-threaded screws resulted in little improvement in the performance. Tension perpendicular to grain splitting and shear failure in connections were observed in this group of connectors. Seismic design detailing was found to be insufficient to resist the applied blast load, with the screws at the top and bottom supports failing near the shear plane between the blocks and the test panel. The brittle failure observed at the connections is hazardous and can lead to potential collapse of the structure. Based on the results it can be concluded that angled screws with sufficient capacity to resist in-plane seismic loading would likely fail in a brittle manner during a blast event. This would completely eliminate the wall's ability to carry loads from floors above and also increase the risk on injuries of the building occupants. Even when the number of screws was increased by a factor of four in specimen DT2 compared to DT1 in order to meet the requirements for connection design following the Canadian blast design code (CSA 2012), brittle tension perpendicular to grain failure was still observed.

In an attempt to improve the behaviour observed with the screw connections, bearing type connections were investigated. This is in principle similar to those used in the context of light-frame wood stud walls (Viau and Doudak 2016b). The choice of bearing type supports was meant to promote compression rather than tension perpendicular to grain failure and thin angles were meant to deform and provide energy dissipation in the system. In order to show consistency in behaviour, multiple tests were conducted with the only difference being the pressure and impulse combinations to which the specimens were subjected. This also allowed to establish the limit to which the connection could be pushed without any catastrophic failure occurring.

In all tests involving the ML24Z ductile angles, significant deformation and crushing in the wood on the tension face was observed. No significant failure was observed in the CLT panel itself. This type of failure mode goes against the design philosophy outlined in the Canadian blast design standard (CSA 2012), where the connections are typically over-designed to allow the member to fully develop its capacity and ductility. For a material like wood, relying on the connection to yield is essential in enhancing the ductility and overall performance of the system. The difficulty, of course, lies in finding the balance between yielding the connections to obtain adequate ductility while preventing complete failure which will compromise the performance of the building and the safety of the occupants. Ideally, the connections would provide the desired ductility ratio and not reach failure before the panel reaches its ultimate flexural failure. In order to investigate and optimize this concept further, more research needs to be done on ductile type connections, where the yielding is isolated in connections with large enough deflection capacity to allow the wood member to fully develop its capacity.

The thick steel angle was another bearing type connection with sufficient capacity to meet the requirements in the blast design standard. This approach could be considered as a viable alternative to that of ductile angles. The screw connection with the stiff steel angle provided some rotational restraint as observed through the small amount of separation between the panel and the block simulating floors. While some translation in the connections was observed, which would imply that some energy was indeed transmitted and absorbed in the connections, the stiff angle connections provided enough capacity that allowed the CLT panels to reach their ultimate capacity in both tests.

4.2 Validity of Simplified Modelling Techniques

Work conducted on simply supported CLT panels identical to those tested in the current study resulted in the development and validation of a material predictive model for CLT panels with simply supported boundary conditions (Poulin et al. 2017). This section aims at investigating the validity of using such a model for CLT members with realistic boundary conditions. Timber systems are generally assumed by designers to have simply supported end conditions because the tension perpendicular to grain capacity is typically insignificant to develop moment resistance at the connection. Some of the test configurations failed prematurely in splitting such as the ones with screws in end grain (EG1 – EG3, and DT1 – DT2). Although designers may still assume that such connections can be idealized as simply supported, the test results clearly show that a model developed for the flexural behaviour of the simply supported CLT panel would likely not be suitable to model such configurations.

Single degree-of-freedom (SDOF) analysis was conducted using the material-predictive model developed and validated by Poulin et al. (2017) to predict the deformation-time response of the CLT panels while using the actual pressure-time histories generated during the tests. SDOF modelling is widely used in blast design and has shown to predict the behaviour of structural elements to blast with good accuracy (Jacques et al. 2013, Jacques et al. 2012, Lacroix and Doudak 2015, Parlin 2010, Viau and Doudak 2016a). This methodology consists of lumping the system in a single mass and spring system, and solving the equation of motion for a given pressure-time history. The main shortcomings of using SDOF analysis are the assumed deflected shape and failure mode that must hold true for the predictions to be accurate. As mentioned earlier, not all connection types are suitable for the comparison with SDOF analysis since it can be confirmed through observation of the failure mode that the shape function for simply supported condition is not applicable. This includes the end-grain screw connections and angled double-threaded screws, which resulted in tension perpendicular to grain failure at the connection.

The tests in which failure did not ultimately occur in the connections were modelled using SDOF analysis in order to verify the validity of using the assumption of simply supported end conditions, which is a typical assumption in wood design. Presented in Figure 4 are the analytical results for two specimens, namely DA4 and SA2. It can be seen from the figure that the SDOF model tends to underpredict the displacement of the system. This is due to the fact that the energy absorbed by the angles is not considered in the model, due to the limitations of assuming simply supported boundary conditions. It can be concluded that the contribution of the connections must be included in the model in order to conduct accurate modelling and obtained reasonable results. One possible way of including the contribution of the end connections would be by the development of composite curves for the purpose of performing SDOF analysis. This entails combining the resistance and stiffness relationships of all the system components into one equivalent resistance curve. Such methodology has been utilized in previous studies on, for example, retrofitted masonry walls subjected to blast (Ciornei 2012, Knox et al. 2000) and has shown to provide accurate numerical predictions. The development of such a model is currently underway. If proven to provide inaccurate predictions, the use of more refined modelling techniques, such as multi degree-of-freedom modelling or finite element analysis, will be explored.

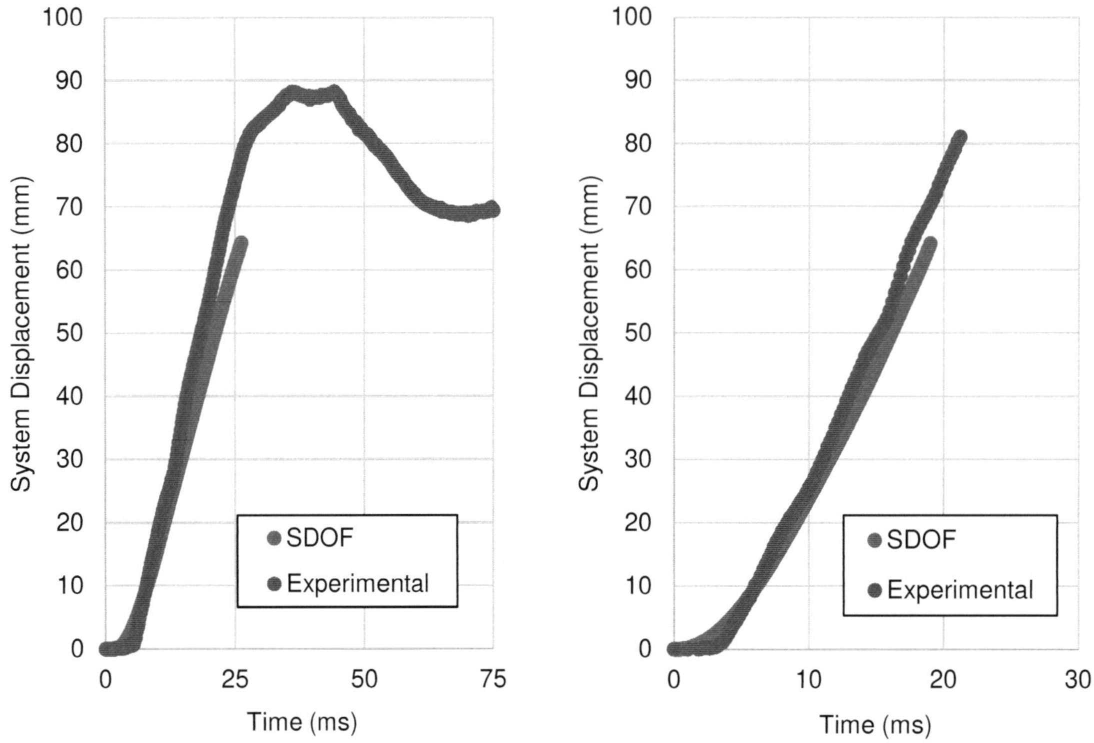

(a) (b)

Figure 4: SDOF modelling results: (a) DA4; (b) SA2

4.3 Code Considerations

The current blast design codes provide simplistic guidelines, which consist of overdesigning the connection based on the capacity of the structural element. The design philosophy aims at allowing the main structural element, in this case the CLT panel, to reach its ultimate capacity prior to failure in the connections. This approach is unsuitable for wood, due to the fact that it fails in a brittle manner in flexure. Additionally, the results presented in this paper show that overdesigning the connections may not be sufficient to prevent the unwanted failure in the boundary conditions. Fuse-based design of connections, similar to what is the design philosophy in seismic engineering, entails allowing only a certain aspect of a structural system to experience inelastic strains. In timber structures, the connections may act as the fuse, since they are typically made of ductile steel mechanical fasteners. Given that the primary objective of blast codes is to protect the lives of building occupants, for which progressive collapse prevention plays an important part, this fuse-based approach may provide an alternative approach to that included in the current blast design standard guidelines.

The results from the testing reported in this paper show that simply relying on the assumption of flexural failure of the CLT slab may not be adequate since unwanted brittle and premature failure in the connection may occur. Controlled failure of ductile connections with high levels of displacement capacity seems promising. The authors are in the process of exploring and optimizing such connection systems.

5 CONCLUSIONS

A total of eleven tests were performed using a shock tube test apparatus on five-ply CLT panels using various realistic connections. An emphasis was placed on damage levels and failure modes in comparison to previous CLT tests which only looked at the behaviour of CLT under idealized boundary conditions. It was found that even when designed according to the specifications of the current blast design standard, connections involving end-grain and double-threaded screws experienced brittle failure that corresponded to a load level inferior to that of the panel flexural capacity. Two bearing type angle connections showed that the higher capacity angle effectively allowed the CLT panel to reach its ultimate capacity, while the more flexible angle acted as an energy dissipater, which allowed the panel to remain elastic. Discussions on the ability of simplified SDOF models to predict the behaviour of wall systems consisting of panels and connections showed poor match between predicted and experimental displacements, due to the fact that the model assumes simply supported end conditions and takes no account for the connections' displacement, nor their ability to dissipate energy. Finally, current design philosophies and code guidelines, where connections are simply required to be overdesigned, may not be adequate. A better understanding of the connection behaviour is required before further guidance can be provided for their design under the effects of blast loading.

References

Ciornei, L. 2012. Performance of Polyurea Retrofitted Unreinforced Concrete Masonry Walls Under Blast Loading. MASc, Civil Engineering, University of Ottawa.

CSA. 2012. Design and assessment of buildings subjected to blast loads. *CSA S850.* Mississauga, ON: CSA Group.

Jacques, E., Lloyd, A., Braimah, A., Saatcioglu, M., Doudak, G., and Abdelalim, O. 2013. Influence of high strain-rates on the dynamic flexural material properties of spruce–pine–fir wood studs. *Canadian Journal of Civil Engineering,* **41**(1): 56-64.

Jacques, E., Lloyd, A., and Saatcioglu, M. 2012. Predicting reinforced concrete response to blast loads. *Canadian Journal of Civil Engineering,* **40**(5): 427-444.

Knox, K. J., Hammons, M. I., Lewis, T. T., and Porter, J. R. 2000. Polymer materials for structural retrofit. Air Force Research Laboratory, Tyndall AFB, FL: Force Protection Branch, Air Expeditionary Forces Technology Division.

Lacroix, D. N., and Doudak, G. 2015. Investigation of Dynamic Increase Factors in Light-Frame Wood Stud Walls Subjected to Out-of-Plane Blast Loading. *Journal of Structural Engineering,* **141**(6): 04014159.

Parlin, N. J. 2010. Behavior of FRP-Coated Wood Panels under Dynamic Loading. MSc, Civil Engineering, University of Maine.

Poulin, M., Viau, C., Lacroix, D. N., and Doudak, G. 2017. Experimental and Analytical Investigation of Cross-Laminated Timber Panels Subjected to Out-of-Plane Blast Loads. *Journal of Structural Engineering,* **144**(2): 04017197.

Trautz, M., and Koj, C. 2009. Self-tapping screws as reinforcement for timber structures. *Evolution and trends in design, analysis and construction of shell and spatial structures : IASS symposium 2009,* Universidad Politécnica de Valencia, Valencia, Spain, 28 September - 2 October 2009.

Viau, C., and Doudak, G. 2016a. Investigating the Behavior of Light-Frame Wood Stud Walls Subjected to Severe Blast Loading. *Journal of Structural Engineering,* **142**(12): 04016138.

Viau, C., and Doudak, G. 2016b. Investigating the behaviour of typical and designed wall-to-floor connections in light-frame wood stud wall structures subjected to blast loading. *Canadian Journal of Civil Engineering,* **43**(6): 562-572.

Building Tomorrow's Society
Bâtir la Société de Demain

Fredericton, Canada
June 13 to 16, 2018 / *13 au 16 juin, 2018*

CALGARY FLOOD MITIGATION RESEARCH

Rennie, Colin D.[1,4], Parsapour-Moghaddam, Parna[1], Perry, Brian[1], Slaney, Jonathan[2], Cornett, Andrew[3], and Knox, Paul[3]
[1] Department of Civil Engineering, University of Ottawa, Ottawa, Canada
[2] Watershed Engineering, City of Calgary, Calgary, Canada
[3] Ocean Coastal and River Engineering, National Research Council of Canada, Ottawa, Canada
[4] crennie@genie.uOttawa.ca

Abstract: The catastrophic 2013 Alberta flood that inundated downtown Calgary was Canada's most costly flood and second most costly natural disaster. The flood caused five deaths, required evacuation of 100,000 people, and resulted in $6B in estimated damages. Subsequently, the Province of Alberta and City of Calgary have undertaken extensive engineered mitigation works to minimize the consequences of future flood events. This paper begins with a brief overview of the flood, followed by research that has informed engineering design of proposed mitigation works; morphodynamic numerical modelling of Bow River, and physical modelling of the proposed Springbank Diversion on the Elbow River.

Morphodynamic models of Bow River were developed using the Delft3D package, including the reach from Bearspaw Dam to the Calgary Weir. Pre- and post-flood surveys of river bed bathymetry were used to calibrate morphodynamic predictions of the model. The model is being used to assess proposed flood mitigation and fish habitat enhancement works, such as manipulation of mid-channel bars.

A large scale (1:16) physical model of the river diversion structure for the proposed Springbank Off-Stream Reservoir Project was constructed at the National Research Council's (NRC) Ocean Coastal and River Engineering (OCRE) Research Centre in Ottawa. The model included structures to divert Elbow River flow from the main river channel to a constructed diversion channel, structures to regulate flow down-river, as well as realistic representation of the approach channel and floodplain. Designs of the diversion and main channel structures were studied and optimized based on model test results to ensure adequate conveyance and control during flood conditions as well as minimal blockage by sediment and woody debris. Natural woody debris was employed during model testing, which proved to be important for realistic model representation of debris jamming.

1 2013 ALBERTA FLOOD

The catastrophic 2013 flood in Alberta resulted in an estimated $6B in damage, Canada's most costly flood and second most costly natural disaster. The flood caused five deaths, required evacuation of 100,000 people, damaged more than 10,000 homes, and resulted in closure of 985 km of degraded roadway, including closure of many bridges (Government of Alberta 2014; Expert Management Panel 2014). The flood arose unexpectedly due to cyclonic orographic precipitation June 19-21 from a large storm event that stalled over the eastern slopes of the Rocky Mountains (Milrad et al. 2015; Pomeroy et al. 2016; Teufel et al. 2017; Milrad et al. 2017). Precipitation was heaviest in the foothills west of Calgary: recorded precipitation totals were 68 mm in Calgary, 200 mm in Canmore, and 345 mm in Burns Creek

(Pomeroy et al. 2016). This precipitation fell on remaining mountain snowpack at elevations above about 2000 m, which contributed to the flooding. The resulting flood zone extended from the US border to the Red Deer River valley north of Calgary (Pomeroy et al. 2016), thus high water in both Elbow River and Bow River contributed to flooding in the City of Calgary.

The Elbow River and Bow River peak discharges and associated return periods were estimated by Golder (2014). Discharge in Elbow River above Glenmore Dam peaked at 1240 m^3/s with an estimated return period 1:200 years. Elbow River flow was regulated through Glenmore Dam, thus discharge downstream of the dam peaked at 700 m^3/s with an estimated return period of 1:90 years. Bow River discharge in Calgary (below Bearspaw Dam but upstream of the confluence with the Elbow River) peaked at 1840 m^3/s. The return period of this flow has been estimated at 1:80 years. However, this return period estimate relies upon historical water marks from three ungauged larger floods in the late 19th and very early 20th centuries.

Calgary, like many other communities in Canada and around the world, has historically developed within a floodplain. As a consequence, portions of the city are inherently vulnerable to flood impacts from extreme events. The downtown core and several communities within the city built in low-lying areas were inundated in 2013 due to their vulnerability to surface flooding, high groundwater levels and/or sewer back-up. The flood zone included portions of Calgary's central business district, the Saddledome sports arena and the Calgary Stampede grounds, and several shoreline communities. The residential neighbourhoods along the Elbow River immediately upstream of the Bow River confluence were particularly affected.

A subsequent flood resiliency action plan for the City of Calgary (Expert Management Panel 2014) identified priority actions, including: assessment of review and update of official flood hazard maps, maintenance of a comprehensive flood risk database, and evaluation of social, economic, and environmental impacts of flood mitigation options. Several flood mitigation solutions are under consideration. For example, in partnership with the Province of Alberta, the off-channel storage of Elbow River flood water at the proposed Springbank Off-Stream Reservoir Project has been deemed a preferred solution to reduce flood flows. Furthermore, within the City of Calgary, critical river channel protection initiatives for the Bow and Elbow Rivers have and are being implemented to protect infrastructure, minimize flood hazards, and safeguard drinking water quality and natural habitat for fluvial species (Klohn Crippen Berger 2016). These channel works have included bank protection, bar reshaping, and channel cutting.

This paper reviews engineering research conducted in support of some of these proposed initiatives. First, three-dimensional morphodynamic numerical modelling of the Bow River is presented. Second, a large scale physical model study of the diversion structure for the proposed Springbank Off-Stream Reservoir Project is reported.

2 MORPHODYNAMIC MODELLING OF BOW RIVER

Central to flood risk mitigation is the need to understand river channel morphodynamics. The spatiotemporal distribution of sediment transfer through a reach of river determines channel morphology, which in turn, dictates flow resistance and river conveyance capacity, and thus has profound implications for flood hazard. Sediments are eroded from scour zones, where flow is converging, and are deposited in locations where flow is diverging. Erosion from the toe of a river bank can lead to bank failure, which can undermine infrastructure located in or near the river. Sediment transport and resulting channel morphology also determine the availability of riverine habitat, and thus have important consequences for water and habitat quality, fluvial biota, and species diversity.

Bed sediments in the Bow and Elbow Rivers were mobilized during the 2013 flood. Erosion was observed particularly at channel banks, and deposition resulted in new mid-channel bars (Klohn Crippen Berger 2016). Notably, deposition occurred upstream of some obstructing bridges. The City has identified increased flood hazard in some locations due to this sediment redistribution, and is currently in the process of designing flood mitigation strategies (Klohn Crippen Berger 2016). At the same time, the City recognizes that flood mitigation works may impact availability of habitat for fish and other aquatic and riparian organisms.

Morphodynamic models can be used to predict dynamic behavior and changes in river channel morphology by considering water and sediment transport, including erosion and deposition of boundary sediments. Morphodynamic models couple hydraulic models of river flow with prediction of sediment transport induced by flow, such that the flow boundary changes over time. Such a model can be used to determine morphological consequences of channel works, and thereby the optimal approaches to mitigate flood risks and enhance habitat.

Delft3D is open-source freeware, developed by Deltares, for 2D or quasi-3D morphodynamic numerical modelling. Given its accessibility, relative ease of implementation, incorporation of several bedload transport prediction equations, handling of bed material sorting in layers, and track record of successful application, Delft3D is widely used for 2D and 3D morphodynamic modelling of rivers, including meandering river bends (e.g., Parsapour-Moghaddam and Rennie 2014; Kasvi et al. 2015) and braided rivers (e.g., Williams et al. 2016; Mineault-Guitard et al. 2017; Mineault-Guitard et al. in review). Delft3D has also been used in combination with a number of fish habitat models to investigate river habitat quality for various aquatic species (e.g., Wang et al. 2012). The morphological module of Delft3D has been validated (Lesser et al. 2004), and is capable of simulating sediment fractions in multiple vertical layers, including non-cohesive and cohesive sediments (e.g. Sdqy et al. in review).

Several in-stream projects proposed by the City of Calgary for flood hazard mitigation and fluvial habitat augmentation have involved complicated 3-D flow fields and bank erosion. In order to understand these flow fields, potential bank erosion, and long-term consequences of in-stream channel works, we have been using the open-source freeware Delft3D software to analyze flooding and morphodynamic factors related to management of Bow River (Parsapour-Moghaddam and Rennie 2016a,b; Parsapour-Moghaddam et al. in review).

2.1 Study Reach and Boundary Conditions

The Bow River has a length of approximately 645 km, extending from Bow Glacier in Banff National Park to the confluence with Oldman River, defined as the start of South Saskatchewan River. The study area was a 50 km reach of Bow River as it passes through Calgary, extending from Bearspaw Dam to Policeman's Flats. The bankfull discharge for this reach is roughly 500 m^3/s, although the contribution of Elbow River flow to the middle portion of the study reach needs to be considered. Following Melethil (2015) we have been modelling the 2013 flood between June 19 and June 26 using 12 hour increments of the measured Bow River and Elbow River flows, as well as associated downstream water levels as determined by the City of Calgary's calibrated 1D HEC-RAS model.

Detailed pre-flood channel bathymetry of Bow River was surveyed by the City of Calgary in 2010. A post-flood survey was also conducted in 2013 (Golder Associates 2015). We interpolated the raw survey data to create pre-and post-flood digital elevation models (DEMs) from which Delft3D boundary meshes were generated. Furthermore, the pre- and post-flood DEMs were compared to determine a DEM of difference to ascertain the degree of morphodynamic channel change during the flood. Based on bathymetric output of 2013 flood model runs starting from the pre-flood DEM, the DEM of difference has been utilized to assess morphodynamic model predictions of channel change during the 2013 flood.

Models have been generated for several different sub-reaches. We conducted both 2D and 3D morphodynamic modelling on a short portion of the downtown core as well as the upper half of the reach between Bearspaw and the Calgary Weir just downstream of the Elbow River confluence (Parsapour-Moghaddam and Rennie 2016a,b). Most recently we have assessed 2D hydrodynamics in greater detail in the short downtown reach (Parsapour-Moghaddam et al., in review).

2.2 Sensitivity Analyses

Sensitivity analyses were performed on a large range of model parameters, including grid resolution, 2D versus 3D modelling, number of vertical layers employed in 3D modelling, eddy diffusivity, horizontal eddy viscosity, roughness, sediment transport formulation, and the influence of explicitly modelling bridge openings (Parsapour-Moghaddam and Rennie 2016a). Many of these analyses focussed on results observed in the Home Road Bend where an initial modelling attempt by Melethil (2015) displayed unrealistic scour in the inner bend and deposition in the outer bend that was not observed during the 2013 flood. The model proved to be most sensitive to roughness, horizontal eddy viscosity, and sediment transport formulation; even small changes in these parameters can impact the simulated results of sedimentation and erosion. It was found that use of spatially distributed roughness and horizontal eddy viscosity allowed for best match with observed erosion and deposition patterns. Lastly, dry points were added at bridge pier locations in the model reach, including at the three bridges in the studied bend. The resulting model suitably predicted erosion at the outer bank and deposition at the inner bank of Home Road Bend (Figure 1). Furthermore, it was found that generation of a better Delft3D model mesh can improve results (Parspapour-Moghaddam and Rennie 2016b). Specifically, grid cells should have aspect ratio close to 1.0 and the model mesh elevations should faithfully represent the surveyed bathymetry data. Finally, 2D modelling of the downtown reach confirmed the model is significantly sensitive to roughness and horizontal eddy viscosity (Parsapour-Moghaddam et al., in review). Furthermore, finer grid resolution with cells averaging 150 m^2 was found to be required to reproduce detailed hydrodynamics in the 2D model.

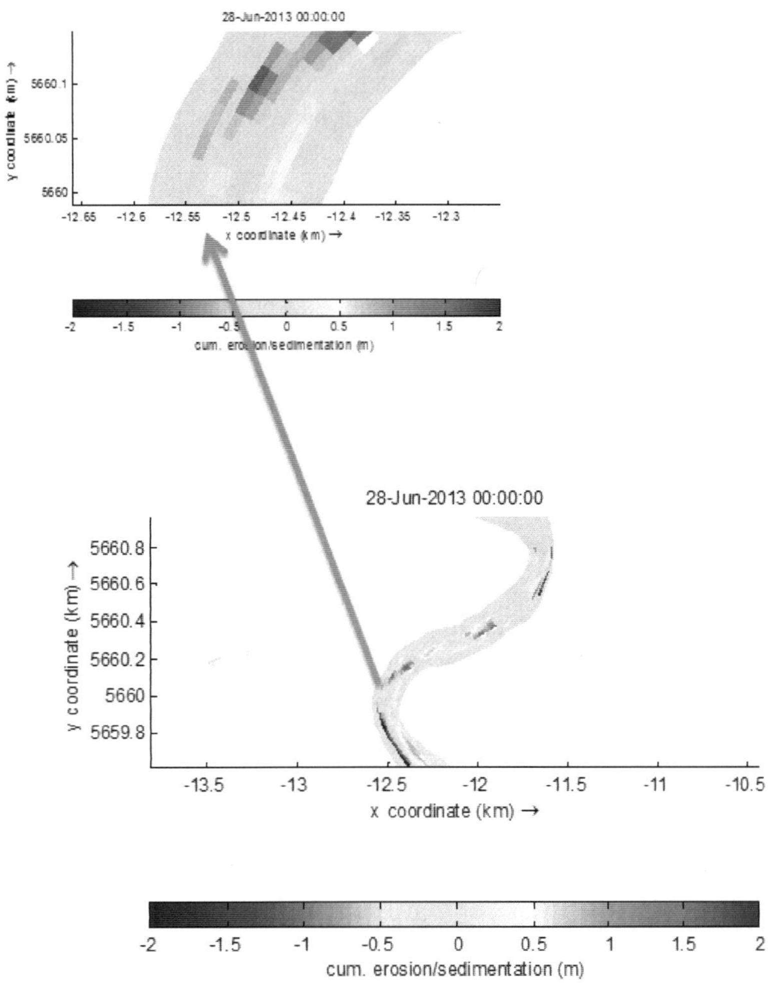

Figure 1: Predicted erosion and deposition in Home Road Bend during 2013 flood using a two-layer 3D Delft3D model of Bow River from Bearspaw Dam to the Calgary Weir with spatially distributed values of

roughness and horizontal eddy viscosity. Flow from north to south (top to bottom) (Parsapour-Moghaddam and Rennie 2016a)

2.3 Model application

The next step will be to develop Delft3D models for key reaches of Bow and Elbow Rivers where the need for management decisions has been identified. The first reach of interest will be the downtown section of the Bow River, between 14th Street and Centre Street. This particular reach has important flood and erosion management issues related to gravel-bar accumulation near bridges and flow bifurcation (Expert Management Panel 2014; Klohn Crippen Berger 2016). Key sub-reaches identified for flood mitigation works that require 3D morphodynamic modelling include the 10th Street reach (between the 14th St Bridge and just downstream of the 10th Street Bridge) and Prince's Island reach (from Peace Bridge to Centre Street Bridge). Gravel bars accumulated at and upstream of the 10th Street Bridge during the 2013 flood. These bars are currently vegetating and pose increased flood and erosion hazards. Potential interventions include gravel bar lowering to reduce flood risk and side channel cutting to improve habitat quality. Morphodynamic modelling is required to guide design of these works. In particular, the potential for long-term erosion and deposition as a result of the proposed works need to be quantified. Prince's Island reach includes a side channel, which is cut through a large meander bend. Extensive flooding occurred in this downtown location in 2013, and highly resolved modelling is required to quantify current and future the flood risks. The reach of interest on the Elbow River is centered at the Mission (4th Street) Bridge (between 26th Ave. SW and Roxborough Ave.). Extensive deposition has led to the growth of Elbow Island below the bridge, such that only one of four bridge spans is open to flow at most river discharges. Again, the proposed solution is to reduce flood risk by lowering the gravel bar and to improve habitat by cutting a channel through the bar. Again, morphodynamic modelling is required to evaluate long-term impacts.

3 SPRINGBANK OFF-STREAM RESERVOIR PROJECT

Some of the most severe flooding in 2013 occurred along the lower reaches of the Elbow River. One of the proposed solutions to reduce flooding in Calgary is to divert a portion of Elbow River flood flow before it reaches the city. The proposed $372M Springbank Off-Stream Reservoir Project is scheduled to begin construction in late 2018, following completion of an ongoing environmental assessment process (Alberta Transportation 2017). The project would be situated approximately 15 km west of Calgary at Springbank Road, and would include construction of a weir structure to divert a portion of high Elbow River flow via a new channel to an off-stream dry storage reservoir (Figure 2). The water stored in this basin would then be routed back to the river after flood recession.

3.1 Diversion Structure Physical Model

The proposed diversion structure would include both an inlet to the diversion channel and a sluice/spillway to pass remaining flow to the Elbow River. Sedimentation and/or obstruction by large woody debris (LWD) are serious concerns for design of any river diversion structure. In order to test and modify proposed designs to prevent undermining erosion or blockage by sediment and/or LWD, a large scale (1:16) physical model was constructed at the National Research Council's (NRC) Ocean Coastal and River Engineering (OCRE) Research Centre in Ottawa. NRC was commissioned to construct and test the model by Stantec for Alberta Transportation. The University of Ottawa assisted NRC with the physical modelling. The physical model has been described previously in Knox et al. (2017), Cornett et al. (2017), and Perry et al. (in review). A brief summary is provided herein.

The physical model was built within a 50 m by 30 m basin. Realistic scaled representations of structures and local topography were constructed, including the diversion inlet structure and the diversion channel, the sluice/spillway leading downstream to the Elbow River main channel, and the Elbow River approach channel (Figure 3). The approach channel included a vegetated mid-channel island. Channel bathymetry was constructed of concrete grout, and vegetation was modelled using natural tree branches inserted into the wet grout. Flow into the system was regulated using a variabl- pitch pump, and downstream water levels in each receiving channel were controlled with individual tail gates.

The diversion inlet and the sluice/spillway were fitted with gates to regulate flow into each channel. Prototype (model) flows tested within the model were 60 m^3/s (0.06 m^3/s), 120 m^3/s (0.12 m^3/s),320 m^3/s (0.31 m^3/s), 760 m^3/s (0.74 m^3/s), and 1240 m^3/s (1.21 m^3/s). Prototype discharge into the downstream Elbow River main channel was maintained at 120 m^3/s for the three lower flow conditions, and 640 m^3/s for the highest flow condition. The original design called for a diversion inlet width of 46 m (2.88 m) divided into four bays and a main channel sluice of 10 m (0.63 m) and spillway of 31 m (1.94 m) width.

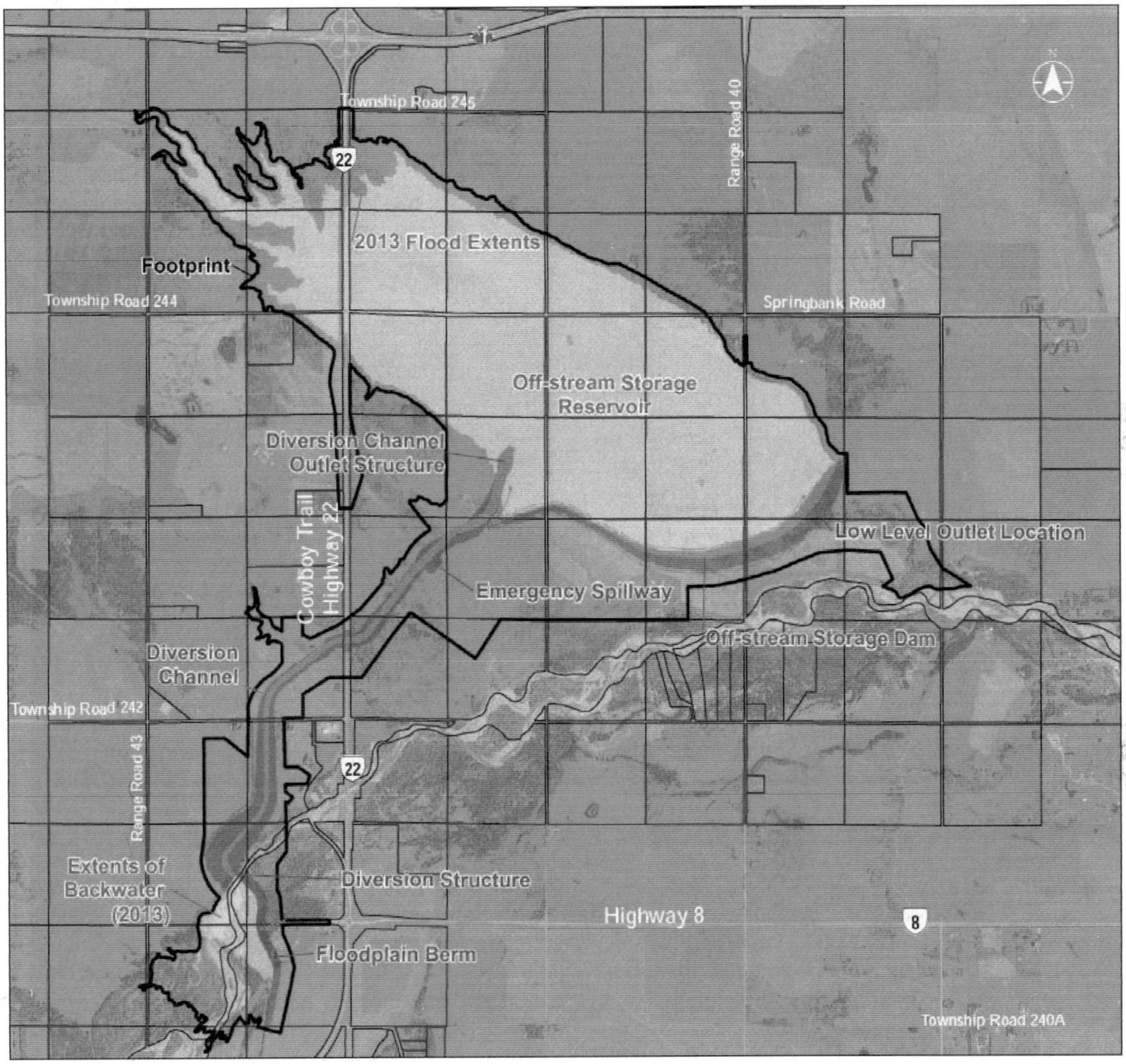

Figure 2: Proposed Springbank Off-Stream Storage Project (Government of Alberta 2017). Flow from west to east.

Figure 3: Undistorted Froude scale (1:16) physical model of modified diversion structure for proposed Springbank Off-Stream Reservoir Project. Facing downstream. Diversion channel to the left, and Elbow River main channel to the right.

3.2 Testing procedure

An extensive series of model tests was performed to evaluate the diversion structure for flow distribution, erosion, sedimentation, and LWD jamming. Initial testing of the four model flows focussed on replication of the desired flow distribution between the diversion channel and Elbow River downstream main channel. Once modelling of the flow distribution was deemed satisfactory, the model was used to test for erosion of riprap protection placed on the upstream edge of the diversion structure. Next, the potential for sedimentation was evaluated. The Elbow River bed sediment distribution was measured and scaled at 1:16. Small adjustments were made to the finer fractions of the model distribution to maintain Shields scaling of equivalent mobility. Regardless, due to modelling constraints, only the upper half the distribution was employed. This sediment was introduced at the upstream end of the model and resulting locations of deposition were identified. Finally, the model was tested for LWD jamming. Model debris was scaled to dimensions of LWD observed at the field site. Three types of model LWD were employed: 1) straight cylindrical dowels, 2) natural LWD cut from tree branches, and 3) natural LWD cut from tree branches augmented with model root wads consisting of thin wood strips attached in an X-pattern to the end of each branch.

3.3 Outcomes

As a result of model testing for flow distribution, erosion, sedimentation, and LWD jamming, the diversion inlet and main channel gate structures were modified. Small adjustments were made to the diversion structure layout to ensure appropriate flow distribution for each model flow. Perhaps most useful were the tests using LWD. It was determined that the initial diversion structure design was prone to debris jamming, largely because the gate widths were less than the LWD length, which allowed LWD to span a gate opening and hence become trapped. The initial trapped piece of LWD was then likely to obstruct and prevent passage of subsequent LWD pieces, leading to a debris jam. Consequently, the diversion structure gates were redesigned with fewer wider gates. It was also found that natural LWD were more prone to jamming than cylindrical dowels (Perry et al., in review), thus natural debris should be employed in physical model testing of LWD jamming.

4 CONCLUSIONS

The 2013 Alberta flood event was unprecedented in Canada. Since the flood, the Province of Alberta and the City of Calgary have proactively pursued flood mitigation measures. This paper briefly describes engineering research support provided to evaluate two such initiatives. First, numerical modelling has been employed to evaluate morphodynamic processes in Bow River. Such models will be used to evaluate in-channel modifications such as bar scalping and channel cutting for flood hazard mitigation and fish habitat enhancement in key reaches. Second, large scale physical modelling was conducted to test and modify a flow diversion structure for the Springbank Off-Stream Reservoir Project to diminish Elbow River flood flows. The physical model tests led to design modifications of the diversion structure to ensure appropriate flow distribution and to minimize potential for large woody debris jamming.

References

Alberta Transportation (2017) Springbank Off-Stream Reservoir. Available at http://www.transportation.alberta.ca/documents/SR1FactSheet.pdf Accessed April 2, 2018.

Cornett, A., Knox, P., Menninger, J., Lux, R., and Provan, M. (2017). Assessment of debris issues impacting design of a flood diversion project in a large scale physical model. *37th IAHR World Congress*, Kuala Lumpur, Malaysia.

Expert Management Panel on River Flood Mitigation (2014) *Calgary's Flood Resilient Future*, Report to City of Calgary. Available at http://www.calgary.ca/_layouts/cocis/DirectDownload.aspx?target=http%3a%2f%2fwww.calgary.ca%2fUEP%2fWater%2fDocuments%2fWater-Documents%2fFlood-Info-Documents%2fExpert-Management-Panel-Report-to-Council.pdf&noredirect=1&sf=1. Accessed February 15, 2018.

Golder Associates (2015) Bow and Elbow River Updated Hydraulic Model Project, Survey Data Collection and Digital Elevation Model Creation. Prepared for the City of Calgary.

Golder Associates (2014) Bow River and Elbow River: Basin Wide Hydrology Assessment and 2013 Flood Documentation. Report prepared for Alberta Environment and Sustainable Resource Development and the City of Calgary.

Government of Alberta (2014) June 2013 Southern Alberta Floods, One Year Report. Available at https://www.alberta.ca/AlbertaCode/images/Flood-Recovery-One-Year-Report.pdf Accessed April 2, 2018

Government of Alberta (2017) Springbank Off-Stream Reservoir Open House. Available at http://www.transportation.alberta.ca/documents/SpringbankReservoirOpenHouse-Aug15-2017.pdf, Accessed April 2, 2018.

Kasvi, E, Alho, P, Lotsari, E, Wang, Y, Kukko, A, Hyyppä, H, & Hyyppä, J. (2015) Two-dimensional and three-dimensional computational models in hydrodynamic and morphodynamic reconstructions of a river bend: sensitivity and functionality. *Hydrological Processes*, **29**(6): 1604-1629.

Klohn Crippen Berger (2016) *Calgary Rivers Morphology and Fish Habitat Study*. Report prepared for City of Calgary.

Knox, P., Cornett, A., Provan, M., Menninger, J., Lux, R. and Abbas, S. (2017). Assessment of debris issues impacting design of a flood diversion project in a large scale physical model. *2017 CSCE Conference*, Vancouver.

Lesser, GR, Roelvink, JA, Van Kester, JATM, Stelling, GS. (2004) Development and validation of a three-dimensional morphological model. *Coastal Engineering*, **51**(8): 883-915.

Melethil, A, (2015) City of Calgary River Morphology Simulation Study, Supported by City of Calgary,University of Calgary, Deltares, Report No: CofC-UofC-RMS.

Milrad, S. M., J. R. Gyakum, and Atallah, E. H. 2015. A meteorological analysis of the 2013 Alberta Flood: Antecedent largescale flow pattern and synoptic-dynamic characteristics. *Mon. Wea. Rev.*, **143**: 2817–2841, doi:10.1175/MWR-D-14-00236.1.

Milrad, S. M., Lombardo, K., Atallah, E. H., and Gyakum, J.R. 2017. Numerical Simulations of the 2013 Alberta Flood: Dynamics, Thermodynamics, and the Role of Orography, *Mon. Wea. Rev.*, **145**: 3049-3072, doi:10.1175/MWR-D-16-0336.1.

Mineault-Guitard, A, Rennie, CD, Williams, R (2016) Validation of observed bedload transport pathways using morphodynamic modeling. *River Flow 2016*, St Louis.

Mineault-Guitard, A, Rennie, CD, Williams, R. (in review) Verification of morphodynamic modelling using observed bedload transport pathways. *J. Geophys. Res. – Earth Surface.*

Parsapour-Moghaddam, P, Rennie, CD. (2014) Morphodynamic modelling of a tortuous meandering clay bed river using Delft3D: Stillwater Creek, Ottawa, *River Flow 2014*, Lausanne, Switzerland.

Parsapour-Moghaddam, P, Rennie, CD. (2016a) *Evaluation of Delft3D Numerical Model of Bow River*, Report prepared for City of Calgary, March 1, 2016.

Parsapour-Moghaddam, P, Rennie, CD. (2016b) *Development of an improved Delft3D Numerical Model of Upper Bow River: Influence of Model Grid and Bathymetry*, Report prepared for City of Calgary, October 25, 2016.

Parsapour-Moghaddam, P, Rennie, CD, Slaney, J. (in review) Hydrodynamic Simulation of an Irregularly Meandering Gravel-Bed River: Application of MIKE 21 FM and Delft3D Flow models, *RiverFlow 2018,* Lyon, France.

Perry, B., Rennie, C., Cornett, A. and Knox, P. (in review) Comparison of large woody debris within a large scale physical model. *RiverFlow 2018,* Lyon, France.

Pomeroy, J. W., R. E. Stewart, and Whitfield, P. H. 2016. The 2013 flood event in the South Saskatchewan and Elk River basins: Causes, assessment and damages. *Can. Water Resour. J.*, **41**: 105–117, doi:10.1080/07011784.2015.1089190.

Sdqy, Z, Rennie, CD, Clemens, D. (in review) Impact of reservoir flushing on modeling morphology of its upstream meandering reach. *International Journal of Sediment Research.*

Teufel, B., and Coauthors, 2017. Investigation of the 2013 Alberta flood from weather and climate perspectives. *Climate Dyn.*,**48**: 2881–2899, doi:10.1007/s00382-016-3239-8.

Wang, Y, Xia, Z, & Wang, D. (2012) Characterization of hydraulic suitability of Chinese sturgeon (*Acipenser sinensis*) spawning habitat in the Yangtze River. *Hydrological Processes*, **26**(23): 3489-3498.

Williams, RD, Measures, R, Hicks, DM, Brasington, J. (2016) Assessment of a numerical model to reproduce event-scale erosion and deposition distributions in a braided river. *Water Resources Research*, **52**(8): 6621-6642.

Building Tomorrow's Society
Bâtir la Société de Demain

Fredericton, Canada
June 13 to 16, 2018 / *13 au 16 juin, 2018*

COMPUTER SOFTWARE FOR THE DESIGN OF BLAST RESISTANT WINDOW RETENTION ANCHORS

Jacques, E.[1,3], and Saatcioglu, M.[2]
[1] Assistant Professor, Virginia Polytechnic Institute and State University, USA.
[2] Distinguished University Professor, University of Ottawa, Canada.
[3] ejacques@vt.edu (corresponding author)

Abstract: Broken glass fragments and flying shards are a significant source of secondary injuries caused by window failure during accidental explosions and terrorist attacks. A comprehensive review of available design procedures used in the construction industry for window retention anchors, as well as related design standards, indicated that most of the previous research and development concentrated on the protective glazing of windows, rather than the anchorage of windows to substrates. This paper introduces BRADS (Blast Resistant Window Anchor Design Software), a computer program for the analysis and design of anchoring systems used in blast-resistant window glazing. The software is intended to aid engineers, architects, and manufacturers enhance the resilience of critical infrastructure against explosive events. The tool may be used to design new glazing anchorages or assess the vulnerability of existing anchors. The innovative design algorithm incorporated in the software can compute the required configuration of window retention anchors based on the major parameters known to influence anchorage capacity, including: substrate deformability, glazing response, window aspect ratio, frame rigidity, the distribution of dynamic reactions, combined stresses, and anchorage boundary conditions. The algorithm, developed and validated by the authors, satisfies the requirements of the new national standard on blast-resistant window retention anchors, CSA S852. The results generated by BRADS were compared against experimental data obtained from shock wave testing of blast-resistant window systems using the University of Ottawa Shock Tube and the University of Kentucky Blast Simulator. The comparison indicated that BRADS can predict the magnitude and distribution of anchor forces of the tested windows with a level of accuracy suitable for the purposes of design and assessment.

1 INTRODUCTION

New materials, construction technologies, and test methods have been developed to address the increased security of structures threatened by terrorist bombing and accidental explosions. A considerable effort has been directed toward increasing the level of protection of windows as secondary blast injuries caused by flying debris, such as glass shards, account for much of the trauma resulting from explosive events (Fryberg, 2002). Examples of technologies which are commonly used to improve the level of protection of blast resistant glazing include the use of anti-shatter security film, laminated glass, cable catch systems, and blast curtains. However, the ability of these systems to fulfill their intended function depends on a continuous load path to transfer applied forces through to the supporting structure. For the case of windows subjected to blast loads, this necessitates continuity of the glazing, frame, anchorage, and substrate.

CSA S852 (CSA, 2018) is a recently developed Canadian standard for the design of blast-resistant window anchor systems. The standard provides guidance for determining the required number, size, and spacing of window retention anchors to ensure sufficient safety against premature anchorage failure. The scope includes blast-resistant punched windows having single pane, multi-pane insulated glass (IG), and laminated or filmed windows. The provisions address different types of substrates and window frame types. It provides detailed design procedures based on two approaches; (i) equivalent static force approach and (ii) dynamic two-degree-of-freedom (TDOF) analysis.

Computer software BRADS (Jacques, 2017) – Blast Resistant Anchor Design Software – has been developed to conduct analysis and design of anchoring systems used in blast-resistant punched windows in accordance with the provisions of CSA S852 (CSA, 2018). The tool was developed to facilitate the design of new window anchorages, or to assess the vulnerability of existing anchors by engineers, security personnel, and manufacturers. This program can account for the major parameters known to influence the performance of anchors in blast resistant windows, including substrate flexibility, glazing response, aspect ratio, window frame rigidity, dynamic reactions, anchor fixity, and combined stresses. This paper is intended to provide an overview of the software and illustrate its capabilities. In addition, experimental data from blast testing of windows is compared against results generated by the software with the purpose of demonstrating its suitability for design and assessment.

2 KEY ELEMENTS OF THE DESIGN SOFTWARE

The design software BRADS was developed to satisfy the Tier II: Dynamic Analysis Procedure permitted by CSA S852 (CSA, 2018) for design of blast-resistant window anchorages. The following discussion provides a general overview of the key elements of the design software. The reader is directed to the references by Jacques and Saatcioglu (2018) and Jacques (2017) for more in-depth discussion.

2.1 Behaviour of Windows Subjected to Blast Loads

The load-deformation characteristics of typical blast-resistant glazed and laminated windows are idealized in Figure 1. These types of windows typically exhibit two phases of response: elastic window response until the glass breaks (known as the "pre-break phase"), and tension membrane response of the broken window (known as the "post-break phase"). However, post-break response is only possible if laminated or glazed windows are sufficiently fastened to the window frame to develop tension membrane forces in the protective film or interlayer.

For a glass window, glazed by a protective film or laminated with an interlayer, the initial pre-break stiffness can be computed using the elastic modulus of glass, the moment of inertia of glass and the appropriate support conditions. The window during the pre-break phase develops small deflections, resulting in anchor shear forces in the direction of blast loading and associated bending moments due to the eccentricity of shear forces relative to the critical anchor section at or near the substrate interface. The mass of the window during the pre-break phase is generally assumed to be the total mass of the window.

Upon failure of glass, the resistance drops, and the protective film or interlayer starts dominating response. This is the post-break phase within which the film or interlayer initially behaves elastically, followed by yielding, exhibiting elasto-plastic response. The effect of broken glass pieces on strength and stiffness can be ignored at this stage, though their contribution to mass must be accounted for. The mass of the window in the post-break phase can be taken as the mass of the protective film or interlayer, and any residual glass pieces still bonded to these elements after glass breakage. As displacements increase, the window develops membrane forces under the action of axial stretching of the protective film or interlayer. The window retention anchors during this phase are subjected to a combination of shear force, axial tension force and bending moment. For anchor design, it is important to check the anchor capacity against two sets of design forces; shear force and bending moment resulting from the pre-break phase; and shear force, axial tension and bending moment resulting from the post-break phase.

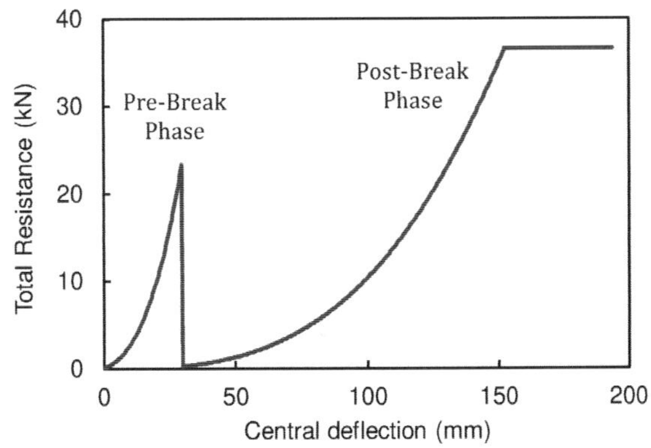

Figure 1: Typical load-deflection curve for a window exhibiting flexural and tension membrane response in the pre- and post-break stages of response, respectively.

2.2 Two-degree-of-freedom (TDOF) Dynamic Analysis

The design of blast-resistant window retention anchors requires the consideration of the effect of dynamic blast loads on the dynamic response of the window-substrate assembly. Exact analytical solutions through Finite Element Analysis, involving the modelling of each component of a window system for two-way plate behaviour can be rather complex and time consuming. This complication cannot be justified for most window anchor designs. The process can be simplified without significantly affecting the accuracy of design. This implies that the complex multiple-degree of freedom analysis of a multi-pane window with insulating air gaps and surface-mounted protective films or interlaminates, developing two-way plate action with distributed mass, interacting with the flexibility of window frames and window substrates in the elastic and plastic range of deformations, can be reduced to a single-degree-of-freedom (SDOF) system, or a two-degree-of-freedom system (TDOF).

The dynamic response of a window mounted into a flexible substrate can be accurately modelled as a TDOF system. A suitable TDOF mass-spring system is illustrated in Figure 2. The equivalent concentrated mass of the substrate is represented by m_2, while the window mass is represented by m_1. The window has an equivalent stiffness k_1 which can be defined from the its resistance function (i.e., Figure 1). The substrate may be assumed to be rigidly supported by the attached structural elements with substrate stiffness represented by spring constant k_2. The entire window-substrate assembly is subjected to a uniform blast pressure with triangular impulsive forcing function $P_r(t)$ with a maximum reflected pressure P_r and positive phase duration t_d. The blast pressure $P_r(t)$ acts on the net area of the window A_w and substrate A_s to develop equivalent time-varying forcing functions F_w and F_S applied to the window and substrate mass-spring systems, respectively. The central deflection of the window is denoted by u_1 and the mid-height deflection of the substrate is denoted by u_2, with respective accelerations of $\ddot{u}_1$and $\ddot{u}_2$.

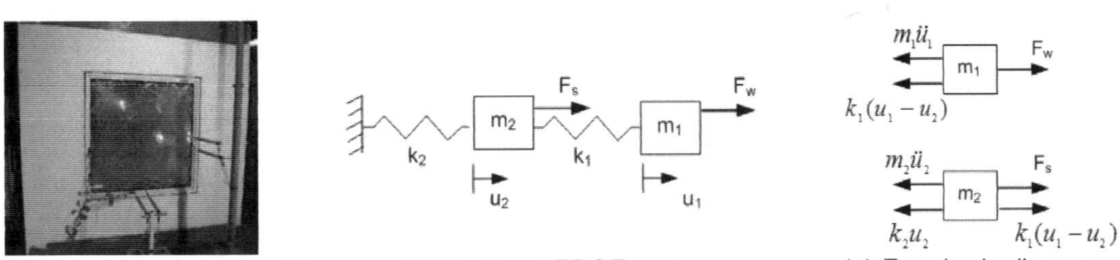

(a) Window-substrate assembly (b) Idealized TDOF system (c) Free-body diagrams

Figure 2: Two-degree-of-freedom (TDOF) idealization of a typical window-substrate assembly.

The following expressions show the equations of motion for a TDOF system without damping, which can best be solved by a numerical procedure:

[1] $m_1\ddot{u}_1 + k_1u_1 - k_1u_2 = F_w$

[2] $m_2\ddot{u}_2 - k_1u_1 + (k_1 + k_2)u_2 = F_s$

The numerical solution employed to solve the TDOF equation of motion should be able to account for variable mass and stiffness of the window system due to pre-break and post-break inelastic response, and potential nonlinearity of the substrate during blast-induced deformations. Once a solution to the TDOF system has been obtained, the dynamic support reactions of the window system can be established using appropriate expressions (i.e., Biggs, 1964, or Morrison, 2007). These support reactions provide design shear forces for the window retention anchors.

2.3 Anchorage Design

Anchorage design conducted in accordance with CSA S852 (CSA, 2018) is the process of selecting the type, size, spacing, and material properties of the fasteners based on a determination of the glazing and frame reactions. The design process also includes ensuring the anchorage has sufficient strength to resist disengagement from the substrate, and that the substrate is designed to resist local failure near the anchor. Given that there are a myriad of different proprietary window retention anchors available on the market, CSA S852 (CSA, 2018) provides explicit guidance on selecting and proportioning fasteners, while ensuring sufficient anchorage strength developed in the fasteners within the substrate. This is achieved in accordance with the manufacturer's product specifications.

The anchor design procedure adopted in CS S852 (CSA, 2018) is based on the von Mises stress condition for plane stresses given in Eq. [3]:

[3] $$\frac{f_{dy}}{\sqrt{\sigma_n^2 + 3\tau_{max}^2}} \geq 1.0$$

where f_{dy} is the dynamic material strength of the fastener, σ_n is the maximum factored normal stress in the anchor at the critical section, and τ_{max} is the maximum factored shear stress in the anchor. Note that Eq. [3] conservatively assumes that the maximum normal and shear stresses at the critical section act simultaneously.

The factored normal stress σ_n and maximum factored shear stress τ_{max} can be calculated following:

[4] $$\sigma_n = \phi\left(\frac{F_T}{A} + \frac{M}{Z}\right)$$

[5] $$\tau_{max} = \phi\left(\frac{4F_V}{3A}\right)$$

where ϕ is the safety factor required for design by CSA S852 (CSA, 2018), A is the anchor cross-sectional area, Z is the plastic section modulus of the anchor, and F_V, F_T, and M are the design shear, tension, and moment in the anchor at the critical section. As will be subsequently discussed, the design moment M is computed by considering the eccentricity of F_V and F_T relative to the critical section, while also accounting for the effect of boundary conditions of the anchor.

3 DESIGN SOFTWARE OVERVIEW

The software uses a database for window resistance curves and other dynamic properties. While the database includes the data for typical windows commonly used in practice, it can also be used for other types and sizes of windows. This requires the addition of new information to the database. User friendly "Add New Window" function enables the user to enter this information to the database. This gives the user the flexibility to expand the window types available for design. The user can extend the functionality of the software by importing the dynamic properties for any window obtained from specialized window analysis software, such as Wingard (ARA, 2005), SBEDS-W (US Army Corps of Engineers, 2012), finite element analysis, or from experimental tests.

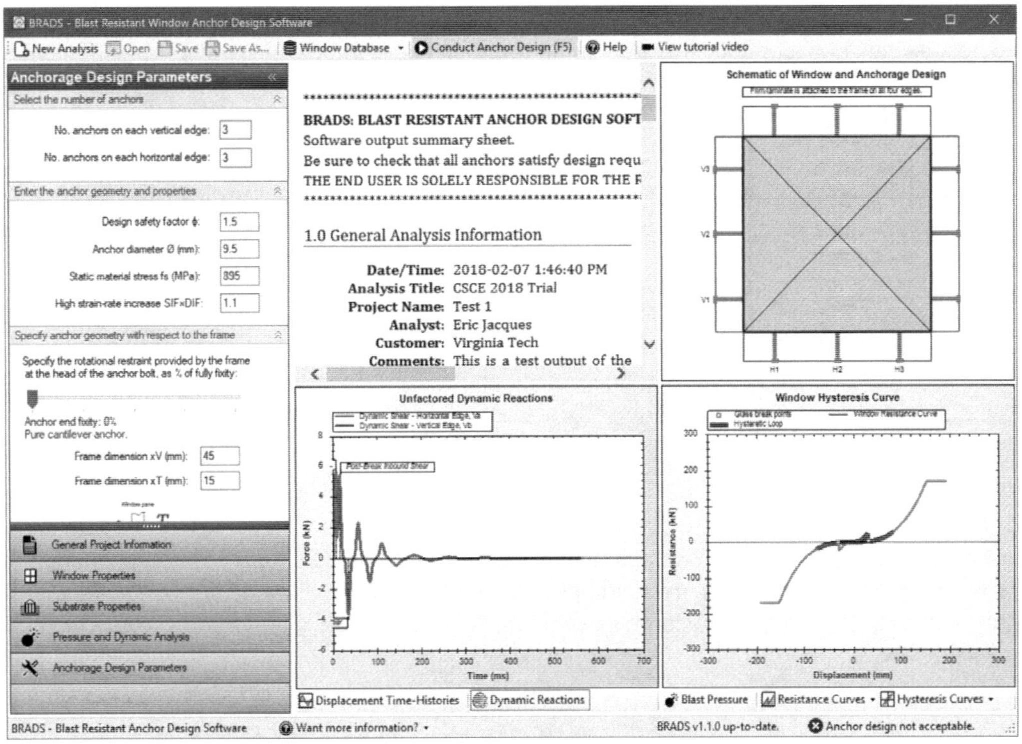

Figure 3: Screen capture of the BRADS analysis output screen.

3.1 Input Data

The graphical user interface (GUI) is structured to accept user input information under five main headings, each relevant to a specific aspect of window anchorage design. These headings are:

1. General Project Information to define and catalog project-specific information, such as the project title, analyst, customer name, and any comments related to the analysis.

2. Window Properties to select a window stored in the database to analyze. Windows are categorized according to their construction type, the configuration of the film/laminate edge attachment, and dimensions (Height × Width). Windows having their film or laminate properly attached to an edge will be designed to resist tensile forces due to membrane action during the post-break phase. The user may specify whether the membrane material is attached on all four sides, the vertical edges, the horizontal edges, or no attachment. The "no attachment" option would be used to analyze windows with daylight application of security film.

3. Substrate Properties to define the geometric and material properties of the mass-spring-damper element representing the substrate. The substrates may be modelled as either elastic or inelastic systems depending on anticipated behaviour. The user can select one of three pre-defined substrates for concrete, masonry, or steel structures. For concrete and masonry substrates, the lumped mass system is assumed to remain perfectly elastic and is constructed based on the equivalent SDOF model for one- and two-way members from UFC 03-340-02 (DoD, 2008). The steel substrate is assumed to be perfectly rigid, thus reducing the TDOF system to a SDOF system composed solely of the window. The software also provides the ability for the user to define a custom nonlinear mass-spring-damper for the substrate. The nonlinear substrate is represented by an inelastic resistance function with degrading stiffness characteristics defined by the yield displacement and yield resistance with elastic and inelastic load-mass transformation factors applied to a lumped substrate mass. The Clough degrading stiffness hysteretic model (Clough, 1966) is used to account for the effect of hysteresis on substrate response.

4. Pressure and Dynamic Analysis allows the user to define the design basis threat (DBT) used in anchorage design and control pertinent parameters of the dynamic analysis solver. The DBT can consist of either: (i) a triangular pressure-time history; (ii) an exponential decay pressure history calculated using the Kinergy-Bulmash polynomials (Kinergy and Bulmash, 1984); or, (iii) as a series of data points imported by the user. The user can also modify the default time-step (0.1 ms) and maximum analysis termination time (5000 ms). The TDOF analysis is programmed to terminate if: (i) the predicted window displacement exceeds the ultimate displacement defined in the window response function (indicating a blow-out failure of the window); (ii) a stable equilibrium of the window is achieved; or, (iii) the TDOF analysis run time exceeds the user-specified termination time.

5. Anchorage Design Parameters: The final input heading is used to establish the design parameters for selecting and proportioning fasteners to satisfy anchor design following CSA S852 (CSA, 2018).
   1. Number of anchors: The number of anchors on each of the vertical and horizontal window frame edges are specified by the user. The software uses this information to calculate anchor spacing.
   2. Anchor properties: The diameter Φ of a circular anchor is specified by the user to compute the cross-sectional area A and plastic section modulus Z used in anchorage design.
   3. Safety factor: CSA S850 (CSA, 2017) and CSA S852 (CSA, 2018) both require a balanced design approach where the window retention anchors are required to be stronger than the window by a factor of 1.5, known as the safety factor ϕ. The shear and tension forces acting on the anchors are multiplied by ϕ to obtain the factored design forces. The default value of ϕ (1.5) may be modified as necessary.
   4. Material properties: The static material stress f_y used in design should be determined from the manufacturer's specifications. The dynamic material stress f_{dy} defines the strength of the anchorage at high strain rates. It is taken as the product of $f_y \times SIF \times DIF$, where SIF is the strength increase factor accounting for additional strength present in the material that is typically neglected under normal loading conditions, and DIF accounts for the apparent increase in material strength under dynamic loads. A default value of 1.1 is used for the product $(SIF \times DIF)$ in the software.
   5. Frame geometry and anchor boundary conditions: The construction of the frame and the anchor boundary conditions are critical for design. The design shear F_V and tension F_T for each anchor are calculated from the tributary area under the edge distribution. Figure 4 (a) shows the assumed point of application of F_V and F_T at the location where the window bites into the glazing stop. This defines the eccentricities of shear force x_V and tension force x_T, respectively, relative to the critical section.

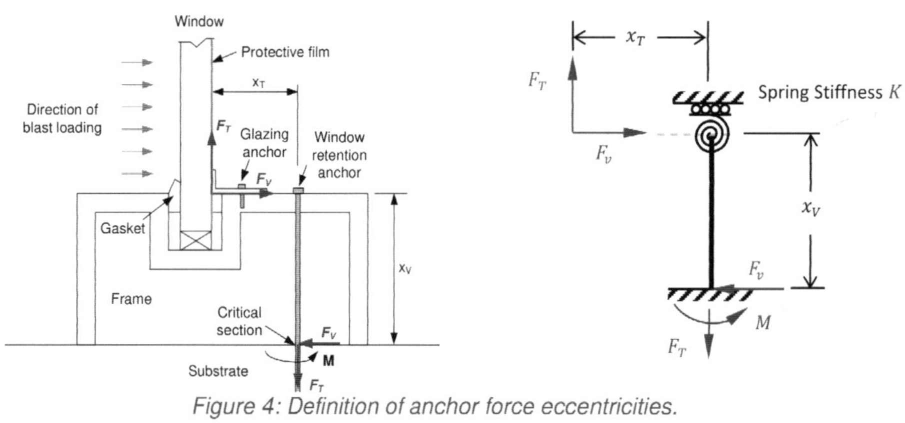

Figure 4: Definition of anchor force eccentricities.

6. The software assumes that the critical section for anchorage design is at the location of the anchor-substrate interface. Anchors are assumed to be fully fixed at the substrate, while the fixity of the anchor head depends on the construction of the frame. The anchor is assumed to behave as a

cantilever when the frame is not able to provide support along the length of the anchor. When the restraint provided by the frame is sufficient to prevent the anchor head from rotating, the anchor behaves as a cantilever with the rotation fixed and translation free at the anchor head. However, the true level of fixity is likely to fall somewhere between these two extremes. Therefore, the software computes the internal forces at the critical section assuming the case of a cantilever with a rotational spring located at the free end, as shown in Figure 4 (b). By varying the relative stiffness K of the spring, the tension F_T, shear F_V, and moment M at the critical section are found from first principles.

7. *Flexural and torsional frame rigidity*: It is well known that the torsional and flexural rigidity of the window frame influences the magnitude of the window deflection and distribution of edge reactions (Morison, 2007). More rigid steel frames tend to produce a more uniform distribution of anchorage forces, while more flexible extruded aluminum frames develop higher anchorage forces near the centre of the frame. The software accounts for the effects of frame rigidity through the empirical edge distributions illustrated in Figure 5. Rigid, semi-rigid, and flexible frames were observed to promote a uniform, trapezoidal, and sinusoidal distribution of edge forces, respectively. The design shear F_V and tension F_T for each anchor is then proportioned from the total shear and tension acting on a window edge based on its tributary area with respect to the distributions illustrated in Figure 5.

8. The potential for reverse corner reactions, known to develop in fully anchored windows subjected to blast loads (Morison, 2007; Department of Defense, 2008), are ignored by the software as their magnitude is negligible compared with the total force developed on an edge.

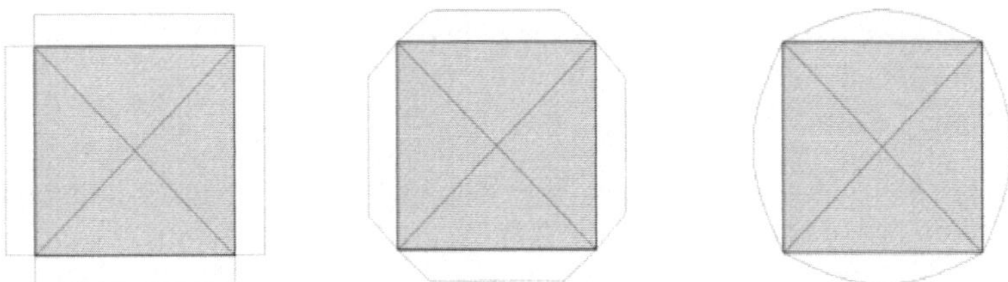

(a) Rigid frame (b) Semi-rigid frame (c) Flexible frame

Figure 5: Effect of frame rigidity on distribution of anchorage forces.

3.2 Analysis Output

The software automatically validates the program inputs once the user has selected a window to analyze, defined the properties of the substrate, established a DBT, and specified the anchorage parameters. If no errors are detected, the software then conducts the anchorage design beginning with the TDOF dynamic analysis to construct the dynamic window reaction time-history. The peak pre-break and post-break reactions are then identified from the dynamic reaction time-history for inbound and rebound response to account for the potential for load reversal during rebound response. The results of this stage of calculations are output to the screen as illustrated in the screen capture shown in Figure 3.

3.3 Design Output

After the software has conducted the TDOF analysis, the capacity of the anchors is verified for pre-break and post-break demands during the inbound and rebound phases. The dynamic reactions are distributed to the anchors in proportion to their tributary areas under the frame rigidity distribution (as shown in Figure 5). These forces are then factored and used to evaluate the anchorage for combined stresses indicated in Eq. [3]. The results of the design calculation are output as illustrated in Figure 6. If the required capacity is higher than the applied design forces, then a green anchor symbol is displayed at the specific location for which the design calculations are done. Otherwise a red anchor symbol is displayed. The user can click on an individual anchor to obtain more detailed information on the applied forces, stresses, and capacities. The user may also vary the anchor size, material, strength, number of anchors, and other design parameters to iteratively obtain an anchorage configuration which satisfies CSA S852

(CSA, 2018) design requirements. This approach provides the designer with complete control of the design process.

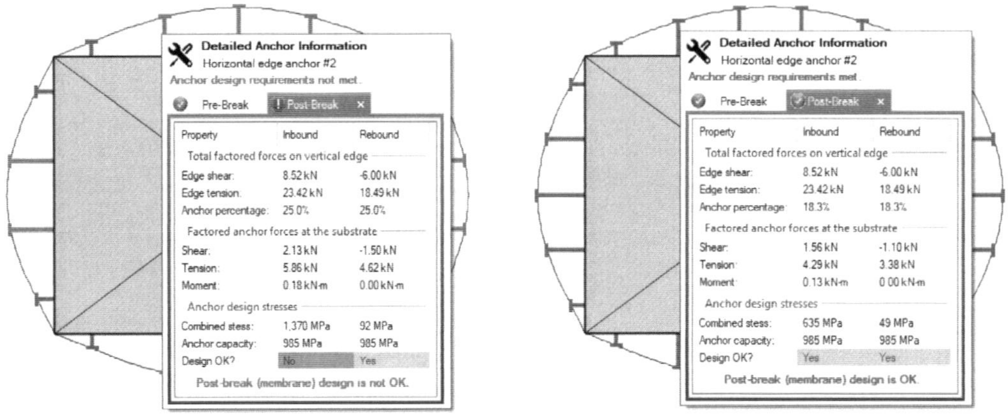

(a) Unsatisfactory anchor design (b) Satisfactory anchor design

Figure 6: Screen capture of the BRADS anchorage design popup window.

4 EXAMPLES OF ANCHORAGE DESIGN USING BRADS

4.1 University of Ottawa Shock Tube Tests

The University of Ottawa Shock Tube (Lloyd et al., 2011) was used to test blast-resistant window systems subjected to shock wave loading (Alameer, et al., 2017a, 2017b). The samples were composed of 25.4 mm thick IG windows with 6 mm annealed glass panes mounted into light aluminum frames. Various thicknesses of security film, anchored to the frame with structural silicone or mechanical attachment, were used to retrofit the windows for improved blast resistance. The punched window samples were installed into structural steel, reinforced concrete, block masonry, and stone masonry substrates using dia. 9.5 mm high-strength (Grade 8) window anchors. The spacing and arrangement of anchors varied depending on the substrate and the applied pressure intensity. Strain gauges were placed on the anchors such that the strain profile in the bolt could be decomposed into flexural and axial strain components, from which moments, shear forces, and tension forces could be computed. In the interest of brevity, the results of test C11, conducted on a reinforced concrete substrate, will be examined here. However, the reader is directed to Alameer, et al. (2017a, 2017b) for more information on the experiments and test results.

Test C11 was conducted on a 1676 mm × 560 mm window (aspect ratio of 3.0) retrofitted with 0.36 mm polyester film mechanically attached to the frame on all four sides. The window anchor configuration consisted of four anchors on each of the vertical edges and three along each horizontal edge, for a total of fourteen anchors. The eccentricities of the anchors, measured prior to testing, were found to be $x_V \approx 45\ mm$ and $x_T \approx 15\ mm$. The reinforced concrete substrate was 150 mm thick $(f_c' = 30\ MPa)$ with a clear span of 2232 mm and width of 2440 mm. The doubly reinforced cross-section had 12 – 10M rebars on each face distributed around the centrally-located window opening. The anchor bolts were fastened into drop-in anchors that were drilled into the concrete opening. Test C11 was subjected to an applied pressure-time history with $P_r = 27.5\ kPa$ and $I_r = 207\ kPa \cdot ms$. This pressure-impulse combination did not cause any damage to the glass, which remained uncracked, or to the framing system. Inspection of the anchor bolts indicated they remained elastic. Experimental anchorage forces F_V and F_T, shown in Table 1, were computed by decomposing strain gauge readings for three fasteners located in the middle and near the corner on the vertical edge, and located in the corner on the horizontal edge.

BRADS was used to predict the experimental anchorage forces shown in Table 1. The dynamic properties of the window were evaluated using other specialized software (ARA, 2005) and imported into BRADS. The forces were predicted using $\phi = 1.0$, semi-rigid frames, and assuming a level of fixity of 25%

at the free end on the anchorage. The Lumped Inelasticity Approach (Jacques et al., 2015; Jacques, 2014) was used to compute the dynamic properties of the substrate using a bi-linear resistance function, giving a yield deflection of 11.1 mm, yield resistance of 268 kN, effective loaded area of 4.5 m^2, mass of 1510 kg, and damping equal to 5% of critical. Table 1 shows that the experimental anchor forces on the vertical edge were predicted with a reasonable level of accuracy, while the software under-predicted F_V and F_T on the horizontal edge. One possible explanation for this discrepancy was a misalignment of the strain gauges on the anchor which might not have been properly oriented in the plane of bending. As a result, strain readings did not accurately record the influence of applied anchor forces. When the anchorage design procedure was re-run using a safety factor $\phi = 1.5$, a total of fourteen 12.7 mm diameter Grade 5 fasteners (five on the vertical edge, and two on the horizontal edge) would be required to satisfy the design requirements of CSA S852 (CSA, 2018).

Table 1: Comparison of anchorage forces on concrete substrates for test C11 (Alameer et al., 2017a).

Anchor Location	**Experiment**		**Predicted**	
	F_V **(kN)**	F_T **(kN)**	F_V **(kN)**	F_T **(kN)**
Middle anchor on vertical edge	3.7	4.5	2.7	5.8
Corner anchor on vertical edge	2.5	4.4	1.8	3.9
Corner anchor on horizontal edge	3.1	3.2	0.6	1.3

4.2 University of Kentucky Shock Tube Tests

Explosive-driven shock tube trials were conducted at the University of Kentucky on 1680 mm × 1220 mm laminated glass windows (Wedding, 2010). The objective of the study was to record the reaction forces and displacements of blast-resistant windows subjected to explosive loading. The windows were composed of a 1.5 mm Uvekol A interlayer sandwiched between two 3 mm thick heat strengthened glass panes and wet glazed to the surrounding extruded aluminum frame. These were anchored to a rigid structural steel substrate at a total of 26 locations around the perimeter, with 8 anchors on each of the vertical edges and 5 anchors on the horizontal edges. The magnitude and distribution of dynamic window reaction shear forces were measured using force sensors installed at the locations of the anchor points. Dynamic tension reaction forces, generated due to membrane action of the laminate, were not reported.

The majority of the tests conducted by Wedding (2010) were for identical windows subjected to an average positive pressure-impulse of $34.0\,kPa$ and $112.1\,kPa \cdot ms$, respectively, and negative pressure-impulse of $-6.1\,kPa$ and $-18.6\,kPa \cdot ms$, respectively. Table 2 summarizes the experimental edge shears, alongside corresponding edge shears predicted using BRADS. The dynamic properties of the window (resistance, mass, damping, etc.) were evaluated using other software (ARA, 2005) and imported into BRADS for analysis. The substrate was set as rigid and a trapezoidal distribution of edge forces was selected based on the reported distribution of forces. In general, the experimental and predicted design shear forces are comparable, although the predicted quantitates are markedly conservative. This is attributed to inaccuracies in constructing the window resistance curve based on the assumption that the Uvekol A interlayer is equivalent to PVB [an assumption made by Wedding (2010) and adopted in this study]. Nonetheless, the predicted shear is sufficiently conservative for the purposes of design.

Table 2: Comparison of experimental and predicted edge forces for tests conducted by Wedding (2010).

Quantity	**Experiment**	**Predicted**
Total peak dynamic shear (kN)	35.0	44.0
Long edge, average shear reaction (kN)	10.9	12.9
Short edge, average shear reaction (kN)	6.6	9.1

SUMMARY

A design procedure was developed for the analysis and design of anchoring systems used in blast-resistant punched windows. The procedure satisfies the requirements of CSA S852 (CSA, 2018) for the design of blast-resistant window anchor systems following a dynamic analysis methodology. The design procedure was implemented into a computer program to compute the required configuration of window retention anchors based on the major parameters known to influence anchorage capacity, including: substrate deformability, glazing response, window aspect ratio, frame rigidity, the distribution of dynamic reactions, combined stresses, and anchorage boundary conditions. The computed performance of anchors was verified extensively against data obtained from shock tube testing of blast-resistant punched windows installed into steel, concrete, block masonry, and stone masonry substrates. The comparison indicated that the design procedure can predict the magnitude and distribution of anchor forces of the tested windows with a level of accuracy suitable for the purposes of design and assessment. Although the design procedure does not explicitly assess the capacity of the fastener to resist disengagement from the substrate, the anchorage forces output by the tool may be used for this calculation.

Acknowledgements

This research was funded under Canadian Safety and Security Program project CSSP-2014-CP-2011.

References

Alameer, A., Elnabelsy, G., Jacques, E., Saatcioglu, M. and Foo, S. 2017a. Tests of blast-resistant windows. Research report, Hazard Mitigation and Disaster Management Research Centre, University of Ottawa, Ottawa, Canada.

Alameer, A., Elnabelsy, G., Jacques, E., Saatcioglu, M. and Foo, S. 2017b. Performance of window retention anchors during shock tube tests. International Journal of Protective Structures. Submitted for publication, 2017.

ARA. 2005. WINGARD PE 5.5 User Guide. Albuquerque, USA: Applied Research Associates, Inc.

Biggs, J.M. 1964. Introduction to structural dynamics, McGraw-Hill, New York.

Clough, R. W. 1966. Effect of stiffness degradation on earthquake ductility requirements. Structural and Materials Research No. 66-16, Structural Engineering Laboratory, Univ. of California, Berkeley, CA.

CSA 2018. "Blast Resistant Window Anchor Systems", S852-18, Canadian Standards Association, Mississauga, Ontario, Canada.

CSA. 2017. Design and Assessment of Buildings Subjected to Blast Loads", S850-12 (R2017) Canadian Standards Association, Mississauga, Ontario, Canada.

DoD. 2008. Structures to resist the effects of accidental explosions. Unified Facilities Code(UFC) 03-340-02,U.S. Department of Defense, Washington, DC.

Frykberg, E.R. 2002 "Medical Management of Disasters and Mass Casualties from Terrorist Bombings: How can we cope?" Journal of Trauma, Injury, Infection, and Critical Care, 53(2), pp. 201–212.

Jacques, E. 2014. RCBlast (version. 0.5.1) [Computer software], ⟨http://www.rcblast.ca/⟩.

Jacques, E. 2017. BRADS (Blast Resistant Anchor Design Software). Software Technical Manual and User's Guide, ⟨http://www.ericjacques.com/BRADS/⟩.

Jacques, E., and Saatcioglu, M. 2018. Computer software for blast-resistant window anchor design. Manuscript submitted to the ASCE Journal of Performance of Constructed Facilities, CFENG-2317.

Jacques, E., Lloyd, A., Imbeau, P., Palermo, D., Quek, J. 2015. GFRP Retrofitted Reinforced Concrete Columns Subjected to Simulated Blast Loading. ASCE Journal of Structural Engineering, 141:11.

Kingery, C.N., and Bulmash, G. 1984. Airblast Parameters From TNT Spherical Air Bursts and Hemispherical Surface Bursts, ARBRL-TR-02555, Army Ballistic Research Laboratory, Aberdeen Proving Ground, MD, USA.

Lloyd, A., Jacques, E., Saatcioglu, M., Palermo, D., Nistor, I., and Tikka, T. 2011. Capabilities and effectiveness of using a shock tube to simulate blast loading on structures and structural components. Behavior of Concrete Structures Subjected to Blast and Impact Loadings, ACI-SP 281, American Concrete Institute, Farmington Hills, MI.

Morison, C. 2007. The resistance of laminated glass to blast pressure loading and the coefficients for single degree of freedom analysis of laminated glass. PhD thesis, Defence College of Management and Technology, 549 pages.

US Army Corps of Engineers 2012. Methodology manual for the single-degree-of-freedom blast effects design spreadsheet for windows (SBEDS-W). Protective Design Center Report PDC TR-12-01.
Wedding, W.C. 2010. Experimental study of blast resistant glazing system response to explosive loading. Master's thesis, University of Kentucky, 114 pages.

Building Tomorrow's Society
Bâtir la Société de Demain

Fredericton, Canada
June 13 to 16, 2018 / *13 au 16 juin, 2018*

CONTACT EXPLOSION RESPONSE OF REINFORCED CONCRETE COLUMNS: EXPERIMENTAL AND VALIDATION OF NUMERICAL MODEL

Dua, Alok[1,3], Braimah, Abass[1] and Kumar, Manish[2]
[1] Carleton University, Canada
[2] Indian Institute of Technology Bombay (IITB), India
[3] alok.dua@carleton.ca

Abstract: Protective design of critical infrastructure and iconic buildings and monuments against terrorist attacks requires the incorporation of blast-resistant design to ensure adequate safety against explosive threats in today's environment. Consequently, the response analysis of structures subjected to blast loading has gained importance with structural engineers in the recent years. Past research efforts have mostly focussed on the far-field blast response of structural components with limited emphasis on the near-field and contact explosion response. Thus, there is a lack of knowledge on the response of structural components to the effects of near-field and contact explosion. The behavior of reinforced concrete components have been studied numerically however there is limited experimental data for validation of these numerical models. The experimental data is scarcest in case of concrete columns. Columns are the most vulnerable and critical member in any structure and are prone to contact explosion attacks using small quantities of explosives. This paper presents a benchmark study for the response of full scale reinforced concrete columns with pad footing when subjected to contact explosions of varying charge mass. The quantitative and qualitative assessments of damage to concrete columns, including spalling and scabbing are presented. The experimental data was used to validate a set of numerical models in LS-DYNA. These models were used in further parametric studies to establish the effects of various design parameters on the response of concrete columns to contact explosions. Preliminary results of the parametric studies show that the response of the concrete columns subject to contact explosion can be reliably predicted using the validated numerical models.

1 INTRODUCTION

Critical infrastructures such as major bridges, petrochemical refineries, thermal or nuclear power plants, stock exchanges, military or political headquarters and iconic monuments have been the targets for terrorist attacks in the past. The past attacks show that the terrorist threats to critical infrastructure are evolving and need an equally evolved and developed protective design capability. There is an ever more stringent control over energetic materials required for large scale bombings. The scenario of critical reinforced concrete (RC) columns being subjected to contact explosion of small charge weights is a credible threat perception. A strategically planned contact explosion on a critical RC column can severely compromise the axial load carrying capacity leading to partial or a complete collapse of the building. Currently, studies on response of RC members subjected to contact explosions are limited to numerical analyses and data from experimental work is scarce. Access to experimental facilities to conduct blast tests is restricted to a select few government organizations and research institutions. Additionally, the data acquisition systems for contact explosion events are not well established in the literature. It is difficult to acquire experimental data such as pressure and shock wave properties due to very high temperature and pressure in the vicinity of the fireball of contact explosion. The lack of quantitative experimental data for validation of numerical models impedes the research on response of RC members subjected to contact explosion. Response of RC slab subjected to contact explosion can be approximated as a 1-D wave propagation phenomenon influenced by limited parameters like charge mass to slab thickness ratio and compressive strength of concrete and yield strength reinforcement. On the other hand response of RC columns subjected to contact explosion can be approximated as a 2-D wave propagation phenomenon influenced by additional parameters like aspect ratio of the column cross-section, axial load level, transverse reinforcement spacing, infill boundary condition etc (Dua & Braimah, 2016). The response of RC columns to contact explosions is hence difficult to predict and requires further investigation.

2 LITERATURE REVIEW

The blast loading behavior on structural elements differs considerably for contact and far-field explosion events (Dua & Braimah, 2016). Contact explosions yield a spatially and temporally non-uniform loading causing predominantly local shear response, while far-field explosions yield spatially uniform loading on the structure causing global flexural response. The material response to contact explosion is highly non-linear and often leads to inelastic deformations in the material. As a result, the empirical methods applicable to far-field blast response analysis are not valid for contact explosion response.

Prior to NCHRP Report No. 645 (2009), no experimental data existed in the public domain documenting the response of RC concrete columns subjected to near-field and contact explosions. The report documented response of RC bridge columns of different cross-sections due to explosions at far-field and near-field scaled distances and also in contact. Numerical modelling guidelines for LS-DYNA were presented and code-ready blast load on RC bridge columns were predicted which correlated well with the experimental results (Williams, 2009).

Experimentally validated empirical relations for residual axial compression capacity of localized blast-damaged RC columns and steel composite RC columns were reported by Wu et al. (2011a) and Wu et al. (2011b). A term 'explosive mass ratio' was defined as the ratio of explosive charge mass to the mass of 1-m high column and used in parametric analysis to study the effect of contact explosions. It was reported that for an explosive mass ratio of 0.04, 60% residual capacity remained due to contact explosion at the bottom of the RC column as compared to 90% residual capacity due to contact explosion at a height of 1.5 m from the bottom of the RC column. Parametric analysis for effect of column height was presented and it was concluded that column height does not affect the residual capacity of blast-damaged columns. An element size of 50-mm was used to model the column as well as the Arbitrary Lagrangian-Eulerian (ALE) domains.

Close-in explosion yields flexural response associated with cover spalling as in most cases the shear capacity exceeds the shear demand from the loading (Cui et al., 2015). Cui et al. (2015) implemented a damage criteria defined as the ratio of the depth of column and relative residual deflection to perform parametric analysis. It was reported that a larger cross-section with higher transverse reinforcement ratio and thinner cover improved the blast response of the columns by reducing the spall depth. Recently Yuan et al. (2017) presented a study on response of RC bridge columns to contact explosion. Experimental tests were conducted on square and circular cross-section columns subjected to contact explosion of 1 kg of TNT. The columns were tested without axial loads. LS-DYNA numerical models captured the experimental behavior appropriately except for the back-face damage. The comparative damage profiles of experimental and numerical results were reported for qualitative validation and comparative accelerometer readings were reported for quantitative validation (Yuan et al., 2017).

In design practice, it is often assumed that the column failure is caused due to transverse loading rather than the axial load. All the studies reviewed in the literature ignored axial load effects citing reasons like high cost prohibitions for inclusion in experimental tests and negligible effect as compared to transverse load. Krauthammer et al. (2013) however reported that axial load can have significant effect on the overall response of RC columns subjected to blast loading. The authors conducted a single degree of freedom (SDOF) analysis and reported that both the quasi-static and impulsive regime response of RC columns subjected to blast loading are affected by the axial load. The flexural strength of the column improved with higher axial load up to the balanced eccentricity. Beyond the balanced eccentricity of the column, the flexural strength started to reduce when subjected to blast loads as the failure mode shifted from ductile tension failure to brittle compression failure. On the other hand, even an axial load of less than 50% of the balanced axial load significantly reduced the ability to withstand impulsive loading. Contrarily to this, Wu et al. (2011a) reported improvement in shear strength to resist localized damage at higher axial load.

Researchers in the past have used multi-material ALE (MMALE) numerical formulation, which is an appropriate method for blast wave generation however its main disadvantage is its mesh dependency (Hanssen et al., 2004; Børvik et al., 2009; Kakogiannis et al., 2010). The accuracy of numerical results is highly sensitive to the mesh properties and can lead to large differences in the results. Zukas and Scheffler (2000) reported that numerical results of a same problem as analysed by different LS-DYNA users can vary by up to 80% due to different mesh properties (Trajkovski et al., 2014). In particular, the mesh dependency is directly related to the computational requirement for simulating a near-field or contact explosion. Hence, mesh sensitivity analysis is important for such studies in order to produce accurate results from a computationally cost effective model.

The literature review presented in the preceding section suggests scarcity of experimental data pertaining to contact explosion on RC columns. Experimental tests reviewed above were conducted on reduced scale columns without realistic boundary conditions. The NCHRP Report No. 645 (2009) presented experimental tests on full scale columns however the specific details of charge mass, cross-sectional details and reinforcement percentage were not made available in the public domain. Numerical studies have concluded that contact explosion induces local response but the effect of boundary conditions at the column's top and bottom for a given location of explosive are yet to be experimentally studied. No experimental data could be found on the effect of service axial loads on response of RC columns subjected to contact explosion placed at the bottom of the column which is a credible threat scenario. The numerical models in the literature reviewed above implemented the detonation process and fluid structure interaction (FSI) with element size of 5 – 50 mm. Thus the empirical expression for residual capacity of columns obtained using these numerical models might not be accurate due to the coarse mesh sizes.

This paper presents a benchmark study for the response of full scale reinforced concrete columns subjected to contact explosions of varying charge mass. The quantitative and qualitative assessments of damage to concrete columns, including spalling and scabbing are presented. The experimental data was used to validate a set of numerical models in LS-DYNA. These models were used in further parametric studies to establish the effects of various design parameters on the response of concrete columns to contact explosions. Preliminary results of the parametric studies show that the response of the concrete columns subject to contact explosion can be reliably predicted using the validated numerical models.

3 EXPERIMENTAL TESTS

3.1 Experimental Setup

Three RC columns were studied for the experimental phase. The material properties, cross-sectional details and reinforcement percentage were same for the three columns (Figure 1(a)) and each column was subjected to contact explosion of different charge mass. Boundary conditions and gravity loads were not provided for the column top for simplification purpose. The effect of axial load and boundary condition at the top of the concrete column will be investigated as part of a future research program. The compressive strength of the concrete was attained through compressive testing of 150-mm cubes. The average compressive strength and density of three 75-mm diameter cores extracted from the test column were determined as 25 MPa and 2635 kg/m^3 respectively. Mat_CSCM concrete constitutive model in LS-DYNA was implemented with 25 MPa unconfined compressive strength. The concrete was poured in two lifts and no splicing was provided in the longitudinal reinforcements. The columns were cured by covering them with wet burlap blankets for 14 days and the testing was carried out after 28 days of casting. The backfill in the foundation pit was allowed to consolidate naturally during this period after initial manual compaction.

The pre-blast experimental setup for 1000-g and 115-g TNT charge are presented as Figure 1(b) and (c). The explosive was placed on the ground in contact with the column face emulating an explosive device placed in a possible terrorist threat scenario.

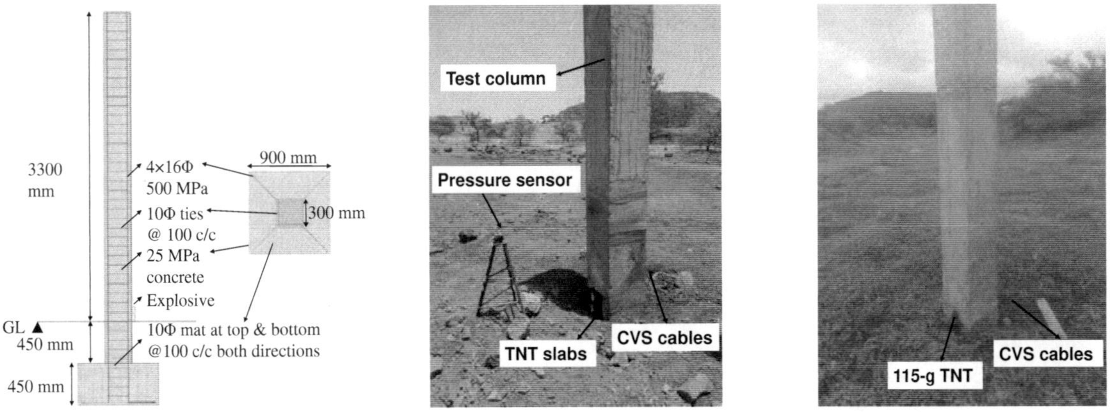

(a) (b) (c)

Figure 1: Experimental setup (a) Column detailing; Setup for (b) 1000-g TNT (c) 115-g TNT equivalent

Standard demolition explosives were used for the tests and the quality controlled chemical and material properties reported by the factory are presented in Table 1. In case of the 115-g explosive charge mass, plastic explosive (PEK) was used because smaller TNT slabs were not available.

Table 1: Chemical Properties for TNT and PEK

Property	Detonation Velocity, m/s	Density, kg/m^3	Mass of one unit, kg	TNT equivalent	Means of Initiation
Plastic Explosive (PEK1)	8750	1549	0.1	1.15	Lead Azide electric detonator
TNT Slab	6930	1550	0.5	1	Lead Azide electric detonator

3.2 Results and Discussion

The damage profiles of RC columns (300mmx300mm) due to contact explosion of different charge mass are presented in Figure 2. For the column subjected to 1000-g of TNT explosive, the concrete core was completely damaged with zero residual axial strength as evident from Figure 2(b). The 500-g TNT explosive also resulted in complete damage to the cover and the concrete core with zero residual capacity (Figure 2(c)). On the other hand, the 115-g TNT equivalent charge caused full-width front face spalling in the concrete cover zone with a height of 180-mm at the center and 400 mm at the edges(Figure 2(d)).

From the experimental observations, the breaching charge for the presented RC column is between the range of 115 g - 500 g. Moreover, the experimentally determined range is consistent with the explosive factor reported by Fujikake and Aemlaor (2013). The columns failed due to local shear and the concrete core was completely crushed during the contact explosion of the 1000-g TNT and 500-g TNT explosive charge masses. The phenomenon leading to this failure mode is discussed later in this paper. The impulse generated due to a contact explosion is transferred to the RC column inducing a compressive shock wave that propagates outwards from the contact point (Puryear et al., 2012). The shock wave attenuates as it propagates through the concrete core towards the side and back faces compressing the material along its path. In case the compressive force generated by the shock wave exceeds the compressive strength of the concrete, it leads to crushing of the material. On arriving at the side and back faces the compressive wave is reflected or transmitted depending on impedance values of the concrete and air surrounding it. The reflected wave is a tensile wave that propagates back into the core of the concrete element. The scabbing on the side and back faces is caused when the tensile strength of concrete is less than the tensile stress caused due to the reflected tensile shock wave. The reinforcement ties confines the concrete core of the column providing enhanced compressive and tensile strengths. This damage mechanism resulted in spalling of the side face corners whereas due to the attenuation and confinement effects, there was no back face scabbing observed in the RC column subjected to 115-g TNT equivalent explosive. The amount of concrete spalling and scabbing has a direct bearing on the residual axial load carrying capacity of the RC columns.

(a) (b)

(c) (d)

Figure 2: Experimental results (a) Fireball after detonation; Post blast damage profiles for (b) 1000-g TNT (c) 500-g TNT and (d) 115-g TNT equivalent

4 NUMERICAL SIMULATIONS

4.1 Mesh Sensitivity Analysis

A mesh sensitivity analysis was performed prior to the numerical simulation of the experimental tests. The mesh sensitivity simulations were compared with experimental results presented by Rigby et al. (2015).

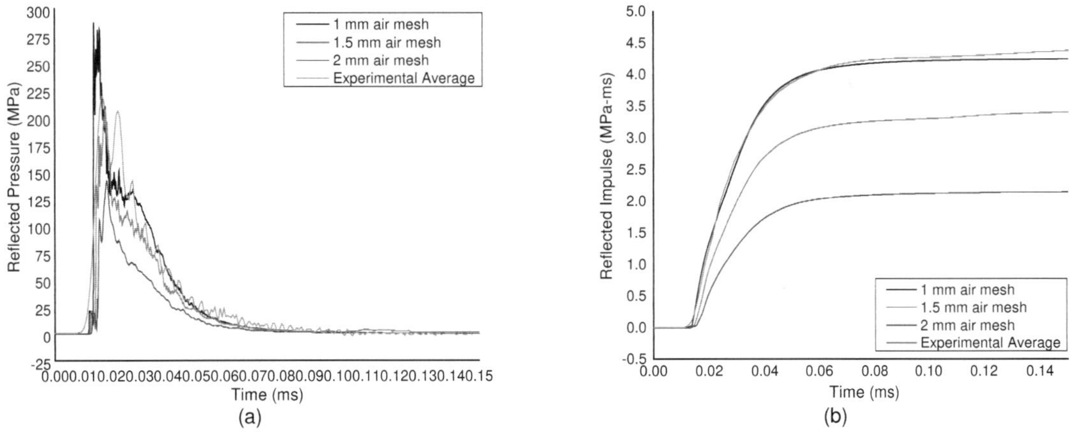

Figure 3: Mesh sensitivity analysis (a) Reflected pressure time-histories for varying element size (b) Reflected impulse time-histories for varying element size

The normal and radial reflected blast parameters acquired via split Hopkinson's pressure bars (SHPB) were reported by Rigby et al. (2015) at a close-in scaled distance of 0.15 m/kg$^{1/3}$. A quarter symmetric model of the experimental setup was implemented with varying element sizes for the air domain. The reflected pressure and impulse histories at 0.15 m/kg$^{1/3}$ scaled distance were plotted with the average of six tests reported by Rigby et al. (2015) and are presented as Figure 3(a) and 3(b), respectively.

In near-field or contact explosions, the duration of blast loading is shorter than the time to maximum response of the member. In such cases the structural members are analyzed for impulse loading and not the blast pressure loading, (Dua & Braimah, 2016). As presented in Table 3, the peak impulse is highly sensitive to the air domain element size. 1-mm element size yielded least error for peak reflected impulse and the error increased to 51% with an element size of 2-mm. This is in agreement with the results reported by Trajkovski et al. (2014) and Cormie et al. (2014). It can be concluded from the forgoing results that 1-mm element size is appropriate to generate accurate peak impulse value which is critical for modelling near-field or contact explosion events. The modeling complexities and computational challenges associated with simulating response of concrete columns subject to contact explosion are discussed in the following section.

Table 2: Mesh Sensitivity Analysis for Reflected Pressure and Impulse

	Experimental average	1 mm air mesh	Error %	1.5 mm air mesh	Error %	2 mm air mesh	Error %
Reflected peak pressure (MPa)	220	287.4	30	204.2	7.2	140.4	36.2
Reflected peak impulse (MPa.ms)	4.45	4.25	4.5	3.41	23.4	2.15	51.7

4.2 Modelling Technique

There are three prevalent techniques used for blast analysis of structures: (a) Simplified analytical methods such as Pressure-Impulse (PI) charts, single degree of freedom (SDOF) or multi degree of freedom (MDOF) dynamic analysis using blast load parameters generated from empirical charts provided by UFC-3-340-02 (2008), Kingery and Bulmash (1984), TM5-855-1 (1987) or other such codes, (b) non-linear finite element analysis using blast load obtained from underlying empirical equations and charts by Kingery and Bulmash (1984), UFC-3-340-02 (2008), TM5-855-1 (1987) and (c) high fidelity finite element analysis programs which use explicit modelling of the detonation process, blast wave propagation and FSI. While the first two methods are appropriate to predict far-field blast response, MMALE formulation is better suited to model contact explosion as it provides a complete description of the blast wave parameters (Børvik et al., 2009). The technique involves modelling the chemical reaction during detonation and consequent blast wave propagation. However, the primary disadvantage of modelling detonation and propagation is mesh sensitivity (Kakogiannis et al., 2010) and hence a higher computational cost due to a fine mesh requirement. Kalra et al. (2014) presented a 2D to 3D MMALE mapping technique in LS-DYNA for reducing the computational cost. The explosive detonation and blast wave propagation was simulated in 2D domain and then mapped to a 3D domain to simulate the FSI between the generated blast wave and the Lagrangian entity. This technique however cannot be implemented for modelling contact explosions as the explosive detonation and FSI occurs simultaneously. Hence, a 3D domain with a fine mesh is required from the beginning ($t = 0$) of the contact explosion simulation. Lately, massive parallel processing (MPP) abilities have been introduced by LS-DYNA that allows parallel processing and availability of large memory resources. Albeit these resources are difficult to access and demands additional skills for setup.

In this paper chained mapping technique were implemented on a single system to accurately predict the impulse generated due to the contact explosion and couple the load with the Lagrangian entity. It involved initially modelling the detonation in a 3D fine mesh while FSI was implemented during the detonation process. The chained simulations were started with a smaller ALE domain of 1-mm element size to ensure accurate prediction of the pressure and impulse values as deduced from the mesh sensitivity analysis. The first simulation was terminated just before the blast wave arrived at the boundaries of the air domain to avoid reflection. Although LS-DYNA provides a keyword BOUNDARY_NON_REFLECTING, this keyword is applicable to Lagrangian entities and not to the ALE domains. Hence each simulation has to be terminated prior to the reflection. A 3D-3D ALE mapping was implemented in the next simulation wherein the underlying algorithm initializes the current ALE domain from the state at the end of the previous simulation. The Lagrangian column and the soil were initialized with the stress at the end of the previous run. These steps were performed with a full deck restart that enables addition or deletion of Lagrangian entities. The ALE domain was increased for subsequent simulations until the generated blast wave cleared away from the column. The material parameters for concrete, reinforcing bars, air and TNT entities were implemented from the data presented by Dua and Braimah (2017); Dua et al. (2017).

4.3 Numerical Results

The multistage simulations presented in this section represent the simulation for 500-g TNT explosive. The final response due to 115-g TNT equivalent explosion are presented subsequently.

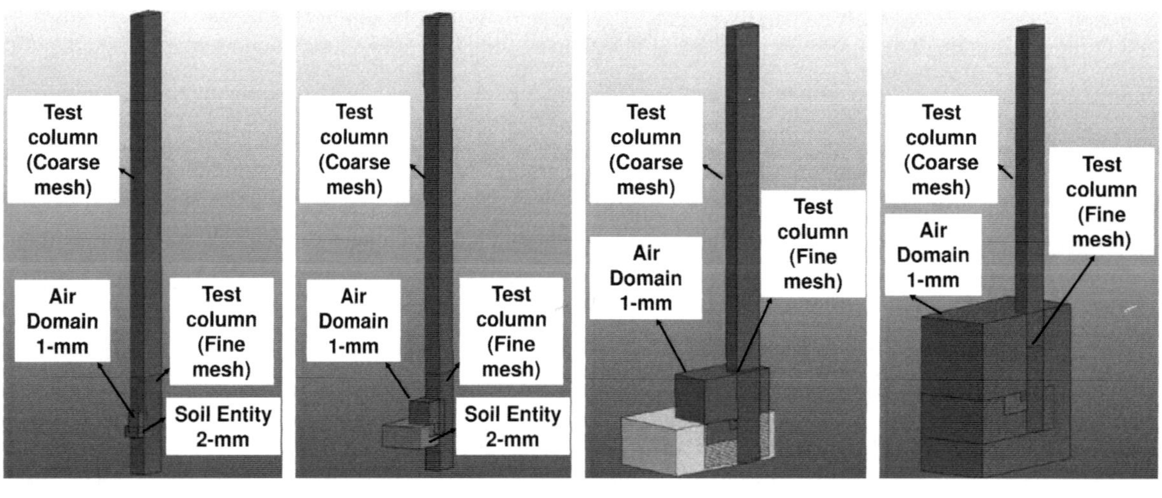

Figure 4: Geometrical models of the chained simulations (a) 1-mm (b) 2-mm (c) 5-mm (d) 10-mm

The first chained simulation was initialized with 1-mm element size air domain and 2-mm element size for the Lagrangian entities of concrete column and soil. The FSI between the detonation products and Lagrangian entities was implemented by the constrained Lagrange in solid (CLIS) keyword with coupling type (CTYPE) equals to 5. L.S.T.C (2015) recommends a smaller element size for the Lagrangian entities being coupled in a FSI problem as compared to the element size of the ALE domain. Alternatively a larger NQUAD (number of coupling points distributed per Lagrangian surface segment) value is recommended that implies longer computational time. Consequently the value for NQUAD parameter was adopted as 2 in the first simulation and subsequently reduced to 1 for subsequent simulations wherein the element size of air domain was larger than the Lagrangian entities (Figure 4).

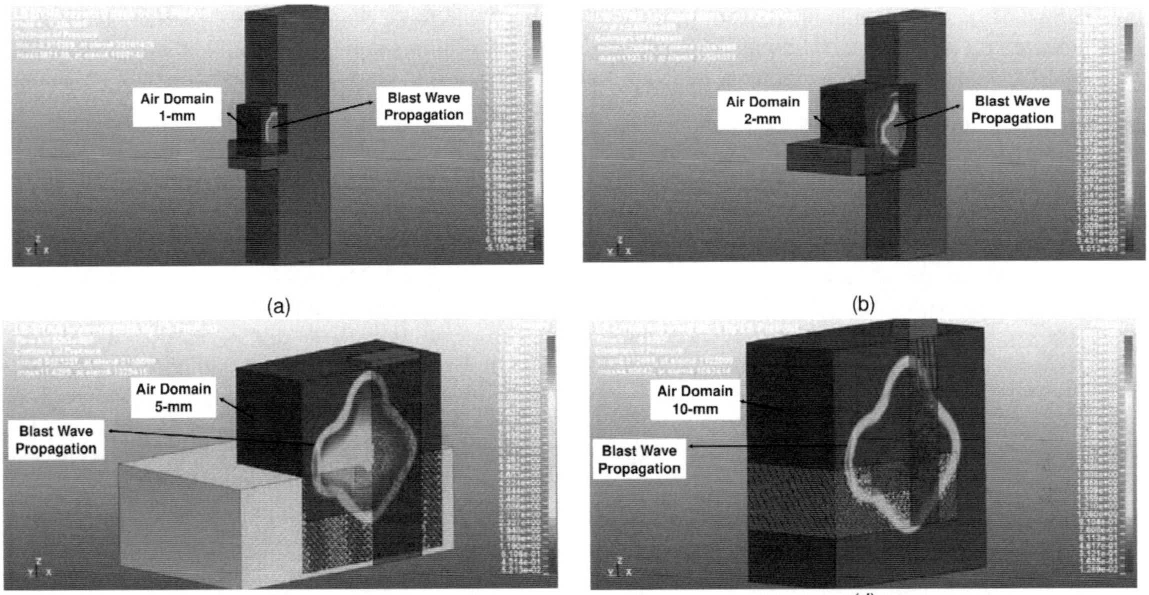

Figure 5: Multistage pressure fringe plots (a) last cycle for 1-mm element size (b) last cycle for 2-mm element size (c) last cycle for 5-mm element size (d) last cycle for 10-mm element size

The first simulation was terminated at 0.0125 ms and the pressure fringe plots at the last cycle of each phase is presented in Figure 5. The range of pressure fringe plot is customized to highlight the extent of blast wave. The pressure during the last cycle of this simulation was mapped to a larger ALE domain with 2-mm element size during the second simulation. The stress time histories of the Lagrangian entities were initialized from the previous run and the simulation was run to 0.0275 ms.

The chained simulations were implemented four times during which the blast wave cleared the column and the blast pressure near the point of detonation reduced to zero. The concrete column takes longer to respond than the impulse uptake from the contact explosion and hence the last simulation was continued up to 30-ms without the ALE domain to capture the response of the column. The concrete part was checked to have near-zero velocity at the end of the simulation to ensure that peak response is captured.

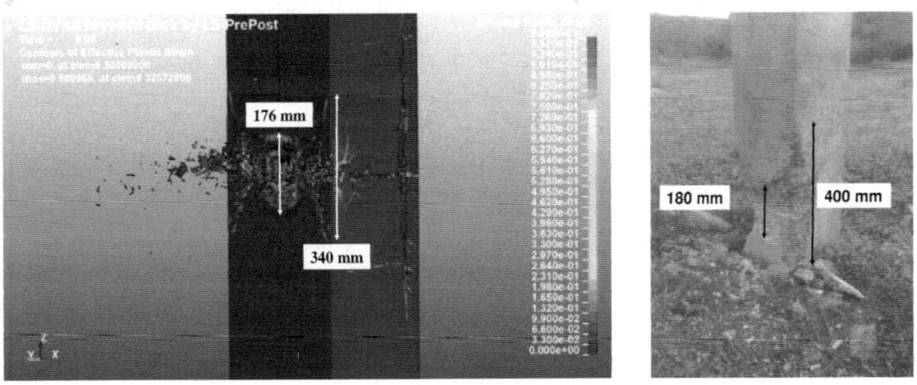

Figure 6: Comparison of Numerical and Experimental results-Front and side face damage profiles

The damage profile of the test column subjected to 115-g TNT equivalent explosive charge mass at the last cycle of the multistage simulation sequence is presented in Figure 6(a). Material erosion criteria was adopted for the Lagrangian entities to address numerical instabilities due to severe distortion of the elements during high strain-rate response in the non-linear regime. The MAT_159 concrete constitutive model allows erosion at accumulated damage of 0.99 which represents elements with no residual strength hence the erosion is linked to material failure. The damage profile presented in Figure 6 was determined from the erosion extent which represents the failed elements in the Lagrangian entities.

5 CONCLUSIONS

Three RC columns of same cross-section and reinforcements percentage were subjected to 1000-g, 500-g and 115-g of TNT equivalent explosion in contact at ground level. The boundary condition are the cottom of the column were provided with a pad footing. No restraint was provided at the top of the column for simplification of the experimental setup and the same is being considered for future experimental programs. The charge mass that could render the RC column with zero axial load carrying was determined to be in the range of 115-500 g which is in agreement with the range reported in the literature. An appropriate mesh size for the numerical model was selected using mesh sensitivity analysis performed at a scaled distance of 0.15 m/kg$^{1/3}$ and validated with experimental results. An element size of 1-mm for the ALE domain predicted the reflected impulse with least error. Furthermore, a multistage simulation technique was presented in the paper to reduce computation demand due to fine mesh requirement for the ALE domain to accurately predict the generated impulse. The column damage predicted with the multistage simulation were qualitatively compared to the experimental results. The damage profile of RC column subjected to 115-g TNT equivalent explosion in contact correlated well with the experimental observations establishing the qualitative efficacy of the numerical technique. The numerical parameters and multistage technique reported in the paper can be reliably implemented for further parametric studies. For further studies, the residual axial load carrying capacities of blast damaged columns will be numerically investigated and used in developing an empirical equation incorporating the concrete compressive strength, reinforcement percentage and the charge mass.

References

Børvik, T., Hanssen, A., Langseth, M., & Olovsson, L. 2009. Response of Structures to Planar Blast Loads - A Finite Element Engineering Approach. *Computers & Structures,* **87**: 507-520.

Cormie, D., Shin, J., Whittaker, A. S., & Wilkinson, W. 2014. Numerical Modelling of Close-in Detonations of High Explosives. *Engineering Structures,* **81**: 88-97.

Cui, J., Shi, Y., Li, Z.-X., & Chen, L. 2015. Failure analysis and damage assessment of RC-columns under close-in explosions. *Journal of Performance of Constructed Facilities,* **29**(5): B4015003-4015001:4015011.

Dua, A., & Braimah, A. 2016. *State-of-the-Art in Near-Field and Contact Explosion Effects on Reinforced Concrete Columns.* 5th International Structural Specialty Conference: Canadian Society for Civil Engineering London, Canada. pp. 1:12 (STR-836).

Dua, A., & Braimah, A. 2017. *Efficacy of Concrete Constitutive Models for Bullet Impact Tests.* 6th International Engineering Mechanics and Materials Conference: Canadian Society of Civil Engineers Vancouver, Canada. pp. 521 521-510.

Dua, A., Braimah, A., & Matsagar, V. 2017. *Understanding the Response of Reinforced Concrete Slabs due to Contact explosion of TNT.* 6th International Engineering Mechanics and Materials Conference: Canadian Society for Civil Engineering (CSCE) Vancouver, Canada. pp. 522 521-510.

Fujikake, K., & Aemlaor, P. 2013. Damage of Reinforced Concrete Columns Under Demolition Blasting. *Engineering Structures,* **55**: 116-125.

Hanssen, A., Enstock, L. K., & Langseth, M. 2004. Close-Range Blast Loading of Aluminum Foam Panels. *International Journal of Impact Engineering,* **27**: 593-618.

Kakogiannis, D., Hemelrijck, D. V., Wastiels, J., Palanivelu, S., Paepegem, W. V., Vantomme, J., . . . Kostopoulos, V. 2010. Assessment of Pressure Waves Generated by Explosive Loading. *CMES-Computer Modelling in Engineering & Sciences,* **65**(1): 75-93.

Kalra, A., Zhu, F., Yang, K. H., & King, A. I. 2014. *Key Parameters in Blast Modeling Using 2D to 3D ALE Mapping Technique.* 13th International LS-DYNA Users Conference: LSTC Detroit, USA.

Kingery, C. N., & Bulmash, G. 1984. *Air blast parameters from TNT spherical air burst and hemishperical surface burst.* US Army Armament Research and Development Center (BRL Technical Report, ARBRL-TR-02555), Aberdeen, MD.

Krauthammer, T., Astarlioglu, S., Morency, D., & Tran, T. P. 2013. Behavior of Reinforced Concrete Columns Under Combined Effects of Axial and Blast-induced Transverse Loads. *Engineering Structures,* **55**: 26-34.

L.S.T.C. 2015. LS-DYNA Keyword User's Manual, Volume I. Livermore, California, USA: LSTC.

NCHRP Report No. 645. 2009. *Blast-Resistant Highway Bridges: Design and Detailing Guidlines.* Transportation Research Board Washington, USA.

Puryear, J. M. H., Stevens, D. J., Marchand, K. A., Williamson, E. B., Sammarco, E. L., & Crane, C. K. 2012. Predicting Spall and Breach for Blast-Loaded Reinforced-Concrete Columns. *Texas ASCE Centennial Conference*: ASCE Dallas, USA

Rigby, S. E., Tyas, A., Clarke, S. D., Fay, S. D., Reay, J. J., Warren, J. A., . . . Elgy, I. 2015. Observations From Preliminary Experiments on Spatial and Temporal Pressure Measurements from Near-Field Free Air Explosions. *International Journal of protective structures,* **6**(2): 175-190.

TM5-855-1. 1987. *Fundamentals of Protective Design for Conventional Weapons.* Washington DC, USA.

Trajkovski, J., Kunc, R., Perenda, J., & Prebil, I. 2014. Minimum mesh design criteria for blast wave development and structural response - MMALE method. *Latin American Journal of Solids and Structures,* **11**: 1999-2017.

UFC-3-340-02. 2008. *Design of structures to resist the effects of accidental explosions.* United States Department of Defense Washington, D.C.

Williams, G. D. 2009. *Analysis and Response Mechanisms of Blast-Loaded Reinforced Concrete Columns.* (Doctor of Philosophy), University of Texas, Austin, USA.

Wu, K.-C., Li, B., & Tsai, K.-C. 2011a. Residual Axial Compression Capacity of Localized Blast-Damaged RC Columns. *International Journal of Impact Engineering,* **38**: 29-40.

Wu, K.-C., Li, B., & Tsai, K.-C. 2011b. The Effects of Explosive Mass Ratio on Residual Compressive Capacity of Contact Blast Damaged Composite Columns. *Journal of Constructional Steel Research,* **67**: 602-612.

Yuan, S., Hao, H., Zong, Z., & Li, J. 2017. A Study of RC Bridge Columns Under Contact Explosion. *International Journal of protective structures,* **109**: 378-390.

Zabihi, N., & Eren, O. 2014. Compressive Strength Conversion Factors of Concrete as Affected by Specimen Shape and Size. *Research Journal of Applied Sciences, Engineering and Technology,* **7**(20): 4251-4257.

Zukas, J. A., & Scheffler, D. R. 2000. Practical Aspects of Numerical Simulations of Dynamic Events: Effects of Meshing. *International Journal of Impact Engineering,* **24**(9): 925-945.

Building Tomorrow's Society
Bâtir la Société de Demain

Fredericton, Canada
June 13 to 16, 2018 / *13 au 16 juin, 2018*

DEBRIS DAMMING LOADS AND EFFECT IN TSUNAMI-LIKE EVENTS

Stolle, Jacob[1,5], Nistor, Ioan[1], Takabatake, Tomoyuki[2], Goseberg, Nils[3], Petriu, Emil[4], Shibayama, Tomoya[2]
[1] Department of Civil Engineering, University of Ottawa, Canada
[2] Department of Civil and Environmental Engineering, Waseda University, Japan
[3] Leichtweiß-Institute for Hydraulic Engineering and Water Resources, Technische Universität Braunschweig, Germany
[4] School of Electrical Engineering and Computer Science, University of Ottawa, Canada
[5] jstol065@uottawa.ca

Abstract: Debris loading during extreme flooding events has been documented by many post-tsunami field surveys of disaster-stricken communities and, as such, it is now considered and accounted for as a critical design consideration in building resilient infrastructure. Debris damming is one of the debris loads of concern, occurring when solid objects entrained within the inundating flow accumulate at the face of a structure or a structural element. The presence of the debris dam results in increased drag loads on the structure and can have other associated effects, such as flow runup and flow accelerations, that can cause increased scouring at foundations. The focus of debris damming studies has been within river engineering: therefore, previous studies have been performed in steady-state conditions. The study presented here is the first to examine debris damming in transient, supercritical flow conditions. The study uses a modified dam-break wave as the hydrodynamic forcing condition and the debris are scaled down debris types common in coastal areas (shipping containers, hydro poles, and boards). The study qualitatively examines the difference in the debris damming mechanisms as a result of distinct flow conditions associated with a dam-break wave interacting with a surface-piercing obstacle. Additionally, the study aimed to determine the influence of the debris dam. It was found that maximum loading condition occurred earlier and of greater magnitude than for the clear water case.

1 INTRODUCTION

The recent devastation generated by major flooding events, such as the 2004 Indian Ocean Tsunami, captured the attention of regulatory agencies and research bodies worldwide due to the failure of coastal structures and critical infrastructure that had been designed to withstand these events (Yeh et al., 2013). One of the primary explanations for these failures was that building standards and design codes did not properly address the variety of extreme loads associated with a tsunami event (Esteban et al., 2015; Yeh et al., 2014). To address this concern in North America, particularly along the West Coast, where the region is at-risk of earthquakes along the Cascadia subduction zone (CREW, 2013), the American Society of Civil Engineering (ASCE) has developed a chapter (ASCE7 Chapter 6) around the design of critical infrastructure exposed to tsunami events (Chock, 2016).

Debris loading was identified within the standard as one of the critical loads. On the one hand, debris impact is of major concern for short-period high-strength forces on structures (Nistor et al. 2017). One the other hand, debris damming occurs when waterborne debris accumulates at the face of a structure or in between adjacent structures. This accumulation can obstruct openings in buildings where water previously could flow freely. The obstruction results in changing hydrostatic and hydrodynamic loads exerted on the structures.

Within standards worldwide, debris damming loads are generally accounted for as hydrodynamic loads acting on a structure (FEMA, 2012; CSA, 2006; AASHTO, 2012), using the drag equation (Bremm et al., 2015):

[1] $F_D = \frac{1}{2} C_D \rho b h u^2$

where C_D is the drag coefficient, ρ is the fluid density, b is the breadth of the debris dam transverse to the flow direction, h is the flow depth, and u is the flow velocity. For bridge piers, C_D is generally taken as 1.4, corresponding to a square-ended pier. In the FEMA P646 Guidelines for the Design for Vertical Evacuation from Tsunamis (FEMA, 2012), to conservatively estimate the hydrodynamic loads, a C_D value of 2.0 for tsunami events was recommended.

The recommended breadth of the debris dam has a variety of definitions. The Canadian Bridge Design Guidelines (CSA, 2006) suggests a breadth equal to that of the bridge pier. The FEMA P646 (2012) recommends a minimum breadth equal to the largest length of the available debris at the site. The ASCE7 Chapter 6 (Chock, 2016) considers any structure, regardless of mitigating breakaway walls, should consider a minimum blockage of 50% for an at-risk exposed section. Field surveys have noted the site-specific nature of debris dam formation (Parola, 2000) and therefore, where possible, site evaluations are recommended across the various standards.

While debris damming has been observed in many extreme flooding events, such as the 2005 Hurricane Katrina, 2006 Switzerland Flood, and the 2010 Chilean Tsunami (Robertson et al., 2007; Schmocker et al., 2011; Takahashi et al., 2010), the study of debris damming has primarily been performed within the context of river engineering. Bocchiola et al., (2006) examined the spatial distribution of debris damming through randomly placed obstacles. The study found that the amount of debris captured increased with the length of the debris and decreased with the flow velocity. Increased length of the debris resulted in the debris being more often captured by a "bridging" mechanism, where the debris would be caught on multiple obstacles. The "bridging" was more stable than the alternative "leaning" mechanism, where the debris would be captured on a single obstacle. Therefore "bridging" was considered more likely to form a debris dam. Similarly, the higher flow velocities resulted in less stable capture of the debris resulting in the accumulation being washed away prior to a stable dam being formed. In this work, a "debris dam" will refer to a stable formation of debris accumulation and "debris accumulation" will refer to the transient features of debris advected within a flow.

A thorough review of debris damming literature leads to the conclusion that the examination of debris damming has been predominantly performed in *steady-state subcritical* flow conditions. However, many flooding events, particularly tsunamis and flash-floods, have *transient* flow properties (Ghobarah et al., 2006; Ioualalen et al., 2007; Saatcioglu et al., 2005) and can enter into trans-critical/supercritical ($Fr > 1$) flow regimes (Fritz et al., 2006; Matsutomi et al., 2010; Titov et al., 1997). Previous research into the hydraulic loads and flow-structure interaction associated with unsteady tsunami-like waves has shown distinctive wave profiles of transient nature that significantly differ from steady-state conditions (Arnason et al., 2009; Goseberg et al., 2014; St-Germain et al., 2013). Therefore, this study examines the formation and loading conditions associated with a debris dam in unsteady supercritical flow conditions with the objectives of:

- Studying the formation of debris dams in unsteady flow conditions.
- Contrast the transient dam formation to the formation mechanisms described by Schmocker et al., (2013).
- Examining the influence of debris shape on the formation and stability of the debris dam.
- Evaluating the loading conditions on the obstacle as a result of the debris dam formation.

This paper is separated into the following sections: "Methodology" details the facilities and instrumentation used in this study as well as the experimental protocol; "Results" details the key results of the study emphasizing the above stated objectives of the paper; "Discussion" describes key findings in the context of the wider research and discusses potential scale effects; "Conclusions" outlines the results of this study and identifies key conclusions related to the objectives.

2 METHODOLOGY

2.1 Facilities

The experiments were performed in the High-Discharge Flume at Waseda University, Tokyo (Japan). The flume shown in Figure 1 was 14.0 m long x 0.40 m wide x 0.80 m high. The flume was separated into two sections: the reservoir (light blue) and the propagation section. The floor of both sections was stainless steel. The propagation section consisted of a 2.0 m long 1:10 slope followed by a 4.0 m flat section. The 1:10 slope was used to provide adequate storage from the high-discharge flow to avoid any backwater effects caused by the flow exceeding the outlet capacity. Rapid opening of a partial lift gate resulted in a dam-break wave which propagated along the flume axis as a hydraulic bore (Chanson, 2006). A full description of the lift gate can be found in Esteban et al., (2017). The debris used in this study were scaled-down versions of common coastal debris: shipping containers (SC), hydro poles (HP), and damaged drywall panels (referred to herein as boards) (B). The debris (discussed further in Section 2.2) were placed 1.0 m from the edge of the slope downstream to allow for adequate time for the bore to develop once it had passed the sloped section.

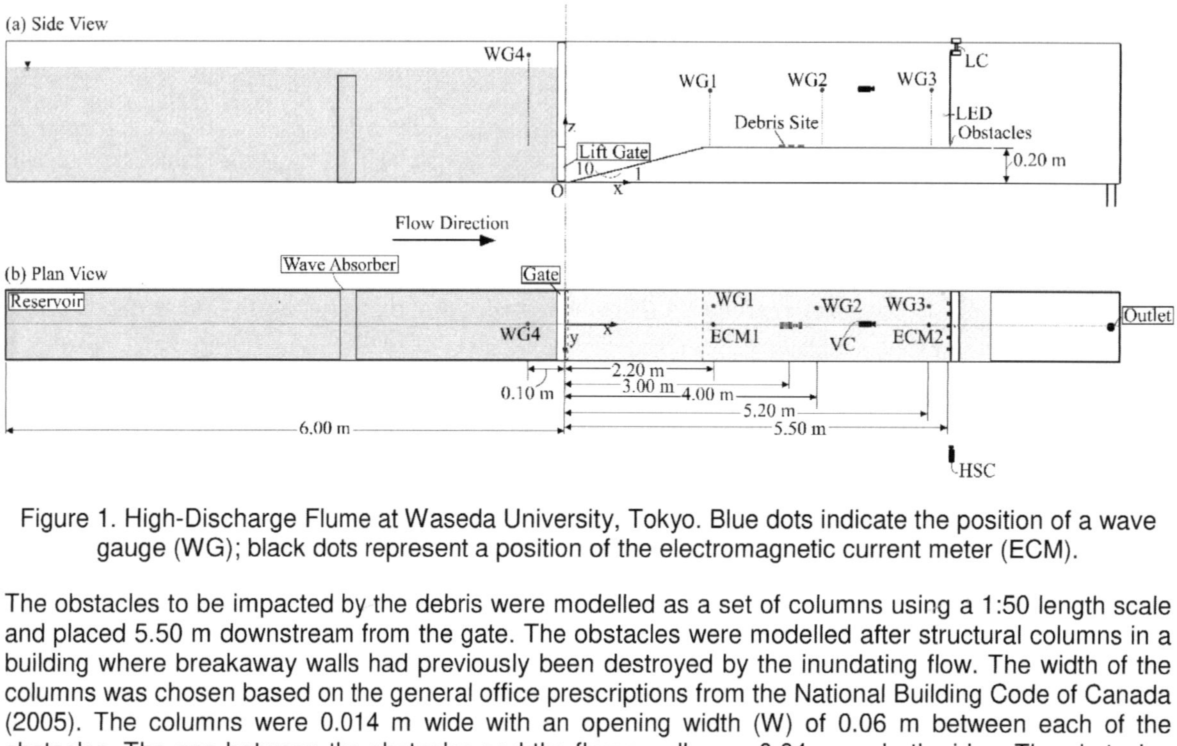

Figure 1. High-Discharge Flume at Waseda University, Tokyo. Blue dots indicate the position of a wave gauge (WG); black dots represent a position of the electromagnetic current meter (ECM).

The obstacles to be impacted by the debris were modelled as a set of columns using a 1:50 length scale and placed 5.50 m downstream from the gate. The obstacles were modelled after structural columns in a building where breakaway walls had previously been destroyed by the inundating flow. The width of the columns was chosen based on the general office prescriptions from the National Building Code of Canada (2005). The columns were 0.014 m wide with an opening width (W) of 0.06 m between each of the obstacles. The gap between the obstacles and the flume wall were 0.04 m on both sides. The obstacles were 0.40 m high. The obstacles were placed 0.005 m above the bed surface to prevent them from touching the bed surface and thus biasing the force measurements from the load cell (LC).

Table 1 lists the instrumentation used in this experiment. The positions of the instruments are shown in Figure 1. Wave gauge (WG) 4, placed in the reservoir section, was used as the reference WG for each of the experimental trials. For the remainder of this study, time = 0.00 s (zero time) refers to the time where the water depth at WG4 drops 0.002 m. The instrumentation placed downstream of the debris site (except the video camera (VC)) was removed for the cases where debris was present to avoid accidental damage. The VC and high-speed camera (HSC) were used to monitor the evolution of the debris dam. To synchronize the cameras with the hydrodynamic data, a LED was placed in the field-of-vision of both cameras. An algorithm was written post-processing to monitor the LED and the time within the video file when the LED turned on could be determined. When the LED turned on, the voltage would be recorded by the data acquisition system (DAQ) to have a reference time between the hydrodynamics recorded by the DAQ and the video files from the cameras.

Table 1. Experimental Instrumentation.

Instrumentation	Model	Instruments	Sampling Rate
Wave Gauge (WG)	KENEK CH-601	WG1, WG2, WG3	100 Hz
Electro-current Meter (ECM)	KENEK MT2-200	ECM1, ECM2	100 Hz
Video Camera (VC)	JVC Everio GZ-HM440		60 FPS
High-Speed Camera (HSC)	KATO KOKEN k4		100 FPS
Load Cell (LC)	SSK LB120-50		100 Hz

2.2 Debris

The experiments were performed considering Froude similitude (1:50 geometric length scale). Three different types of debris were selected based on the debris indicated in the ASCE7 Chapter 6 (Chock, 2016) and the FEMA P646 (FEMA, 2012) (Figure 2). All the debris were built from pine wood (specific gravity (SG) = 0.47) and were therefore positively buoyant. The shipping containers (SC) were modelled as an International Organization of Standardization (ISO) 6.1 m shipping container (0.12 m x 0.045 m x 0.045 m) (Goseberg et al., 2016; Nistor et al., 2016; Stolle et al., 2016). The hydro poles (HP) were modelled as standard 6.1 m hydro poles (0.12 m x 0.005 m diameter). The board (B) was designed considering an arbitrary shape of a piece of drywall washed away from another structure (0.06 m x 0.04 m x 0.002 m). Each debris was weighed for an average weight of 0.111 kg (prototype scale = 13 875 kg), 0.002 kg (250 kg), and 0.004 kg (500 kg) for the SC, B, and HP, respectively. The standard deviation of the weight between debris was 0.001 kg.

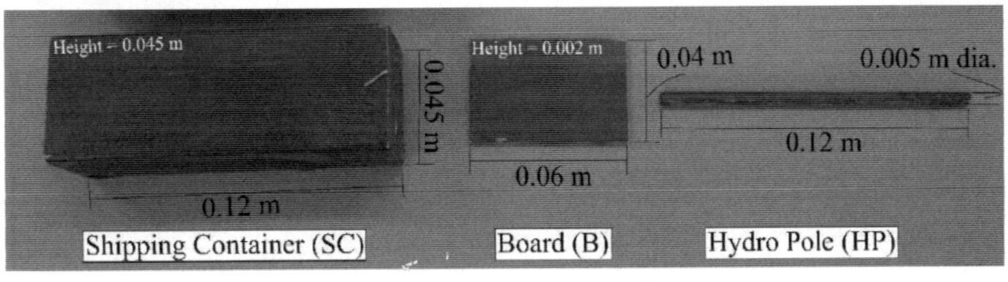

Figure 2. Type of debris modelling in tsunami-like conditions: Shipping containers (SC), board (B), and hydro poles (HP)

2.3 Experimental Protocol

The experiments can be separated into three categories: same volume cases (SV), same plan area cases (PA), and the debris mixture cases (DM). An additional set of tests were performed to examine the repeatability of the hydrodynamics with nod debris and without (no obstacles) the structure. The same volume cases (equally combined volume of all debris in a test) were performed to examine the influence of the individual debris properties on the capture efficiency and associated loads. The plan area experiments refer to the plan area of the debris when observing the debris from a top view. The ASCE7 Chapter 6 assessment of debris impact potential relies upon the plan area of the debris (Chock, 2016; Naito et al., 2014; Naito et al., 2016), therefore the amount of debris in the same plan area category was selected based on the plan area. The debris mixtures were selected to examine the influence of the debris types in combination on the dam formation. For each category, the experiments were performed with three different impoundment water depths: 0.40 m, 0.50 m, and 0.60 m. Each experiment was performed with 3 repetitions for a total of 125 experimental trials.

Table 2. Experimental Protocol.

Impoundment Water Depth [m]	Debris Mixture [SC, HP, B]	Experimental Category
0.4	0,0,0	
0.5	9,0,0	PA
0.6	0,81,0	PA
	0,0,20	PA
	3,27,7	PA/DM
	7,9,2	PA/DM
	1,63,2	PA/DM
	1,9,16	PA/DM
	6,54,14	DM
	9,81,20	DM
	1,0,0	SV
	0,103,0	SV
	0,0,51	SV

3 RESULTS

3.1 Hydrodynamics

Figure 3 shows the clear-water (no debris) conditions at the obstacles position. Figure 3a-b show the water depth and the flow velocity at the obstacle site (WG3 and ECM2). The wave front velocities at the obstacle site are plotted as open diamonds in Figure 3b. These were estimated using the difference in the wave arrival time between WG2 and WG3. Flow velocities were measured using ECM2; however, due to air entrainment and cavitation occurring in the initial wave front, data could not be accurately captured for the first ~2 s of the wave front.

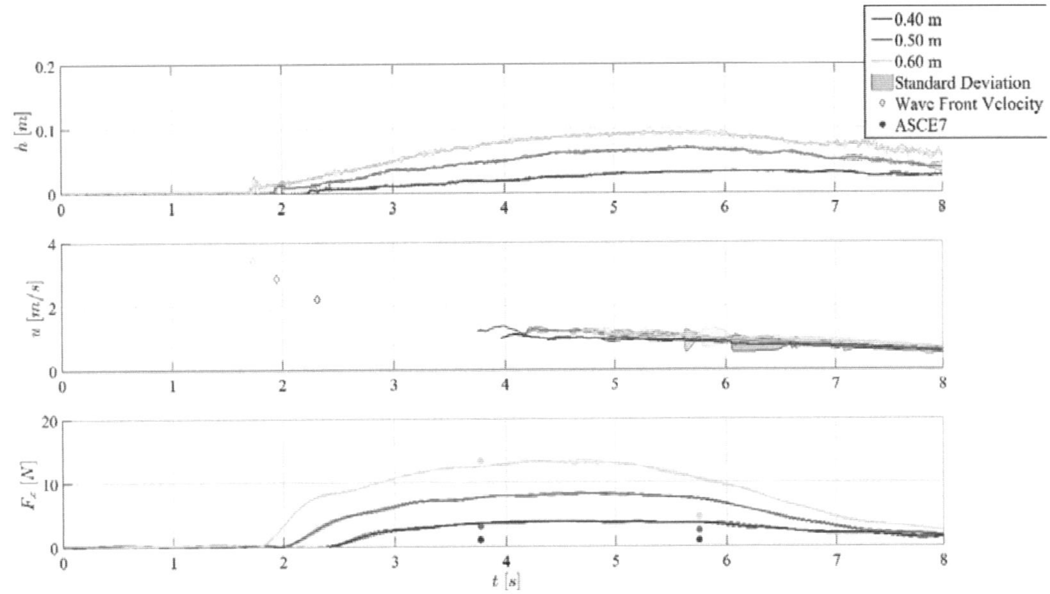

Figure 3. Hydrodynamic conditions at the site of the obstacle (no debris case). The water surface elevation and flow velocity taken with the obstacles in place. The filled circular markers represent the calculated drag forces from the ASCE7 Chapter 6.

The clear-water force profile in the flow direction (Figure 3c) was measured with the LC connected to the obstacles. The force profiles are compared to the ASCE7 standard (Eq. 1) at the point of 2/3 the wave front velocity and the maximum water depth (filled circles). These two time instances were chosen based on the critical design loads outlined in the ASCE7 standard (Chock, 2016). Generally, the ASCE7 standard overpredicted the hydrodynamic forces for the clear-water case. The Froude number from field investigations of tsunami events have ranged from 0.6 – 1.4 (Fritz et al., 2006; Fritz et al., 2012; Matsutomi et al., 2010). The Froude number in these experiments ranged from 1.2 – 1.9, exceeding the expected range in a tsunami-like event. Bricker et al., (2015) in an examination of Manning's *n* values used in tsunami modelling, emphasized the necessity of properly scaling experiments, particularly related to the Reynolds (Re) and Weber (We) numbers. The Re and We number were calculated at the point where the ECM first recorded consistent velocity values. Te Chow, (1959) noted that flow conditions should be fully turbulent (Re > 1.25×10^4) and Peakall et al., (1996) indicated that We > 120 - both of these conditions are achieved in these experiments.

3.2 Debris Dam Formation

The formation of a debris dam in steady-state subcritical flow conditions has been well-established by Schmocker et al., (2013). According to these authors, the debris initially passed through the obstacles until the first debris was caught in the obstacle, commonly referred to as the "key log." Once the "key log" had formed, the debris were pushed by the streamlines surrounding the "key log" to fill the width of the obstacle. As the width was filled, the debris would begin to compact those at the front of the structure, forcing them towards the bed and obstructing thus the channel cross-section. As a result, the water depth upstream of the dam began to rise and the flow velocity decreased. Once the flow velocity had decreased, compaction no longer occurred and a debris carpet on the water surface begun forming upstream of the obstacle. The extent of each of these steps could be dependent on the flume dimensions and initial flow conditions. Equivalently, the companion study to the work performed here (Stolle et al., 2017) with the same debris under steady-state flown conditions displayed a similar behavior.

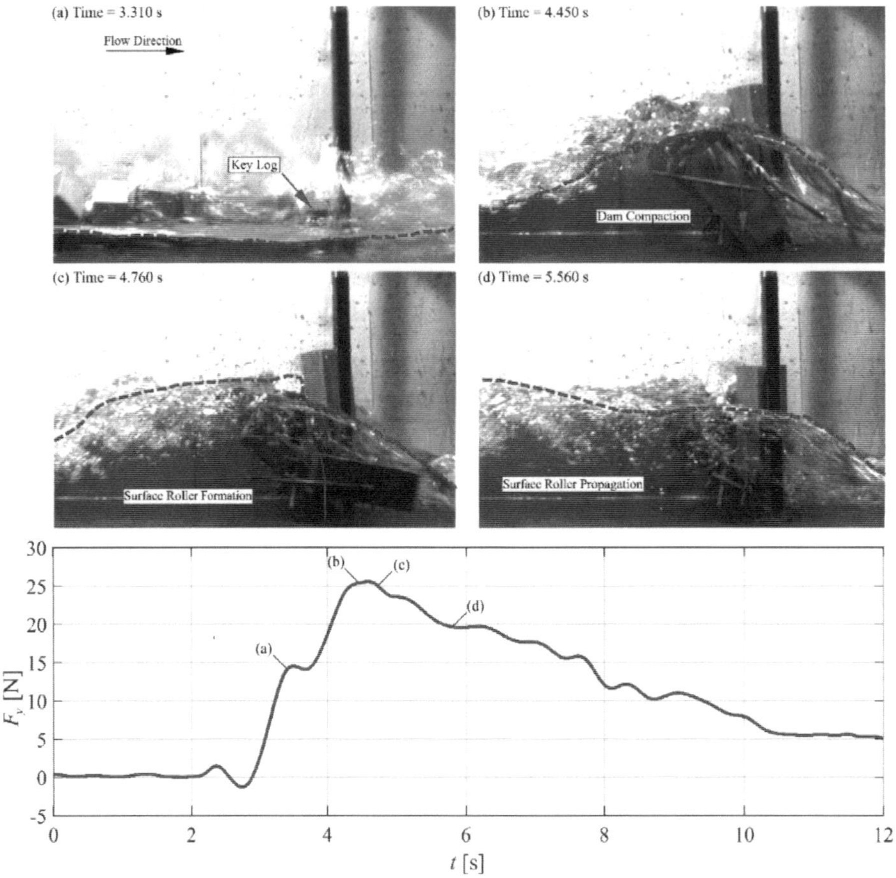

Figure 4. Debris dam formation for the case with 9 SC, 81 HP, 20 B and 0.40 m impoundment depth. Panels (a)-(d) show still images from the HS at various key point. The red dashed line is marking the water surface elevation. Panel (e) shows the force-time history for the experiment, the times corresponding to the above images are indicated within the figure.

Figure 4 shows the formation of the debris dam in unsteady supercritical flow conditions. Due to the supercritical flow conditions and the high velocity flow, stream lines did not fully form (Soares-Frazão et al., 2007) and debris were not pushed to obstruct the full width of the obstacle. Unlike the steady-state cases where the stagnation zone had adequate time to form before the debris arrived. The transient flow case instead, caused by the high-velocity flow, moved into the rapid compaction of the dam causing a constriction of the cross-section. As the constriction formed, the water would runup the structure formed a surface roller directly upstream of the structure (Figure 4b), similar to the surface piercing columns in dam-break flows (St-Germain et al., 2013). Due to the supercritical flow conditions, the rise in water surface level did not propagate upstream as was observed in the steady-state case. Instead, the surface roller expanded in cross-stream direction until it reached the walls of the flume; at that point, the surface roller propagated upstream. St-Germain et al., (2013) also noted that this was the point of maximum runup. Similarly, the maximum force and runup corresponded with the instant just before the surface roller began to propagate back upstream (Figure 4e). The vortices formed by the surface roller destabilised the dam formation causing its constituting parts to break apart and be washed downstream, past the obstacles (Figure 4c). As the surface roller propagated upstream, the reduction of flow velocity and return of predominantly unidirectional flow resulted in the dam stabilizing (Figure 4d).

3.3 Debris Dam Loading

In Equation 1, the drag coefficient (C_D) was used to calculate the drag force acting on a structure. However, the classical definition of the drag coefficient refers to the case when the obstacle is fully submerged in the flow (Qi et al., 2014). In the experiments presented here, due to the unsteady, energetic nature of the flow, a wake formed behind the structure resulting in an unbalanced hydrostatic force acting on the structure (Figure 8). As such, the drag coefficient cannot be determined within this study. Arnason et al., (2009), in a study of bore impingement on a vertical column, instead used a similar resistance coefficient (C_R) which manifests as a time history due to the fluctuation in the hydrostatic component.

[2] $C_R = \frac{2F_x}{\rho hbu^2}$

The resistance coefficient C_R was taken as mean resistance coefficient in the first 5 seconds after the wave arrival to avoid any influence from the reservoir. The characteristic length used in the calculation of the Fr_d and Re_d was the dam width at the time of occurrence of the maximum force. Due to the unsteady conditions and formation of the surface roller, a reference unperturbed upstream state could not be achieved. Therefore, the velocity and water depth used in the calculation of C_R was taken from the clear-water condition (Bremm et al., 2015) at the same instant. A log-log transformation was applied to the dataset to achieve an approximately linear regression (R^2 = 0.821). The resulting coefficients of the regression were $a = 3.58 \times 10^5$ and $b = -1.089$. The focus of this study addressed the forces in supercritical flow conditions: it is therefore unclear if the same trend could be extended to subcritical flow conditions, rendering thus necessary future research on transient subcritical conditions.

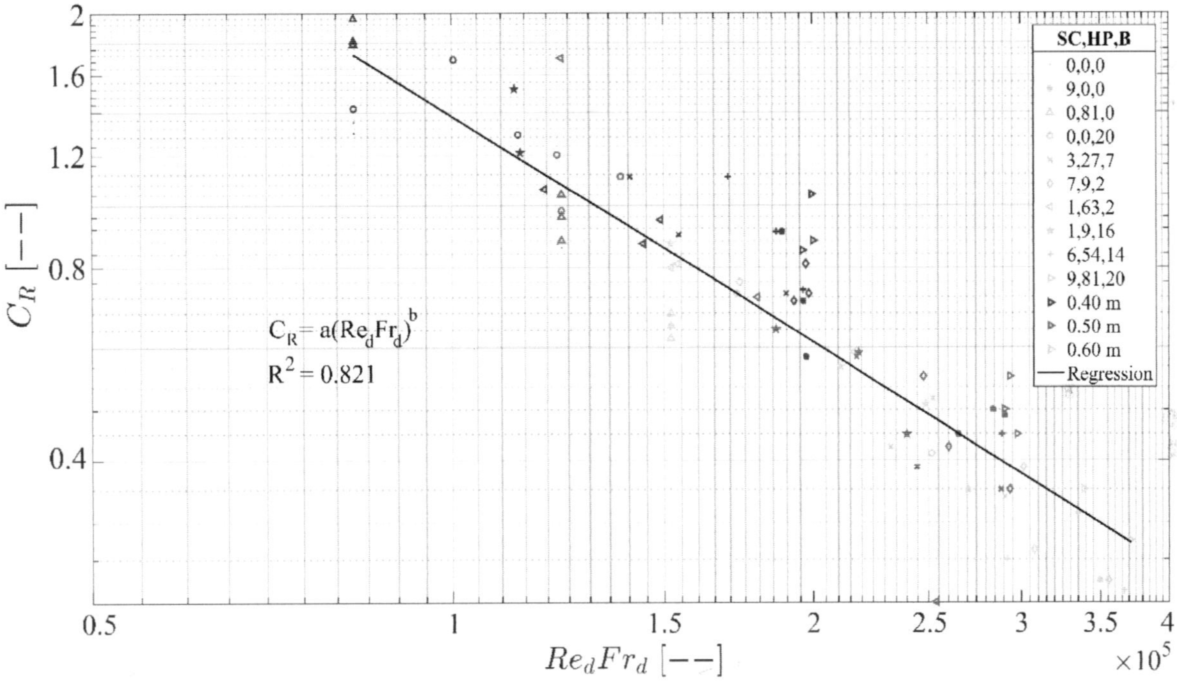

Figure 5. Resistance coefficient, C_R as a function of the product of the Reynolds and Froude number. The different debris mixtures are indicated by the marker type and the initial impoundment depth by the marker color. The log-log regression line is displayed as a solid black line.

4 CONCLUSIONS

The study discussed herein examined the formation and loads associated with debris dams in supercritical, unsteady flow conditions. The study focused on the debris damming in a coastal setting by modelling debris common to a coastal environment and by using a hydrodynamic forcing condition similar to a tsunami wave. The paper investigated the influence of the debris mixtures on the size of the debris dam, the runup at the structure face, and the loads exerted on the structure.

Based on the results, the following conclusions can be drawn:

- The debris dam formation was initiated similarly to the steady-state cases, as described in Schmocker et al. (2013). However, due to the supercritical flow conditions, a debris carpet was not formed in front of the dam.
- Due to the formation of the surface roller at the structure face, the dam would break apart as the surface roller propagated upstream, resulting in the dam being less stable than the unidirectional cases.
- The resistance coefficient showed an approximately log-log linear relationship with the product of the Reynolds and Froude number.

Debris damming has been shown to be a major concern in flooding events, and therefore needs to be carefully considered in the design process of infrastructure exposed to such effects. The study presented herein was the first to examine debris damming in unsteady flow conditions. Based on climate change studies on precipitation patters, intensity and possibly frequency of transient flows will likely increase; hence, designing for such extreme flows is of major interest for engineers. However, due to concerns of damage to instrumentation in such conditions, research is further needed into the changing hydrodynamic conditions around the debris dam formation. While the focus of this experimental program was modelling tsunami-like flow conditions, the general conclusions of this study can be applied to rapidly occurring flooding events, such as flash floods and flood waves from breaching dams, where supercritical flow are present.

Acknowledgements

The authors are acknowledging the support of the NSERC CGS-D Scholarship (Jacob Stolle), of the NSERC Discovery Grant (No. 210282) (Ioan Nistor), the Marie Curie International Outgoing Fellowship within the 7th European Community Framework Program (No. 622214) (Nils Goseberg), and the Strategic Research Foundation Grant-aided Project for Private Universities from the Japanese Ministry of Education, Culture, Sports, Science and Technology (No. S1311028) (Tomoya Shibayama). The authors would also like to acknowledge the help of the Waseda University technical staff in the design and implementation of the experimental setup, in particular Mr. Kazuhiro Yamanashi, Mr. Kiyoji Egawa, and Mr. Akihiro Kagami.

References

AASHTO 2012. Bridge design specifications.

Arnason, H., Petroff, C. & Yeh, H. 2009. Tsunami bore impingement onto a vertical column. Journal of Disaster Research 4(6), 391–403.

Bocchiola, D., Rulli, M. & Rosso, R. 2006. Transport of large woody debris in the presence of obstacles. Geomorphology 76(1), 166–178.

Bremm, G.C., Goseberg, N., Schlurmann, T. & Nistor, I. 2015. Long Wave Flow Interaction with a Single Square Structure on a Sloping Beach. Journal of Marine Science and Engineering 3(3), 821.

Bricker, J.D., Gibson, S., Takagi, H. & Imamura, F. 2015. On the need for larger Manning's roughness coefficients in depth-integrated tsunami inundation models. Coastal Engineering Journal 57(02), 1550005.

Chanson, H. 2006. Tsunami surges on dry coastal plains: Application of dam break wave equations. Coastal Engineering Journal 48(04), 355–370.

Chock, G.Y. 2016. Design for tsunami loads and effects in the ASCE 7-16 standard. Journal of Structural Engineering, 04016093.

CREW 2013. Cascadia subduction zone earthquakes: A magnitude 9.0 earthquake scenario, Oregon Department of Geology and Mineral Industries,p.

CSA 2006. Canadian highway bridge design code, CSA,p.

Esteban, M., Takagi, H. & Shibayama, T. 2015. Handbook of Coastal Disaster Mitigation for Engineers and Planners.

Esteban, M., Glasbergen, T., Takabatake, T., Hofland, B., Nishizaki, S., Nishida, Y., Stolle, J., Nistor, I., Bricker, J., Takagi, H. & others 2017. Overtopping of Coastal Structures by Tsunami Waves. Geosciences 7(4), 121.

FEMA 2012. P646 Guidelines for Design of Structure for Vertical Evacuation from Tsunamis. Federal Emergency Management Agency.
Fritz, H.M., Borrero, J.C., Synolakis, C.E. & Yoo, J. 2006. 2004 Indian Ocean tsunami flow velocity measurements from survivor videos. Geophysical Research Letters 33(24).
Fritz, H.M., Phillips, D.A., Okayasu, A., Shimozono, T., Liu, H., Mohammed, F., Skanavis, V., Synolakis, C.E. & Takahashi, T. 2012. The 2011 Japan tsunami current velocity measurements from survivor videos at Kesennuma Bay using LiDAR. Geophysical Research Letters 39(7).
Ghobarah, A., Saatcioglu, M. & Nistor, I. 2006. The impact of the 26 December 2004 earthquake and tsunami on structures and infrastructure. Engineering structures 28(2), 312–326.
Goseberg, N. & Schlurmann, T. 2014. Non-stationary flow around buildings during run-up of tsunami waves on a plain beach. Coastal Engineering Proceedings 1(34), 21.
Goseberg, N., Stolle, J., Nistor, I. & Shibayama, T. 2016. Experimental analysis of debris motion due the obstruction from fixed obstacles in tsunami-like flow conditions. Coastal Engineering 118, 35–49.
Ioualalen, M., Asavanant, J., Kaewbanjak, N., Grilli, S., Kirby, J. & Watts, P. 2007. Modeling the 26 December 2004 Indian Ocean tsunami: Case study of impact in Thailand. Journal of Geophysical Research: Oceans (1978–2012) 112(C7).
Matsutomi, H. & Okamoto, K. 2010. Inundation flow velocity of tsunami on land. Island Arc 19(3), 443–457.
Naito, C., Cercone, C., Riggs, H.R. & Cox, D. 2014. Procedure for site assessment of the potential for tsunami debris impact. Journal of Waterway, Port, Coastal and Ocean Engineering 140(2), 223–232.
Naito, C., Riggs, H., Wei, Y. & Cercone, C. 2016. Shipping-Container Impact Assessment for Tsunamis. Journal of Waterway, Port, Coastal, and Ocean Engineering, 05016003.
Nistor, I., Goseberg, N., Mikami, T., Shibayama, T., Stolle, J., Nakamura, R. & Matsuba, S. 2016. Hydraulic Experiments on Debris Dynamics over a Horizontal Plane. Journal of Waterway, Port, Coastal and Ocean Engineering, 04016022.
Parola, A.C. 2000. Debris forces on highway bridges, Transportation Research Board,p.
Peakall, J. & Warburton, J. 1996. Surface tension in small hydraulic river models-the significance of the Weber number. Journal of Hydrology New Zealand 35, 199–212.
Qi, Z., Eames, I. & Johnson, E. 2014. Force acting on a square cylinder fixed in a free-surface channel flow. Journal of Fluid Mechanics 756, 716–727.
Robertson, I., Riggs, H.R., Yim, S.C. & Young, Y.L. 2007. Lessons from Hurricane Katrina storm surge on bridges and buildings. Journal of Waterway, Port, Coastal, and Ocean Engineering 133(6), 463–483.
Saatcioglu, M., Ghobarah, A. & Nistor, I. 2005. Effects of the December 26, 2004 Sumatra earthquake and tsunami on physical infrastructure. ISET Journal of earthquake technology 42(4), 79–94.
Schmocker, L. & Hager, W.H. 2011. Probability of drift blockage at bridge decks. Journal of Hydraulic Engineering 137(4), 470–479.
Schmocker, L. & Hager, W.H. 2013. Scale modeling of wooden debris accumulation at a debris rack. Journal of Hydraulic Engineering 139(8), 827–836.
Soares-Frazão, S. & Zech, Y. 2007. Experimental study of dam-break flow against an isolated obstacle. Journal of Hydraulic Research 45(sup1), 27–36.
St-Germain, P., Nistor, I., Townsend, R. & Shibayama, T. 2013. Smoothed-particle hydrodynamics numerical modeling of structures impacted by tsunami bores. Journal of Waterway, Port, Coastal, and Ocean Engineering.
Stolle, J., Nistor, I. & Goseberg, N. 2016. Optical Tracking of Floating Shipping Containers in a High-Velocity Flow. Coastal Engineering Journal, 1650005.
Stolle, J., Takabatake, T., Mikami, T., Shibayama, T., Goseberg, N., Nistor, I. & Petriu, E. 2017. Experimental Investigation of Debris-Induced Loading in Tsunami-Like Flood Events. Geosciences 7, 74.
Takahashi, S., Sugano, T., Tomita, T., Arikawa, T., Tatsumi, D., Kashima, H., Murata, S., Matsuoka, Y. & Nakamura, T. 2010. Joint survey for 2010 Chilean earthquake and tsunami disaster in ports and coasts. Port and Airport Res Inst.
Te Chow, V. 1959. Open channel hydraulics, McGraw-Hill Book Company, Inc; New York,p.
Titov, V.V. & Synolakis, C.E. 1997. Extreme inundation flows during the Hokkaido-Nansei-Oki tsunami. Geophysical Research Letters 24(11), 1315–1318.
Yeh, H., Sato, S. & Tajima, Y. 2013. The 11 March 2011 East Japan earthquake and tsunami: tsunami effects on coastal infrastructure and buildings. Pure and Applied Geophysics 170(6-8), 1019–1031.

Yeh, H., Barbosa, A.R., Ko, H. & Cawley, J.G. 2014. Tsunami Loadings on Structures: Review and Analysis. Coastal Engineering Proceedings 1(34), currents–4.

Building Tomorrow's Society
Bâtir la Société de Demain

Fredericton, Canada
June 13 to 16, 2018 / *13 au 16 juin, 2018*

DESIGN CONSIDERATIONS FOR CROSS-LAMINATED TIMBER PANELS SUBJECTED TO SIMULATED BLAST LOADS

Viau, Christian[1,4], Lacroix, Daniel[2] and Doudak, Ghasan[3]
[1] University of Ottawa, Canada
[2] Carleton University, Canada
[3] University of Ottawa, Canada
[4] cviau037@uottawa.ca

Abstract: This paper presents the results from an experimental program investigating the out-of-plane behaviour of cross-laminated timber (CLT) panels under static and simulated blast loading. A total of eighteen CLT panels were investigated with the aim to determine the dynamic behaviour of CLT panels. Dynamic testing was conducted through the use of simulated blast loads produced by a shock tube testing apparatus. An average dynamic increase factor (DIF) on the resistance is developed. Two material predictive models are proposed, which take into consideration high strain rate effects, failure modes, and the experimentally observed post-peak residual properties. A single degree-of-freedom (SDOF) model was validated using full-scale simulated blast load test results, and the predictions were found to match well with the experimental displacement-time histories. Current Canadian Blast Design Standard provisions were found to provide relatively conservative and accurate predictions when using the proposed DIF and analytical model.

1 INTRODUCTION

Cross-laminated timber (CLT) panels consist of lumber boards stacked in layers which run perpendicular to one another. The end product is dimensionally stable with high in- and out-of-plane strength and stiffness. Recent interest in sustainable construction has led to an increase in the use of heavy- and mass-timber elements, particularly CLT. The usage of wood beyond that of low-rise construction has been very limited primarily due to code requirements that are outdated and prescriptive in nature. The use of wood in tall (e.g. UBC Brock Commons) and high profile (e.g. Richmond Olympic Oval) buildings has also meant that such buildings may be exposed to heightened risk, including that stemming from intentional or unintentional blast loading.

Blast loading occurs during an extremely brief time period, generating high strain rates in the materials exposed to the loading. This short duration loading affects the mechanical properties of the material and an apparent increase in strength relative to the material's static strength is typically observed. This increase is quantified as the ratio of the dynamic to the static strength, termed "dynamic increase factor" (DIF). DIFs for different materials can be found in various blast design codes (ASCE 2011, CSA 2012).

Several studies have investigated CLT's material properties, specifically relating to rolling shear (Flores et al. 2016, Li and Lam 2016, Zhou et al. 2014), as well as the behaviour of CLT when subjected to out-of-plane loads (Jacquier 2015, Sikora et al. 2016, Steiger and Gülzow 2009), in-plane loads (Ashtari et al. 2014, Ceccotti et al. 2006, Popovski and Gavric 2016, Yasumura et al. 2016), and fire (Fragiacomo et al.

2013, Frangi et al. 2009, Frangi et al. 2004). However, research is lacking in the area of establishing the behaviour of CLT under hazardous and short-duration loads, such as that arising from blast loading. Understanding the behaviour of CLT when subjected to such loads is imperative in the development of harm-mitigation strategies which focuses on occupant life safety and progressive collapse prevention.

The majority of studies on DIF have been on reinforced concrete (e.g. Barreiro 2016, Burrell 2012, Jacques 2011, 2016, Lloyd 2010) and structural steel (e.g. Jama et al. 2009, Nassr et al. 2012) members, due to their inherent blast-resistant properties. Studies on bare studs (Jacques et al. 2014) and light-frame wood stud walls (e.g. Collins and Kasal 2009, Lacroix and Doudak 2015, Lacroix et al. 2014, Marchand 2002, Viau 2016, Viau and Doudak 2016b) have shown a significant increase in the flexural strength when subjected to high strain rates. A recent project investigated whether CLT buildings would meet the United States Department of Defense level of protection requirements (Senalik and Podesto 2017). The results from these live blast tests show that desired levels of protections can be achieved with CLT, however, little details have been made available to the public, including whether a DIF was observed during testing.

The goal of this paper is to investigate the out-of-plane behaviour of CLT panels under static and blast loading. The development of the material-predictive model will first be summarized and used to develop a generalized code-based design approach of one-way CLT panels. The code-based approach will then be validated using the experimental test results.

2 EXPERIMENTAL PROGRAM AND RESULTS

A recent study conducted at the University of Ottawa on eighteen spruce-pine-fir CLT panels focused on quantifying high strain rate effects in CLT (Poulin et al. 2017). The width and the length of all panels were 445 mm and 2,500 mm, respectively, with thicknesses of 105 mm and 175 mm (i.e. 3-ply and 5-ply). The study consisted of static and dynamic testing phases, during which the properties were experimentally obtained.

2.1 Static Testing Phase

The static tests consisted of subjecting the panels to four-point bending with simply supported boundary conditions until failure was obtained. Through a hydraulic jack, loading was applied at a rate which would induce failure at a time between 2 and 10 minutes. The applied load, mid-span deflection, and compressive and tensile strains were measured at a sampling rate of 10 samples per second. Of the eighteen panels, six were tested to destruction and twelve were tested within their elastic range in order to determine their stiffness. This information was used as input in modelling the blast behaviour of each panel. The results of the static testing phase are summarized in Table 1, which includes the mean and coefficient of variation (CoV) of the maximum resistance, ductility ratio, initial stiffness based on experimental load resistance curve from 10 to 40%, and the experimentally observed post-peak resistance expressed as a percentage of the maximum resistance.

Table 1: Results of Static Testing

Specimen Group	Max. Resistance		Initial Stiffness		Ductility Ratio		Post-Peak Resistance	
	Average (kN)	CoV	Average (kN/mm)	CoV	Average	CoV	Average (%)	CoV
3-ply	68.4	0.08	2.1	0.01	1.8	0.11	20	0.09
5-ply	128.6	0.03	5.9	0.03	2.2	0.19	20	0.15

For both CLT panel groups, a linear relationship up to the maximum resistance was observed at which outer longitudinal lamination failure occurred, as shown in Figure 1 (a). Resistance plateaus were observed past the maximum resistance, which correlated to the resistance of the residual undamaged section of the panel. Superimposing a typical 5-ply panels' resistance curve over that of a 3-ply (Figure 1 (b)), it can be seen that the first drop in resistance of the 5-ply panels corresponds to the ultimate strength of a 3-ply panel.

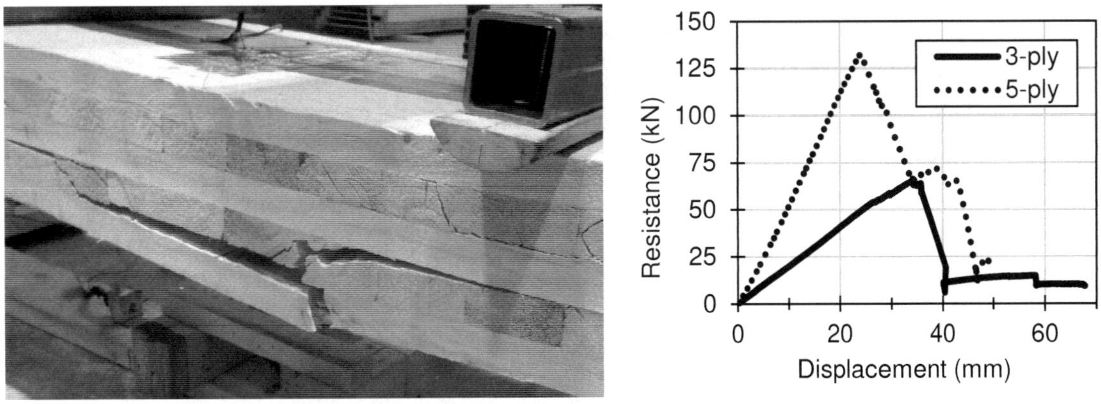

(a) (b)

Figure 1: (a) Typical static flexural failure; (b) Typical experimental static resistance curves

2.2 Dynamic Testing Phase

Through the use of a shock tube, twenty simulated blast loads were applied to the twelve specimens that were tested non-destructively during the static testing phase. This method of generating high strain rates on structural components has been widely used in past studies and accepted as an effective methodology to generate blast loads in timber and reinforced concrete elements (Jacques et al. 2015, Lacroix and Doudak 2015, 2018, Viau and Doudak 2016a) as well as fibre reinforced structural elements (Aoude et al. 2015).

Four-point bending was used in the dynamic testing in order to allow direct comparison of resistance and stiffness with those obtained through static testing. Pressure waves were collected and transferred to the specimens via a load-transfer device (LTD), which converted the blast loading to two concentrated loads at the third points. Load-cells were placed at the specimen support points in order to measure the dynamic reactions. The reactions in addition with the applied loads and inertial forces of both the specimen and LTD were used in the calculation of the dynamic resistance-displacement relationships for each test specimen, according to dynamic equilibrium (Biggs 1964). The dynamic response was documented through the use of a data acquisition system recording 100,000 samples per second, as well as high-speed camera. A summary of the dynamic results for the panels is presented in Table 2.

Table 2: Results of Dynamic Testing

Specimen Group	Max. Resistance		Initial Static Stiffness		Ductility Ratio	
	Average (kN)	CoV	Average (kN/mm)	CoV	Average	CoV
3-ply	83.7	0.04	1.9	0.10	2.7	0.14
5-ply	174.2	0.01	5.4	0.05	2.3	0.11

Whereas flexural failure was the primary failure mode observed during static testing, some specimens under dynamic loading experienced significant rolling shear. An example of such rolling shear failure is shown in Figure 2 (a). Due to the early onset of rolling shear failure, the longitudinal laminates deflected almost as individual elements rather than a composite section, causing a degradation in stiffness. The loss of stiffness meant that the specimen could deflect much more and return to its original position. In cases where dynamic flexural failures were observed, they were similar to those observed during static testing, with little to no significant rolling shear failure, as seen in Figure 2 (b). Panels where rolling shear damage was documented following a shock tube test were later tested statically in order to determine their residual stiffness. It was found that the degraded stiffness of these panels varied between 20 to 35% of that observed in their undamaged condition.

(a) (b)

Figure 2: Representative dynamic: (a) Rolling shear failure; (b) Flexural failure

2.3 Dynamic Increase Factor

Experimentally obtained static and dynamic resistance-displacement relationships showed an increase in strength when subjected to high strain rates, while no increase was observed in the stiffness. An average dynamic increase factor (DIF) of 1.28 was determined from comparing the static and dynamic testing results (Poulin et al. 2017).

3 DEVELOPMENT OF A MODELLING APPROACH FOR CLT

Based on the experimentally obtained resistance curves, a generalized approach for constructing the resistance curve of both 3-ply and 5-ply specimens was developed, for both failure modes observed – namely flexure and rolling shear failure (Poulin et al. 2017).

The flexural behaviour of both specimen groups can be described as initially linear elastic, after which the loss of the bottom longitudinal laminates would cause a sudden drop in resistance. For the 3-ply specimens, this post-peak behaviour was modelled as a sudden drop in resistance to a value equal to 20% of the ultimate resistance. The value of 20 % was determined by experimental testing. For the 5-ply specimens, the loss of the outermost laminates means that the specimen would now behave as a 3-ply specimen. At a deflection equal to that of the elastic limit of the equivalent 3-ply, the resistance curve would drop to a value equal to approximately 20% of the ultimate resistance of the 5-ply specimen, which is consistent with the approach for the 3-ply specimens. The post-peak residual deflections were capped at a maximum ductility ratio of 2.5, based on the experimental findings. Representative flexural resistance curves are shown in Figure 3 (a) for the 3- and 5-ply specimens.

Specimens in which rolling shear failure occurred prior to any flexural failure necessitated a modified approach to capture this behaviour. When rolling failure is fully developed, the section no longer behaves as one rigid component, but rather as a set of (undamaged) longitudinal laminates weakly connected by the transverse laminates which have become heavily damaged by rolling shear. The partial composite action between the laminates caused a significant reduction in stiffness (on the order of 65-80% reduction). An example of a resistance curve modified for rolling shear failure is shown in Figure 3 (b).

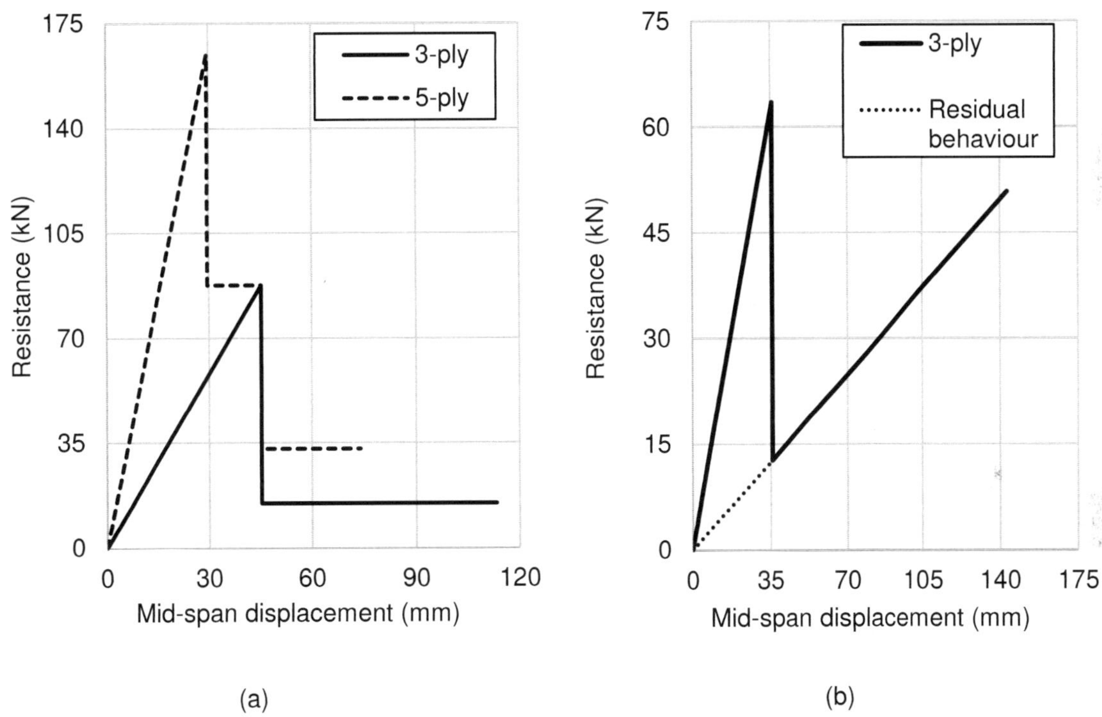

Figure 3: Representative resistance curves: (a) flexural failure; (b) rolling shear failure

4 PROPOSED DESIGN APPROACH

Current blast provisions provide designers with the means to conduct blast analysis and design for various materials, including wood structures (ASCE 2011, CSA 2012). Blast analysis is currently based on establishing a desired level of protection (LOP) to which an element can be assessed. Once established, this methodology is used in determining whether the quantifiable response and overall damage level meet the desired LOP. The basis of blast design consists of increasing design-level capacities to mean values, with additional consideration for high strain rate effects. This adjustment is done through the use of a strength increase factor (SIF) and a dynamic increase factor (DIF). The ultimate dynamic moment capacity and resistance are calculated using Equations 1 and 2, respectively.

[1] $M_{dyn} = \Phi \cdot SIF \cdot DIF \cdot M_{static}$

[2] $R_u = 6 \cdot M_{dyn} \cdot L^{-1}$

where Φ is the material adjustment factor, which is set to unity for blast, SIF is the strength increase factor, DIF is the dynamic increase factor, M_{static} is the specified static moment strength obtained from manufacturer tables (kN·m), and L is the clear span (m).

The Canadian blast standard currently assigns a SIF of 1.2 for glulam and engineered wood products (CSA 2012). This value is consistent with the adjustment factor of 1.25 provided in CSA O86 for stress grade E1 CLT panels (CSA 2014) and is in line with what was observed in this study (1.24). A DIF of 1.28 was found to be appropriate for CLT based on findings from the experimental program. The stiffness of a CLT panel subjected to third point bending is calculated using Equation 3 (ASTM 2015).

[3] $K = (L^3 \cdot (56.4 \cdot EI_{eff})^{-1} + \kappa \cdot L \cdot (5 \cdot GA_{eff})^{-1})^{-1}$

where L is the clear span (m), EI_{eff} is the effective bending stiffness (kN·m²), GA_{eff} is the effective shear rigidity (kN), and κ is the shear coefficient factor, equal to 1.2 for rectangular sections (Timoshenko and Woinowsky-Krieger 1959). M_{static}, EI_{eff}, and GA_{eff} are obtained from published manufacturer tables.

The elastic limit, denoted x_e (m), for both panels is calculated by dividing the resistance by the stiffness obtained from Equations 2 and 3, respectively. The inputs used in calculating the resistance curves are presented in Table 3.

Table 3: Modelling inputs based on published *Nordic Structures®* design data

Panels	SIF	DIF	M_{static} (kN·m)	EI_{eff} (kN·m²)	GA_{eff} (kN)	R_u (kN)	K (kN/m)	x_e (m)
3-ply	1.2	1.28	18.8	481.0	3248.5	77.5	1734	0.045
5-ply	1.2	1.28	43.0	1842.3	6675.0	177.4	5324	0.033

Shown in Figure 4 are the resulting resistance curves for the 3-ply and 5-ply panels based on published data assuming flexure model only. The rolling shear model presented in this paper can be used if a designer is able to predict this failure mode with some level of certainty. Although lower stiffness at the early stages of loading would cause higher displacements, this displacement cannot directly correlate to significant damage when compared to panels that failed in flexure. It was therefore decided that for the purpose of illustrating a simplified design approach, only the flexure model would be considered, as it represents the more critical of the two failure modes.

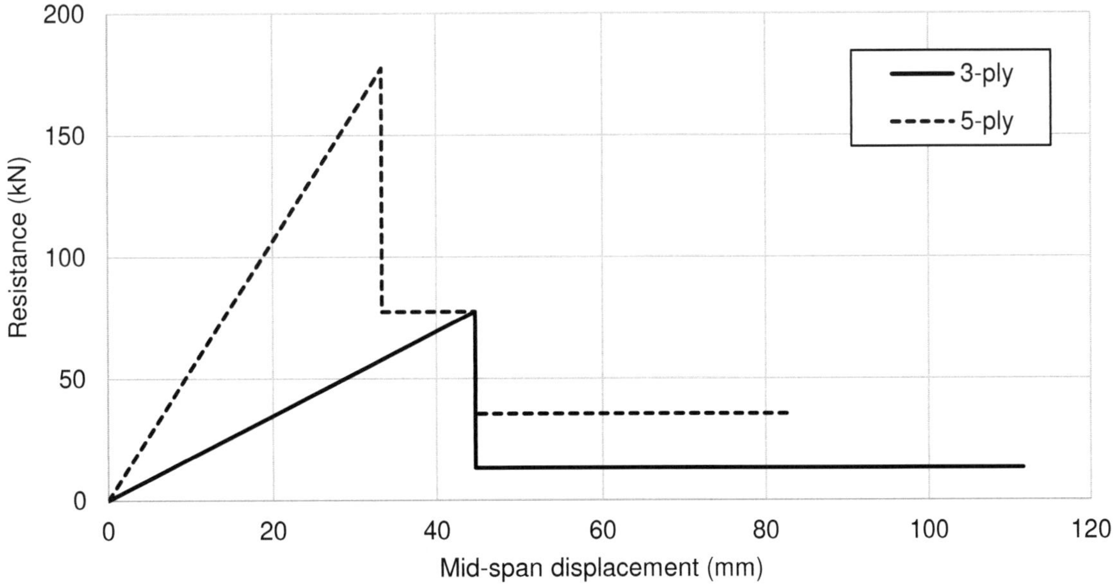

Figure 4: Resistance curves based on published strength modified using code-based SIF and experimental DIF

The analytically predicted displacements against those experimentally obtained is presented in Figure 5. The prediction of the model seems reasonable on average, with the model based on published data over predicting displacements on average by 1 %. Based on these results, it can be concluded that conducting blast analysis based on published static values modified for the published value of SIF (1.2) and the obtained DIF value (1.28), while assuming flexural failure seems to predict the displacements of CLT panels with reasonable accuracy.

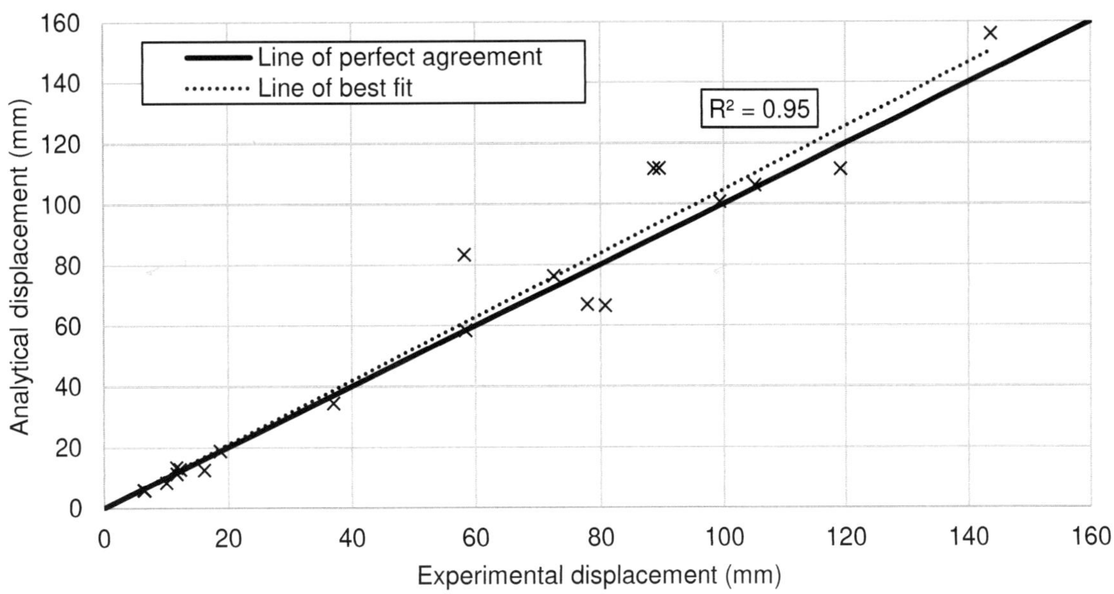

Figure 5: Comparative view of analytical and experimental displacements – code approach

5 CONCLUSIONS

An experimental investigation to determine the static and dynamic behaviour of cross-laminated timber panels was undertaken. An average dynamic increase factor of 1.28 on the resistance was determined, while the dynamic stiffness did not seem to be affected by the increase in strain rate. Residual post-peak resistances were observed in both static and dynamic test, and a modelling approach which incorporates these residual resistances was successfully used. A single degree-of-freedom model utilizing material properties and observed behaviour was developed and validated with experimental dynamic test results. Reasonable correlation between predicted and experimentally obtained panel displacements was found when assuming flexural failure and while modifying the static strength with a SIF and DIF.

References

Aoude, H., Dagenais, F. P., Burrell, R. P., and Saatcioglu, M. 2015. Behavior of ultra-high performance fiber reinforced concrete columns under blast loading. *International Journal of Impact Engineering*, **80**: 185-202.

ASCE. 2011. Blast Protection of Buildings. *ASCE/SEI 59-11*. Reston, VA: American Society of Civil Engineers.

Ashtari, S., Haukaas, T., and Lam, F. 2014. In-Plane Stiffness of Cross-Laminated Timber Floors. *World Conference on Timber Engineering*, Quebec, QC.

ASTM. 2015. Standard Test Methods of Static Tests of Lumber in Structural Sizes. *ASTM D198-15*. West Conshohocken, PA: ASTM International.

Barreiro, J. 2016. Blast Resistance of Non-Composite Tilt-Up Sandwich Panels and their Connections. MASc, Civil Engineering, University of Ottawa.

Biggs, J. M. 1964. *Introduction to Structural Dynamics*. McGraw-Hill, New York, NY.

Burrell, R. P. 2012. Performance of steel fibre reinforced concrete columns under shock tube induced shock wave loading. MASc, Civil Engineering, University of Ottawa.

Ceccotti, A., Follesa, M., Lauriola, M. P., Minowa, C., Sandhaas, C., and Yasumura, M. 2006. SOFIE Project: Test results on the lateral resistance of cross-laminated wooden panels. *First European Conference on Earthquake Engineering and Seismicity,*, Geneva, Switzerland.

Collins, M., and Kasal, B. 2009. Dynamically loaded light-frame wood stud walls: experimental verification of an analytical model. *Materials and Structures*, **43**(9): 1203-1216.

CSA. 2012. Design and assessment of buildings subjected to blast loads. *CSA S850*. Mississauga, ON: CSA Group.

CSA. 2014. Engineering design in wood. *CSA O86*. Mississauga, ON: CSA Group.

Flores, E. I. S., Saavedra, K., Hinojosa, J., Chandra, Y., and Das, R. 2016. Multi-scale modelling of rolling shear failure in cross-laminated timber structures by homogenisation and cohesive zone models. *International Journal of Solids and Structures*, **81**: 219-232.

Fragiacomo, M., Menis, A., Clemente, I., Bochicchio, G., and Ceccotti, A. 2013. Fire Resistance of Cross-Laminated Timber Panels Loaded Out of Plane. *Journal of Structural Engineering*, **139**(12): 04013018.

Frangi, A., Fontana, M., Hugi, E., and Jübstl, R. 2009. Experimental analysis of cross-laminated timber panels in fire. *Fire Safety Journal*, **44**(8): 1078-1087.

Frangi, A., Fontana, M., and Mischler, A. 2004. Shear behaviour of bond lines in glued laminated timber beams at high temperatures. *Wood Science and Technology*, **38**(2): 119-126.

Jacques, E. 2011. Blast Retrofit of Reinforced Concrete Walls and Slabs. MASc, Civil Engineering, University of Ottawa.

Jacques, E. 2016. Characteristics of Reinforced Concrete Bond at High Strain Rates. Ph.D., Civil Engineering, University of Ottawa.

Jacques, E., Lloyd, A., Braimah, A., Saatcioglu, M., Doudak, G., and Abdelalim, O. 2014. Influence of high strain-rates on the dynamic flexural material properties of spruce–pine–fir wood studs. *Canadian Journal of Civil Engineering*, **41**(1): 56-64.

Jacques, E., Lloyd, A., Imbeau, P., Palermo, D., and Quek, J. 2015. GFRP-Retrofitted Reinforced Concrete Columns Subjected to Simulated Blast Loading. *Journal of Structural Engineering*, **141**(11): 04015028.

Jacquier, N. 2015. Development and Evaluation of Mechanical Joints for Composite Floor Elements with Cross Laminated Timber. PhD, Department of Civil, Mining and Environmental Engineering, Luleå University of Technology.
Jama, H. H., Bambach, M. R., Nurick, G. N., Grzebieta, R. H., and Zhao, X. L. 2009. Numerical modelling of square tubular steel beams subjected to transverse blast loads. *Thin-Walled Structures*, **47**(12): 1523-1534.
Lacroix, D. N., and Doudak, G. 2015. Investigation of Dynamic Increase Factors in Light-Frame Wood Stud Walls Subjected to Out-of-Plane Blast Loading. *Journal of Structural Engineering*, **141**(6): 04014159.
Lacroix, D. N., and Doudak, G. 2018. Effects of High Strain Rates on the Response of Glulam Beams and Columns. *Journal of Structural Engineering*, **144**(5): 04018029.
Lacroix, D. N., Doudak, G., and El-Domiaty, K. 2014. Retrofit Options for Light-Frame Wood Stud Walls Subjected to Blast Loading. *Journal of Structural Engineering*, **140**(4): 04013104.
Li, Y., and Lam, F. 2016. Low cycle fatigue tests and damage accumulation models on the rolling shear strength of cross-laminated timber. *Journal of Wood Science*, **62**(3): 251-262.
Lloyd, A. 2010. Performance of Reinforced Concrete Columns Under Shock Tube Induced Shock Wave Loading. MASc, Civil Engineering, University of Ottawa.
Marchand, K. A. 2002. BAIT, BASS & RODS Testing Results. Prepared for the USAF Force Protection Battlelab: Applied Research Associates.
Nassr, A. A., Razaqpur, A. G., Tait, M. J., Campidelli, M., and Foo, S. 2012. Single and multi degree of freedom analysis of steel beams under blast loading. *Nuclear Engineering and Design*, **242**: 63-77.
Popovski, M., and Gavric, I. 2016. Performance of a 2-Story CLT House Subjected to Lateral Loads. *Journal of Structural Engineering*, **142**(4): E4015006.
Poulin, M., Viau, C., Lacroix, D. N., and Doudak, G. 2017. Experimental and Analytical Investigation of Cross-Laminated Timber Panels Subjected to Out-of-Plane Blast Loads. *Journal of Structural Engineering*, **144**(2): 04017197.
Senalik, C. A., and Podesto, L. 2017. Blast-Resistant Testing for Loaded Mass Timber Structures. *Research in Progress*. Madison, Wisconsin: Forest Products Laboratory & Wood Products Council.
Sikora, K. S., McPolin, D. O., and Harte, A. M. 2016. Effects of the thickness of cross-laminated timber (CLT) panels made from Irish Sitka spruce on mechanical performance in bending and shear. *Construction and Building Materials*, **116**: 141-150.
Steiger, R., and Gülzow, A. 2009. Validity of bending tests on strip-shaped specimens to derive bending strength and stiffness properties of cross-laminated solid timber (CLT). Duebendorf, Switzerland: International Council for Research and Innovation in Building and Construction.
Timoshenko, S. P., and Woinowsky-Krieger, S. 1959. *Theory of plates and shells*. McGraw-Hill, New-York, NY.
Viau, C. 2016. Investigating the Response of Light-Frame Wood Stud Walls with and Without Boundary Connections to Blast Loads. MASc, Civil Engineering, University of Ottawa.
Viau, C., and Doudak, G. 2016a. Investigating the Behavior of Light-Frame Wood Stud Walls Subjected to Severe Blast Loading. *Journal of Structural Engineering*, **142**(12): 04016138.
Viau, C., and Doudak, G. 2016b. Investigating the behaviour of typical and designed wall-to-floor connections in light-frame wood stud wall structures subjected to blast loading. *Canadian Journal of Civil Engineering*, **43**(6): 562-572.
Yasumura, M., Kobayashi, K., Okabe, M., Miyake, T., and Matsumoto, K. 2016. Full-Scale Tests and Numerical Analysis of Low-Rise CLT Structures under Lateral Loading. *Journal of Structural Engineering*, **142**(4): E4015007.
Zhou, Q., Gong, M., Chui, Y. H., and Mohammad, M. 2014. Measurement of rolling shear modulus and strength of cross laminated timber fabricated with black spruce. *Construction and Building Materials*, **64**: 379-386.

Building Tomorrow's Society
Bâtir la Société de Demain

Fredericton, Canada
June 13 – June 16, 2018/ *Juin 13 – Juin 16, 2018*June 13 to 16, 2018 / *13 au 16 juin, 2018*

DESIGN CONSIDERATIONS FOR GLULAM BEAMS AND COLUMNS UNDER HIGH STRAIN-RATES

Lacroix, Daniel[1,3], Viau, Christian[2] and Doudak, Ghasan[2]
[1] Carleton University, Canada
[2] University of Ottawa, Canada
[3] Daniel.Lacroix2@Carleton.ca

Abstract: An experimental program investigating the behaviour of glulam beams and columns under high strain rates subjected to simulated blast loading was undertaken. Key observations from the full-scale tests such as the dynamic increase factor (DIF) and failure modes are presented in this paper. A DIF of 1.14 was found to be present only in the absence of continuous or closely aligned finger-joints (FJs) in the outer tension laminations. No increase was observed when continuous or closely aligned FJs were present. The behaviour was observed to be linear-elastic with no significant post-peak capacity. Also, no DIF was observed on the stiffness of the glulam members. A more refined approach using moment-curvature analysis and incorporating second-order effects in the design resistance curve yielded a more reasonable match with the actual behaviour of the glulam beam and column elements.

1 INTRODUCTION

Glulam consists of two or more layers of dimensional lumber joined together with glue and finger-joints (FJs) in the parallel-to-grain direction and is primarily used as beams and columns in the construction of larger wood structures. The layers are stress-rated, thereby allowing for tighter control on the individual lamination properties. A change in regulations permitting the construction of mid- to high-rise structures using wood (NRC 2015) combined with advancement in manufacturing technologies to produce high-performing engineered wood products (EWPs) has allowed wood to be utilized beyond the traditional low-rise light-frame structures and to become a viable material option for much larger structures.

Numerous research studies dealing with the effect of blast loads on the behaviour of light-frame wood structures have been undertaken, where expressions for the dynamic increase factor (DIF) for lumber studs (Jacques et al. 2014) and light-frame wood stud walls (Lacroix and Doudak 2015, Viau and Doudak 2016a, Viau and Doudak 2016b, Viau et al. 2016) were developed. Viau and Doudak (2016b) investigated the effect of realistic boundary conditions on the response of light-frame walls with typical and enhanced connection detailing. Research has also been conducted on the retrofitting of stud walls where enhancements of typical light-frame wall construction and connections were provided (Lacroix et al. 2014, Syron 2010, Viau and Doudak 2016b). While the aforementioned research contributed to the field of blast loading on wood structures, the studies primarily focused on the behaviour of light-frame structures using dimensional lumber. Relatively little research has been conducted on establishing the behaviour of heavy timber and EWPs subjected to blast loading. Thus far, research has been limited to the experimental investigation of strain-rate effects and failure modes of cross-laminated-timber (CLT) panels (Poulin et al. 2018) and glued laminated timber beams and columns (Lacroix 2017).

Taller wood structures are accompanied by an increase in gravity loads acting on the supporting members of lower stories of the building necessitating consideration for the effects of combined lateral-blast and axial gravity loads. To the authors' knowledge, prior to the current research program, no dynamic tests have been conducted on the response of glulam subjected to the effects of the combined lateral-blast and axial gravity loading.

The current paper presents key findings from the experimental program and provides design and analysis methodologies for glulam beam-columns subjected to simulated blast loading. Presented herein are the strain-rate effects, which were quantified by comparing the dynamic to static test results, and the implications of the observed dynamic failure mode on the DIF. More specifically, the focus of this paper is on the effects of variable axial load on the dynamic behaviour. The importance of key parameters and their impact on the response of columns is also discussed.

2 EXPERIMENTAL PROGRAM AND RESULTS

2.1 Experimental Program

Full-scale static and dynamic tests with and without axial load have been conducted with the objective to establish the flexural behaviour of glulam beams and columns as well as to validate a proposed predictive model. A suggested design approach is discussed and validated by comparing the methodology to results obtained from the experimental research program.

2.1.1 Material Description

The experimental program involves the testing of forty-four glulam specimens, statically and dynamically, under four-point bending. The glulam specimens investigated in this research program were obtained from two different manufacturers in which three different cross-sections were considered. The specimens' dimensions and grade consisted of 80 x 228 x 2,500 mm^3 Spruce-Pine (SP), 86 x 318 x 2,500 mm^3 24f-ES SP, and 137 x 222 x 2,500 mm^3 24f-ES SP. The nomenclature of the specimens is such that the letters (B or C) refer to the member type (beam or column) followed by the specimen number. The number in square brackets indicates the width of the narrow face of the cross-section. For example, B3-[86] refers to beam specimen number 3 with dimensions 86 x 318 x 2,500 mm^3 and grade 24f-ES. In addition to having multiple laminations composing the depth of the members, the width of the specimens investigated consisted of either a single laminate (e.g. 80 x 228 mm^2) or multiple laminates (e.g. 86 x 318 mm^2, 137 x 222 mm^2). The specimens with multiple laminates in the beam's width may consist of closely aligning FJs, which, as will be discussed later, has a significant effect on their behaviour. Throughout the paper, the term "beam" is used to describe specimens that have no axial load while being subjected to out-of-plane loading, whereas specimens subjected to combined axial and out-of-plane loads are referred to as "column".

2.1.2 Test Apparatuses and Instrumentation

The static tests were conducted using a four-point bending setup consisting of a reaction frame, hydraulic jack, and steel I-beam, in accordance with ASTM D198 (2015). Using load cells, wire gauges, and strain gauges, the time histories of the static reactions, mid-span displacement, and tensile and compressive wood strains were recorded with a data acquisition system at a sampling rate of 10 samples per second. The dynamic test setup (Fig. 1) closely mimicked the conditions of the static setup. The shock wave resulting from the release of the air in the driver section of the Shock Tube was transferred to the third points of the specimens using a load transfer device (LTD) covering the 2,032 x 2,032 mm^2 end frame opening. The LTD was able to freely displace up to 200 mm due to the slotted hinge system. Axial load was applied using two hydraulic jacks placed between the floor slab and bottom of the specimen whereas load cells measuring the axial load time history were placed at the top end of the specimen. The recorded output for the dynamic tests consisted of time histories for the reflected pressure, displacement, strain, reactions, and axial load. The sensors were connected to a data acquisition system recording at a sampling rate of 100,000 samples per second.

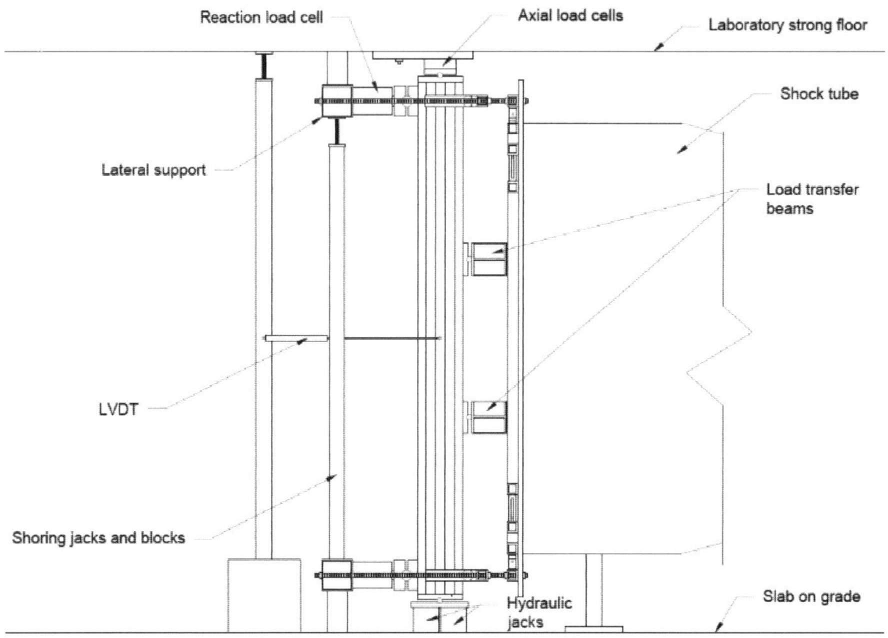

Figure 1: Dynamic test setup for columns

2.2 Full-Scale Test Analysis and Results

2.2.1 Characterization of Failure Modes

Figure 2 shows representative static and dynamic failure modes. During the static testing, the failure modes were limited to simple or splintering tension failure with initial failure occurring either at a knot or FJ (Fig. 2a) followed by propagation of cracks along the glulam layers. For beams with multiple laminates, for which the FJs were not closely aligned, the failure on the tension side appeared to be staggered (Fig. 2b) in comparison to the straight cut across the member width for beams with single laminate or closely aligned FJs across the beam width.

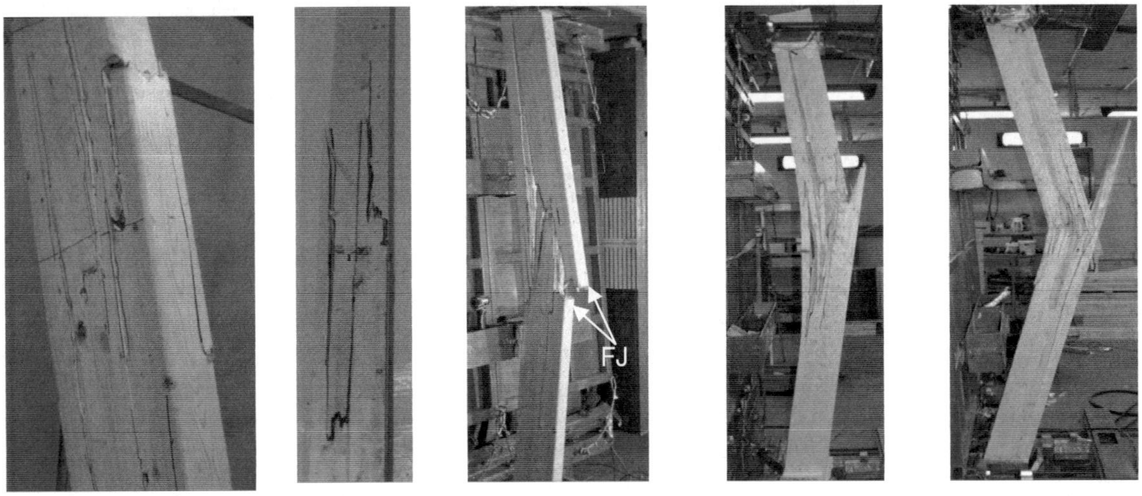

(a) B7-[80] (b) B4-[137] (c) B21.1-[80] (d) B8.2-[137] (e) C4.1-[137]

Figure 2: Representative static and dynamic failure modes

A brash tension failure occurred under dynamic loading for single laminate width beams (Fig. 2c) in comparison to the simple or splintering tension failure under static loading. It was also observed that specimens that had the presence of a FJ in the outer tension lamination within the two load points consistently failed at the FJ (Fig. 2c). Beams with multiple laminates across the member's width generally experienced significantly more crack propagation and splintering (Fig. 2d) than those observed during static testing. For the columns, the damage was concentrated near the member midspan with significant compression failure as evidenced by the wrinkling of the fibres (Fig. 2e).

2.2.2 Strain-Rate Effects – Dynamic Increase Factor (DIF)

Key results describing the average properties for each group of beams investigated including the modulus of elasticity (MOE), modulus of rupture (MOR), and tensile rupture strain ($\varepsilon_{f\text{-}T}$) are presented in Table 1 along with the coefficient of variation (COV). Strain-rates, defined as the strain at yield divided by the time at yield, varied between 0.14 – 0.51s^{-1}.

Table 1: Summary of beams' static and dynamic properties

Specimen group	Property	Static		Dynamic	
		Average	COV	Average	COV
B-[80]	MOE (MPa)	8,368	0.04	9,551	0.12
	MOR (MPa)	51.2	0.09	53.5	0.10
	$\varepsilon_{f\text{-}T} \times 10^{-4}$ (mm/mm)	46.3	0.10	45.2	0.26
B-[86]	MOE (MPa)	8,025	0.04	7,807	0.05
	MOR (MPa)	41.5	0.03	46.8	0.20
	$\varepsilon_{f\text{-}T} \times 10^{-4}$ (mm/mm)	36.0	0.05	35.8	0.19
B-[137]	MOE (MPa)	10,073	0.05	9,313	0.03
	MOR (MPa)	45.6	0.08	48.8	0.11
	$\varepsilon_{f\text{-}T} \times 10^{-4}$ (mm/mm)	40.8	0.20	36.0	0.20

The MOE and MOR presented in Table 1 are "apparent material properties" based on the global response of the beams and were obtained using classical beam equations for the case of a simply supported beam with two concentrated loads at the span's third points (Biggs 1964). Additional steps in obtaining the dynamic MOR and MOE were necessary, where the dynamic resistance was obtained using a single-degree-of-freedom approach considering the inertia effects of both the LTD and glulam specimen in the dynamic equilibrium (Jacques 2016, Lacroix 2017). Determining an average DIF for glulam beams should simply be a matter of relating the dynamic MOR of each glulam beam to the static average for the same type of beam, however, significant difference in the DIF values was observed depending on the failure mode.

Comparisons based on statistical analysis, involving the grouping of specimens by failure (e.g. continuous and staggered) and loading type (e.g. static and dynamic), yielded a DIF of 1.14 for specimens with staggered FJs or no continuous FJ (single width laminate beams) in the outer tension laminations within the maximum stress region (Lacroix 2017). No significant difference was observed between the static and dynamic specimens with a continuous failure indicating that specimens with continuous failure lack the increase associated with high strain-rates. Similar analyses were conducted by the authors for the modulus of elasticity and failure strain, however, the results showed no evidence of a dynamic increase (Lacroix 2017).

2.3 Effect of Axial Load

The effect of axial load was investigated by testing six columns with three different axial load ratios (ALRs). Figures 3a and 3b show representative data for C3-[137] obtained from the column tests. From Figure 3a it can be seen that although the column returned to initial position it experienced some level of damage as indicated by the sudden drop in tensile strain. A comparison of the time at which tensile failure occurred (Fig. 3a) to the peak dynamic reaction (Fig. 3b) shows that both occurred approximately at the same time. A loss in axial load was observed throughout the response of the columns (Fig. 3c), which is consistent with previous findings (e.g. Kadhom 2015, Lloyd 2015). However, the magnitude of the axial load corresponding to peak resistance was more significant (ranging from 50 to 83% of the initial load) than that reported elsewhere for reinforced concrete columns (e.g. Lloyd 2015). Since the span was similar to the aforementioned studies, the difference can be attributed to the compressive stiffness of the glulam columns being significantly less than that of concrete. The dynamic resistance of the column, $R_r(y,t)$, shown in Figure 3c was obtained using Equation 1:

[1] $R_r(y,t) = R(y,t) - \frac{6P_a(t)}{L}y(t)$

where $R(y,t)$ is the dynamic resistance if there were no axial load, $P_a(t)$ is the experimentally measured axial load time history, $y(t)$ is the midspan deflection of the column as a function of time, and L is the clear span. Because the measured dynamic reactions do not account for the secondary moments, an equivalent lateral load (ELL) reflecting the spatial distribution of the blast load (e.g. four-point bending) with the same magnitude as the secondary moments was used.

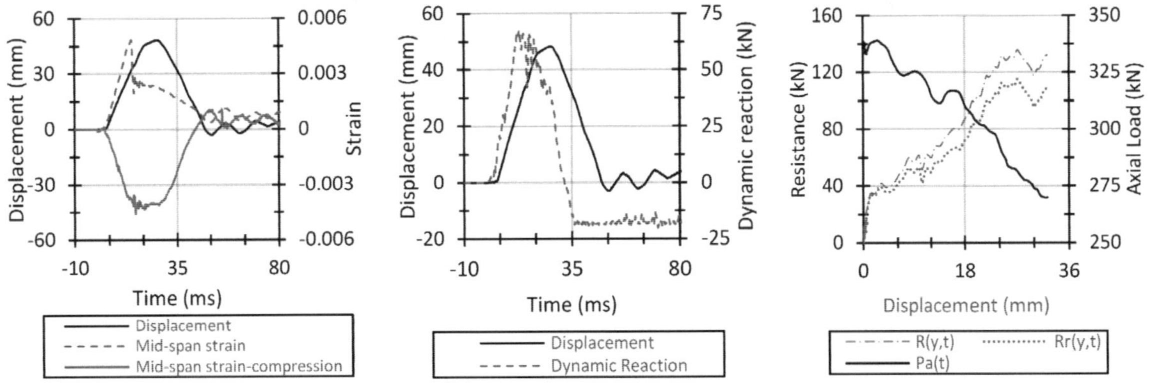

(a) Displacement and strain (b) Displacement and reaction (c) Resistance and axial load

Figure 3: Representative data for C3.1-[137]

It was observed that the addition of axial load has, on average, resulted in a displacement at maximum resistance and a tensile failure strain that are 1.05 and 1.26 times, respectively, greater than those attained in the beam specimens tested dynamically.

3 DESIGN CONSIDERATIONS FOR GLULAM BEAMS AND COLUMNS

3.1 Current CSA S850 Design Approach

Guidance on how to obtain the dynamic design strength of structural components based on the specified static strength is provided in the Canadian blast design standard (CSA 2012). The accuracy of the resistance curve to be used in the dynamic analysis is highly influenced by the assumptions and input parameters for the model. The following investigates the effects of the Canadian blast standard design assumptions on the prediction of the dynamic flexural resistance curves for the B-[137] beams. The resistance and stiffness for a simply supported beam under four-point bending can be found using Equations 2 and 3, respectively:

[2] $R_e = \frac{6S_D}{\mathrm{L}}$

[3] $\mathrm{K} = \frac{56.4\ \mathrm{EI}}{5L^3}$

where R_e is the yield resistance, S_D the dynamic design strength, *K* the stiffness, *E* the elastic modulus, and *I* the moment of inertia. The yield point (x_e) can be found using Equation 4.

[4] $\mathrm{x_e} = \frac{\mathrm{R_e}}{\mathrm{K}}$

The dynamic design strength, shown in Equation 2, is obtained by amplifying the specified static strength (S_S) by a dynamic increase factor (*DIF*) and strength increase factor (*SIF*) as shown in Equation 5.

[5] $\mathrm{S_D} = \mathrm{DIF} * \mathrm{SIF} * \mathrm{S_S}$

The DIF and SIF for glulam suggested by the blast standard (CSA 2012) are 1.4 and 1.2, respectively, whereas the specified static strength, S_s, is determined according to the timber design standard (CSA 2014) as shown in Equation 6.

[6] $S_s = \min\{\varphi[f_b(K_D K_H K_{Sb} K_T)] S K_X K_L;\ \varphi[f_b(K_D K_H K_{Sb} K_T)] S K_X K_{Zbg}\}$

where φ is the material resistance factor, f_b is the bending strength, K_D is the load duration factor, K_H is the system factor, K_{Sb} is the service condition factor for bending at extreme fiber, K_T is the treatment factor, S is the section modulus, K_X is the curvature factor, K_L is the lateral stability factor, and K_{Zbg} is the size factor.

The design assumptions provided by the blast standard (CSA 2012) were investigated and compared to both the experimentally determined dynamic resistance curves and the proposed design approach for glulam beams. The two key parameters investigated are the DIF and response limits. Currently, the blast standard (CSA 2012) assigns a DIF of 1.4 for glulam while the response limits for flexure vary from ductility ratios of 1 through 4 corresponding to increasing damage levels representing elastic to blowout, respectively.

The material resistance factor (φ) is assumed to be equal to unity mainly due to the uncertainties and rarity of occurrence of loading (CSA 2012). The load duration factor K_D is also taken as unity in blast design (CSA 2012). The service factor in bending, treatment factor, curvature factor, and lateral stability factor are all taken as unity to reflect the test conditions, whereas the size factor was taken as 1.3 in conformance with the timber design provisions (CSA 2014). The bending strength was taken as 30.7 MPa based on published data while the modulus of elasticity was taken as the average experimental MOE shown in Table 1. To provide a meaningful comparison with experimental results, the SIF derived from the experimental results was used. The value is reflective of the particular specimens tested and as such, the authors are not suggesting to modify the value existing in the blast design standard.

The flexural resistance resulting from the code design approach is shown in Figure 4 along with the experimental dynamic resistance curves for the B-[137] beams. It can be seen that using a DIF of 1.4 over-predicts the dynamic resistance of the beams. Also, it can be seen that using a perfectly linear elastic-plastic resistance curve with a ductility ratio of 4 is not reflective of the actual glulam beams' behaviour. The beams are observed to behave linearly up to peak resistance followed a significant drop in resistance providing little to no additional energy dissipation.

Also shown in Figure 4 is the flexural resistance curve obtained using the proposed design approach which consists of using a DIF of 1.14 with a linear-elastic resistance curve due to the lack of post-peak resistance (Lacroix 2017). It can be seen that the proposed design approach is more reflective of the actual behaviour of the beams.

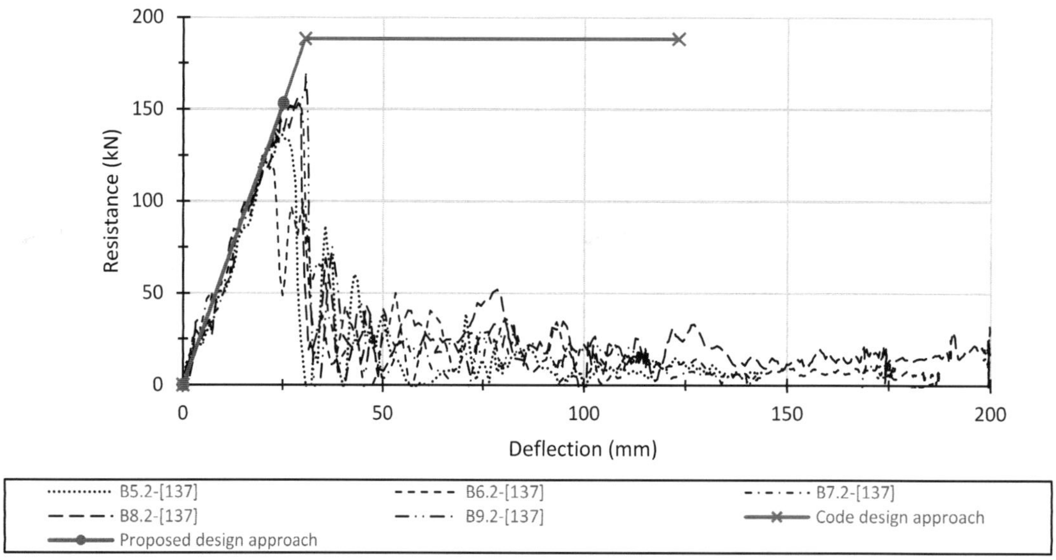

Figure 4: Resistance curves generated from code approach

3.2 Effect of Axial Load

The columns' experimental resistance curves were determined using the ELL approach together with experimentally determined inputs (i.e. dynamic reaction-, applied pressure-, and axial load-time histories) and were shown to corroborate well with the predicted behaviour from the moment-curvature analysis (Lacroix 2017). A material model using tensile and compressive wood stress-strain relationship (determined experimentally) was developed and composite resistance curves were generated to capture the observed loss in axial load. Further details on how the material model and composite resistance curves were developed can be found in Lacroix (2017).

The discussion herein focuses on the effect of varying the axial load with the following hypothetical case. The column is assumed to have the following material properties: ε_{tf} = 0.0045 mm/mm, E_{Wc} = 10,000 MPa, f_{cy} = 35 MPa, and ε_{cult} = 0.014 mm/mm. Figures 5 to 8 present the results from the sensitivity analysis for the following six cases: Case 1 – $E_{Wt} = E_{Wc}$; *m = 0.01*, Case 2 – $E_{Wt} = E_{Wc}$; *m = 0.2*, Case 3 – $E_{Wt} = 2E_{Wc}$; *m = 0.01*, Case 4 – $E_{Wt} = 2E_{Wc}$; *m = 0.2*, Case 5 – $E_{Wt} = 0.5E_{Wc}$; *m = 0.01*, and Case 6 – $E_{Wt} = 0.5E_{Wc}$; m = 0.2. The analyses were conducted for the 137 x 222 mm^2 cross-section with a span length of 2,235 mm. It should be noted that the resistance curves shown include the effect of the secondary moments and would be more prominent with increasing column length.

In general, the analysis revealed that for all six cases increasing the level of axial load resulted in a decrease in maximum moment and resistance, which is consistent with the observations made by Buchanan (1986) for the case $E_{Wt} = E_{Wc}$. This behaviour is at odds with the reported increase in stiffness and resistance observed with increasing compressive axial load for reinforced concrete columns (Burrell et al. 2015, Jacques et al. 2013, Lloyd 2015). It should be noted, however, that this behaviour depends on the stress-strain relationship used as input in the moment-curvature analysis. For example, when the tension MOE (E_{Wt}) is significantly higher than that of the compression MOE (E_{Wc}), a slight reduction in stiffness is observed as the levels of applied axial load increases (Figs. 5c, 6c, 7a, and 8a). This particular behaviour results in an increase in stiffness in the composite resistance curve as there is a loss of applied axial load. It can also be observed that while there is an increase in ultimate curvature due the applied axial load (Figs. 5a, 5b, 5c and 7a) it is not necessarily reflected in the resistance curves due to the secondary moments dominating the overall response (Figs. 6a, 6b, 6c and 8a) resulting in lower failure deflections.

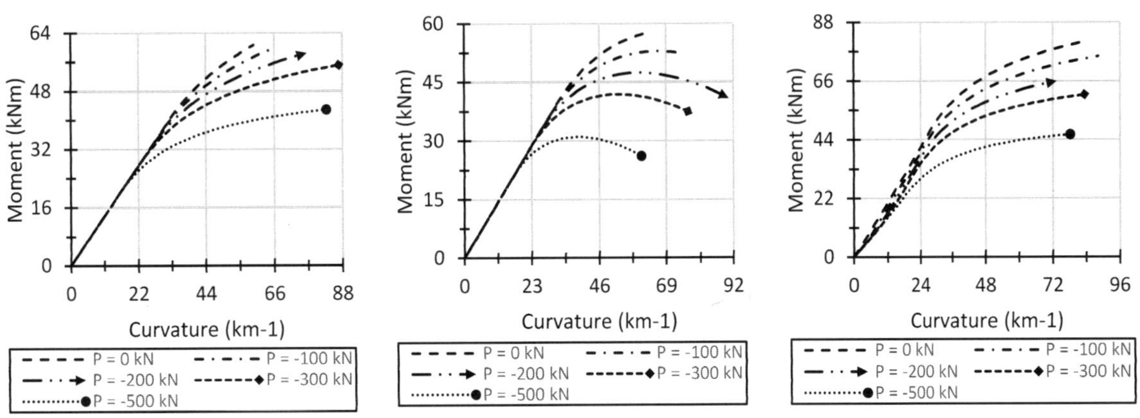

(a) Case 1 – $E_{Wt} = E_{Wc}$; *m = 0.01* (b) Case 2 – $E_{Wt} = E_{Wc}$; *m = 0.2* (c) Case 3 – $E_{Wt} = 2E_{Wc}$; *m = 0.01*

Figure 5: Moment-curvature for Cases 1 to 3

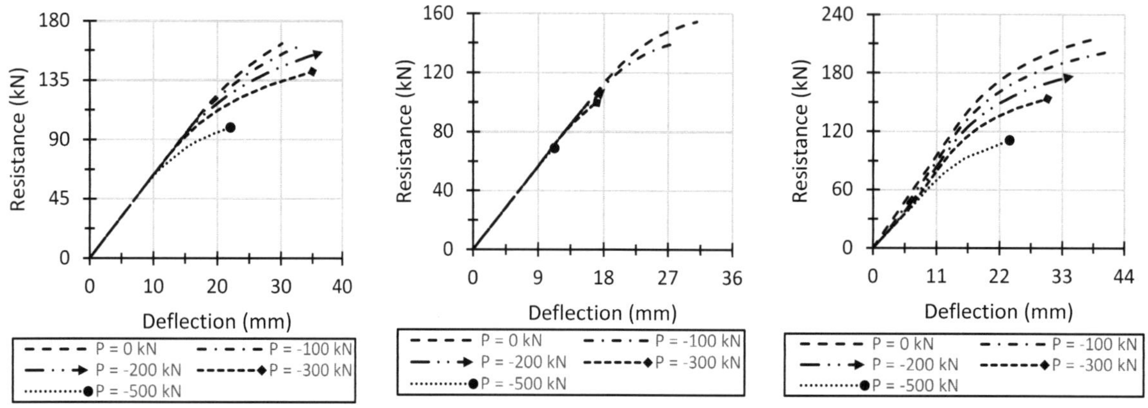

(a) Case 1 – $E_{Wt} = E_{Wc}$; $m = 0.01$ (b) Case 2 – $E_{Wt} = E_{Wc}$; $m = 0.2$ (c) Case 3 – $E_{Wt} = 2E_{Wc}$; $m = 0.01$

Figure 6: Resistance curves for Cases 1 to 3

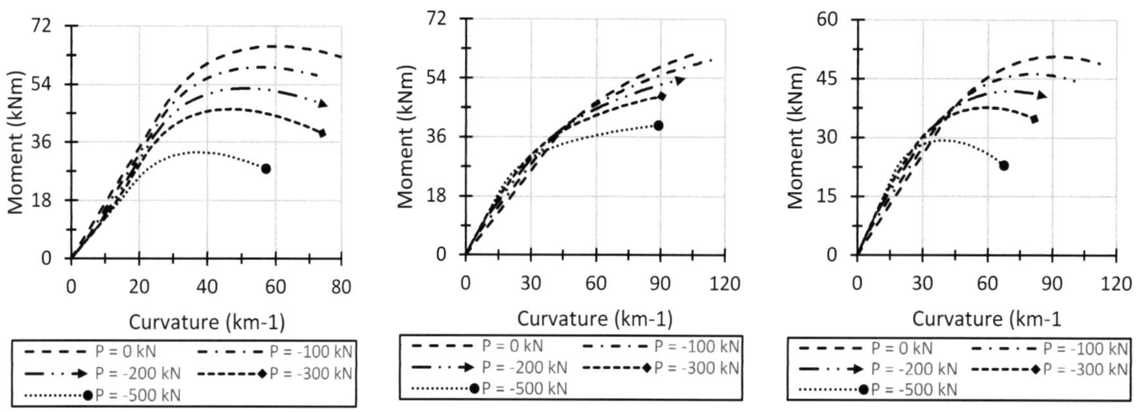

(a) Case 4 – $E_{Wt} = 2E_{Wc}$; $m = 0.2$ (b) Case 5 – $E_{Wt} = 0.5E_{Wc}$; $m = 0.01$ (c) Case 6 – $E_{Wt} = 0.5E_{Wc}$; $m = 0.2$

Figure 7: Moment-curvature for Cases 4 to 6

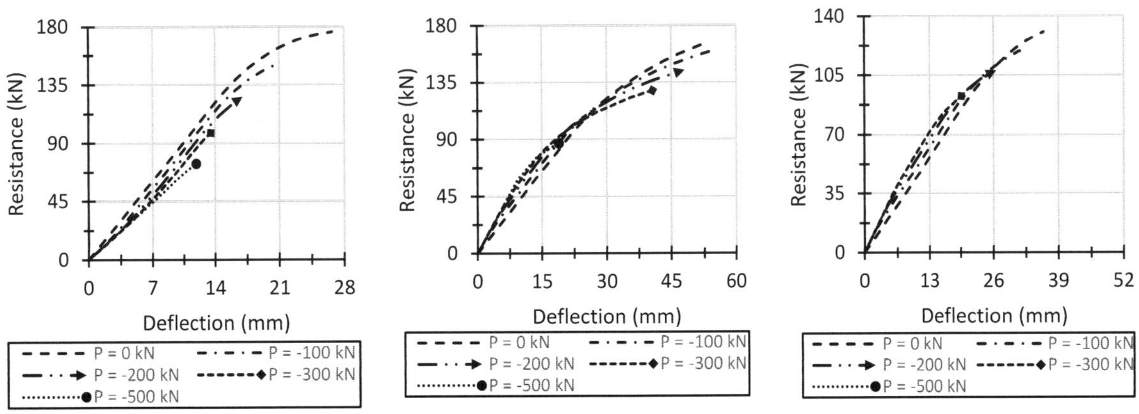

(a) Case 4 – $E_{Wt} = 2E_{Wc}$; $m = 0.2$

(b) Case 5 – $E_{Wt} = 0.5E_{Wc}$; $m = 0.01$

(c) Case 6 – $E_{Wt} = 0.5E_{Wc}$; m = 0.2

Figure 8: Resistance curves for Cases 4 to 6

For the case where E_{Wt} is significantly lower than E_{Wc} (e.g. Cases 5 and 6), there is an initial increase in stiffness with increasing applied axial load, thus resulting in a decrease in stiffness in the composite curve. Although there is an increase in stiffness due to increasing applied axial load, it should be noted that the ultimate moment with axial load is lower than the case of no axial load.

The analysis presented herein highlights the importance of using representative material properties when conducting moment-curvature analysis for wood due to the notable differences discussed. Although the ELL method was used to determine the experimental resistance curves, its use in design without nonlinear reaction-, axial load-, displacement-, and pressure-time histories should be cautioned. Simply applying the ELL method to the resistance curve of the case where there is no axial load may not capture key aspects of the behaviour such as increase in stiffness and increase or decrease in failure displacement. In terms of design approach, it is suggested that a moment-curvature be used to capture the non-linear effects resulting from the material properties and second-order effects. Furthermore, a ductility ratio of 1 should be used due to the lack of post-peak resistance observed in both the beams and columns.

4 CONCLUSIONS

Presented herein are the results of an experimental program investigating the behaviour of glulam beams and columns subjected to blast loading. The results have shown that a DIF of 1.14 is appropriate only when continuous or closely aligned FJs are not found in the outer tension lamination of the glulam specimen. No increase was observed when continuous or closely aligned FJs were present. The post-peak resistance of the glulam beams and columns was not observed to be significant and thus it is suggested that they be designed using a linear-elastic flexural resistance curve, thereby limiting the ductility ratio to unity. Further, no DIF on the stiffness should be applied. In dealing with axial loading, a more refined approach using moment-curvature and incorporating second-order effects should be employed to reflect the actual behaviour of the column.

References

ASTM. 2015. *Standard Test Methods of Static Tests of Lumber in Structural Sizes. ASTM D198-15.* West Conshohocken, PA, American Society for Testing and Materials.

Biggs, J.M. 1964. *Introduction to Structural Dynamics.* New York, NY: McGraw-Hill.

Buchanan, A., H. 1986. Combined bending and axial loading in lumber. *Journal of Structural Engineering*, **112**(12):18.
Burrell, R., Aoude, H., and Saatcioglu, M. 2015. Response of SFRC columns under blast loads. *Journal of Structural Engineering*, **141**(9).
CSA. 2012. Design and assessment of buildings subjected to blast loads. *CSA S850*. Mississauga, ON, Canadian Standards Association Group.
CSA. 2014. Engineering design in wood. *CSA O86*. Mississauga, ON, Canadian Standards Association Group.
Jacques, E. 2016. Characteristics of reinforced concrete bond at high strain rates. Ph.D., Civil Engineering, University of Ottawa.
Jacques, E., Lloyd, A., Braimah, A., Saatcioglu, M., Doudak, G., and Abdelalim, O. 2014. Influence of high strain-rates on the dynamic flexural material properties of spruce–pine–fir wood studs. *Canadian Journal of Civil Engineering*, **41**(1):56-64.
Jacques, E., Lloyd, A., and Saatcioglu, M. 2013. Predicting reinforced concrete response to blast loads. *Canadian Journal of Civil Engineering*, **40**(5):427-444.
Kadhom, B. 2015. Blast performance of reinforced concrete columns protected by FRP laminates. Ph.D., Civil Engineering, University of Ottawa.
Lacroix, D. 2017. Investigating the Behaviour of Glulam Beams and Columns Subjected to Simulated Blast Loading. Ph.D., Civil Engineering, University of Ottawa.
Lacroix, D.N., and Doudak, G. 2015. Investigation of Dynamic Increase Factors in Light-Frame Wood Stud Walls Subjected to Out-of-Plane Blast Loading. *Journal of Structural Engineering*, **141**(6):04014159.
Lacroix, D.N., Doudak, G., and El-Domiaty, K. 2014. Retrofit options for light-frame wood stud walls subjected to blast loading. *Journal of Structural Engineering*, **140**(4):8.
Lloyd, A. 2015. Blast retrofit of reinforced concrete columns. Ph.D., Civil Engineering, University of Ottawa.
NRC. 2015. National Building Code of Canada. Ottawa, National Research Council of Canada,.
Poulin, M., Viau, C., Lacroix, D.N., and Doudak, G. 2018. Experimental and Analytical Investigation of Cross-Laminated Timber Panels Subjected to Out-of-Plane Blast Loads. *Journal of Structural Engineering*, **144**(2):04017197.
Syron, W.D. 2010. Strain Rate-Dependent Behavior of Laminated Strand Lumber. University of Maine.
Viau, C., and Doudak, G. 2016a. Investigating the Behavior of Light-Frame Wood Stud Walls Subjected to Severe Blast Loading. *Journal of Structural Engineering*, **142**(12):04016138.
Viau, C., and Doudak, G. 2016b. Investigating the behaviour of typical and designed wall-to-floor connections in light-frame wood stud wall structures subjected to blast loading. *Canadian Journal of Civil Engineering*, **43**(6):562-572.
Viau, C., Lacroix, D.N., and Doudak, G. 2016. Damage level assessment of response limits in light-frame wood stud walls subjected to blast loading. *Canadian Journal of Civil Engineering*, **44**(2):11.

Building Tomorrow's Society
Bâtir la Société de Demain

Fredericton, Canada
June 13 to 16, 2018 / *13 au 16 juin, 2018*

DEVELOPMENT OF PROBABILISTIC COST FUNCTION FOR FLOOD DAMAGE TO RESIDENTIAL STRUCTURES

McGrath, Heather[1,2], Abo El Ezz, Ahmad[1] and Nastev, Miroslav[1]
[1] Natural Resources Canada, Geological Survey of Canada, Quebec, QC, Canada
[2] heather.mcgrath@canada.ca

Abstract: Risk models, which describe the relationship between hazard intensity and a damage ratio are increasingly used in flood risk management. Direct tangible damage resulting from flooding is typically computed based on the internationally accepted method of depth-damage curves. Depth-damage curves relate absolute damage (in terms of currency) or relative loss (percentage of the estimated total replacement value of property) to a given flood depth. Many depth-damage functions in use today are computed from synthetic data, where data are collected from a representative sample of buildings with similar properties in a floodplain during field surveys. A primary problem when assessing risk at an object-based spatial resolution using depth-damage curves is that these damage functions represent an average structure in the study location. There is great variability across any given structural class as well as variation within individual structures in a structural category (and across communities), for example, not all one-storey residences with basements are the same size, nor constructed of the same quality of materials and workmanship. The variability within a given class of buildings and the resulting depth-damage curve are often not transparent to the end user, thus damage estimates for individual buildings may be over/under estimated. In this paper, synthetic depth-damage curve data from communities in southern Ontario are used to develop probabilistic cost functions such that monetary damage estimates, as spent in Canadian dollars, and their likelihood of being exceeded at any given flood depth are more clearly expressed and communicated to end users.

1 INTRODUCTION

The most common and internationally accepted method for assessment of urban flood damage is through the use of depth-damage curves or stage-damage curves (Plazak1984, Prettenthaler et al. 2010). Depth-damage curves, as the name suggests, relate absolute cost damage (in terms of currency) or relative loss (percentage of the estimated total replacement value of property) to a given flood depth. While many factors can influence the damage incurred from flooding, e.g.: velocity, duration, etc., typically, only inundation depth is considered as a damage causing factor in the formulation of most depth-damage curves (Merz et al. 2004). These curves are developed separately for structural or load-bearing components; for contents (e.g.: interior furniture, art, appliances, etc.,); and for inventory in place (e.g.: commercial stock) (FEMA 2010).

Two common models are used to derive depth-damage curves: empirical and synthetic (Dottori et al. 2016). Empirical models use a data-driven approach, relying on actual damage datasets from past events in order to link building vulnerability and the flood event to damage data, whereas synthetic curves are generated based on a conceptual approach and expert knowledge, hypothesizing and making assumptions about the damage, related to specific components in the building. This approach is thus less dependant on observed datasets for derivation, however data are necessary for validation and calibration (Dottori et al. 2016). Due to the relative scarcity of empirical data for a given community, synthetic depth-damage curves are common.

Depth-damage curves represent the generalized or "average" structure response for the community where they were derived. Within a single building class, e.g.: one-storey single family home with a basement, there is variability in building size, construction materials and quality of finishes. However, depth-damage curves give the impression of a one-to-one relationship, i.e.: for a single given water depth, "x" percent damage or "*y*" dollars of damage will incur. When considered over an entire community, the over and under estimations of damage typically average out (FEMA 2010) - but for a single building there is an unknown uncertainty in the damage estimation results, which is not explicitly stated.

In this paper, synthetic depth-damage curves derived from a 1985 report for the Ministry of Ontario are used as the basis to generate probabilistic cost functions which better represent the variability and variation in the estimated cost of building damages with respect to a range of flood levels. Depth-damage curve data are converted from 1984 dollars to 2016 dollars using indexes from Statistics Canada (https://www.statcan.gc.ca/): Consumer Price Index (CPI), Survey of Household Spending (SHS) and the New Housing Price Index. The resulting probabilistic cost functions are plotted to show the probability of exceeding a specific dollar amount (*"$ XX"*) across a range of flood levels.

2 DATA

The data utilized in this research are from a 1985 report for the Ministry of Ontario (Paragon Engineering Limited et al. 1985), in which synthetic depth-damage curves were created from a sample of seven communities in southern Ontario. The synthetic depth-damage curves were computed via an on-site survey of households of representative structures in each of these communities. The survey results were processed against numerous water depths to generate depth-damage curves. A total of 34 curves were derived for residential structures, including curves for: one-storey, two-storey, split-level residences as well as townhouses and mobile homes. Eleven of these curves are devoted to one-story residences, including variants such as building quality, presence or absence of basement, and construction style (wood frame or masonry foundation). A recommendation in the Paragon Engineering Limited et al. (1985) report suggests mean damage curves for 7 basic types of structures are sufficient. Figure 1 illustrates the aggregation of (a) one-storey structures with basements and (b) two –storey structures with basements with damage cost converted to 2016 dollars and highlights the variability within the Ontario sample datasets over a range of water depths. Figure 1a represents summarized data from a total of 76 structures, while Figure 1b is the result of 99 field surveys on two-storey structures with basements. In both Figures 1a and 1b, the variability in the damage estimate to the structure and contents increases with increasing depth, from approximately $1,000 at -2.4m, to $10,000 at 0 m, up to ~$25,000 difference between different buildings within the same class at 2.4 m flood level.

3 METHOD

The original depth-damage curves were developed in 1984 and reported in 1984 dollars. To convert to present day dollars, a combination of tables and data from the CPI and SHS tables via the procedure outlined by Natural Resources Canada and Public Safety Canada (2017) in the *Canadian Guidelines and database of Flood Vulnerability Functions* were followed. The method to convert contents depth-damage data is via the CPI for weights and classes from the SHS of household spending data tables, Eq. 1 and Eq. 2. While the SHS is not a direct measure of changing prices, spending amounts in each of the categories account for changes in quality and quantity over time (Natural Resources Canada and Public Safety Canada 2017). The CPI is a widely used measure of inflation, published by Statistics Canada (http://www.statcan.gc.ca/). Structure damage is updated to 2016 dollars using the New Housing Price Index (http://www23.statcan.gc.ca/imdb-bmdi/pub/indexN-eng.htm), Eq. 3.

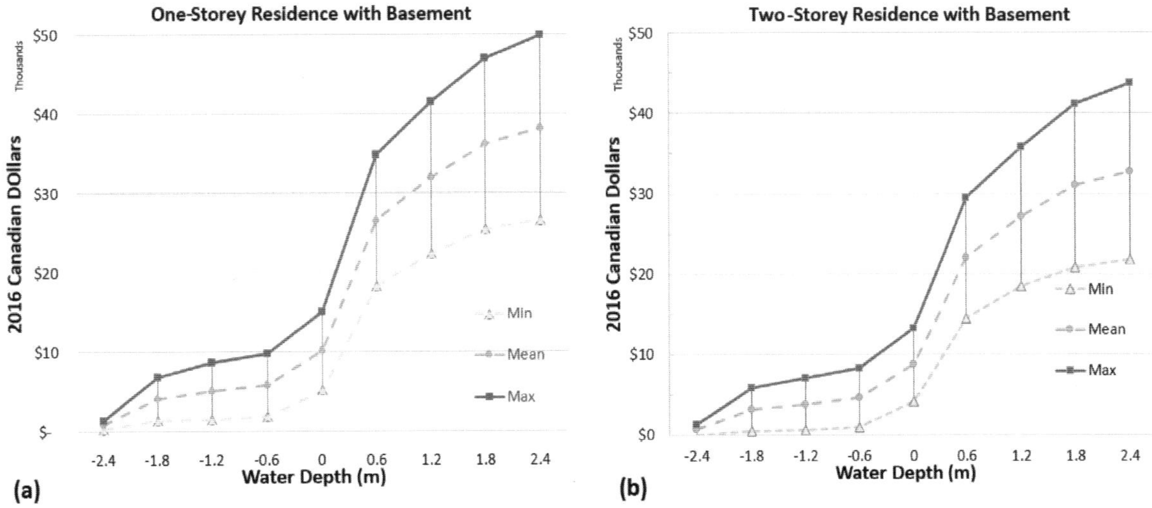

Figure 1: Variability in the total estimated damages for (a) one-storey residences with basement and (b) two-storey residences with basement. The depth-damage curves are based on data from the 1985 Ministry of Ontario, report converted to 2016 dollars

[1] $$Contents\ Flood\ Damage\ Index = \sum(Component\ Index_i * Weight_i)$$

[2] $$Current\ \$\ (Contents) = Base\ Year\ \$\ x\ (Current\ Weighted\ SHS\ /\ Base\ Year\ Weighted\ SHS)$$

[3] $$Current\ \$\ (Structure) = \frac{Current\ New\ Housing\ Index}{Base\ Year\ Housing\ Index} * Base\ Year\ Damages$$

The probabilistic cost functions (as shown in Figure 2) are the result of applying a series of treatments to the depth-damage curve data. The mean (μ) and standard deviation (σ) for each of the 7 structure classes were used to compute the Coefficient of Variation (δ), Eq. 4. Next the lognormal standard deviation (ζ) is calculated using the square of δ, Eq. 5. The median of the lognormal distribution (λ) at each water depth is computed with Eq. 6. Finally, the probabilistic cost functions are derived based on the computed probability of exceedance of a specific dollar loss using the threshold values from Table 1 and Eq. 7. These damage class values (a_i) are based on the Hazus damage classification scheme (FEMA 2010), converted to Canadian dollars and rounded to the nearest hundred for all values other than the minimum.

[4] $$\delta = \frac{\sigma}{\mu}$$

[5] $$\zeta = \sqrt{log(1 + \delta^2)}$$

[6] $$\lambda = log(\mu) - (\tfrac{1}{2}\zeta^2)$$

[7] $$PCF = 1 - NormDist(\log(a_i), \lambda, \zeta, true)$$

Table 1: Damage Class Values

Damage Class Value (a)	Class
$1	Slight
$3,500	Moderate
$6,000	Minor
$10,000	Major
$20,000	Extensive

4 RESULTS

The results for 6 of the 7 structure classes (mobile homes are omitted) are presented in Figure 2. For structures with basements, flood depths ranging from -2.4 m to +2.4 m are considered, to include basement/below grade level flooding, those without basements consider flood depth from -0.6 m to +2.4 m. Zero metres flooding represents the upper plane of the height of the first floor. In all scenarios, there is damage expected at this water level.

In Figure 2a, the one-storey building with basement structure class, the probability of exceeding $1 in damages is lowest at lower water levels, with the inverse being true for the major class (> $20,000) of damage costs. At 0 m water depth, the probability of exceeding $20,000 is negligible, 0.3%. There is a 43% chance of exceeding $10,000 in damages, and 41% of exceeding minor ($6,000) damage at 0 m water level for a one-storey building with basement. At this 0 m water depth there is a 10% probability of minimal damage, less than $3,500. In both one-storey probability cost functions (with and without basements), the probability of exceeding $20,000 in damage rises steeply beyond 0 m water depth, reaching 85% at 0.6 m for structures with and without a basement respectively.

Two-storey structures display a different damage response function to flooding than the one-storey structures. In the two-storey structures, the probability of exceeding $20,000 is lower than the one-storey buildings, with greater than 80% probability at 1.2 m and 1.8 m for basement and no basement respectively. This is, in part, due to the consideration of the second storey. In a two-storey home, the total value of contents (furnishings, furniture, art, etc.) is distributed among multiple levels, whereas a single storey will have its contents on the first floor, thus, the expected total damages incurred, are typically less in a two-storey than a one-storey. It can be observed in Figures 2a and 2c, that the shape of the probabilistic cost functions are quite similar, but shifted toward lower probabilities of exceedance in the two-story building, indicating expected lower costs at the same water levels. At the maximum water depth of 2.4 m the two-storey with no basement shows there is, an 88% probability of exceeding $20,000, which differs from the other one- and two-storey curves which are between 97% and 100% at this flood level.

For a split-level home, which typically has a finished lower level, which is partially below grade, the probability of exceeding $20,000 is the greatest at lower water levels, with a 16% probability of exceeding $20,000 at 0 m water depth and 100% exceedance at 1.2 m water depth.

The computed probability cost function for the townhouse shows a difference damage response than the other classes. In this class, the probability of exceedance of $20,000 in damages is very low, with only a 27% probability at the maximum water depth of 2.4 m. At 0 m water depth, there is a 50% probability of damages being less than $3,500. Thus the cost of expected damages overall is lower; however, as is seen in the other classes, cost of damage shows a steep increase between 0 and 0.6 m of flood water.

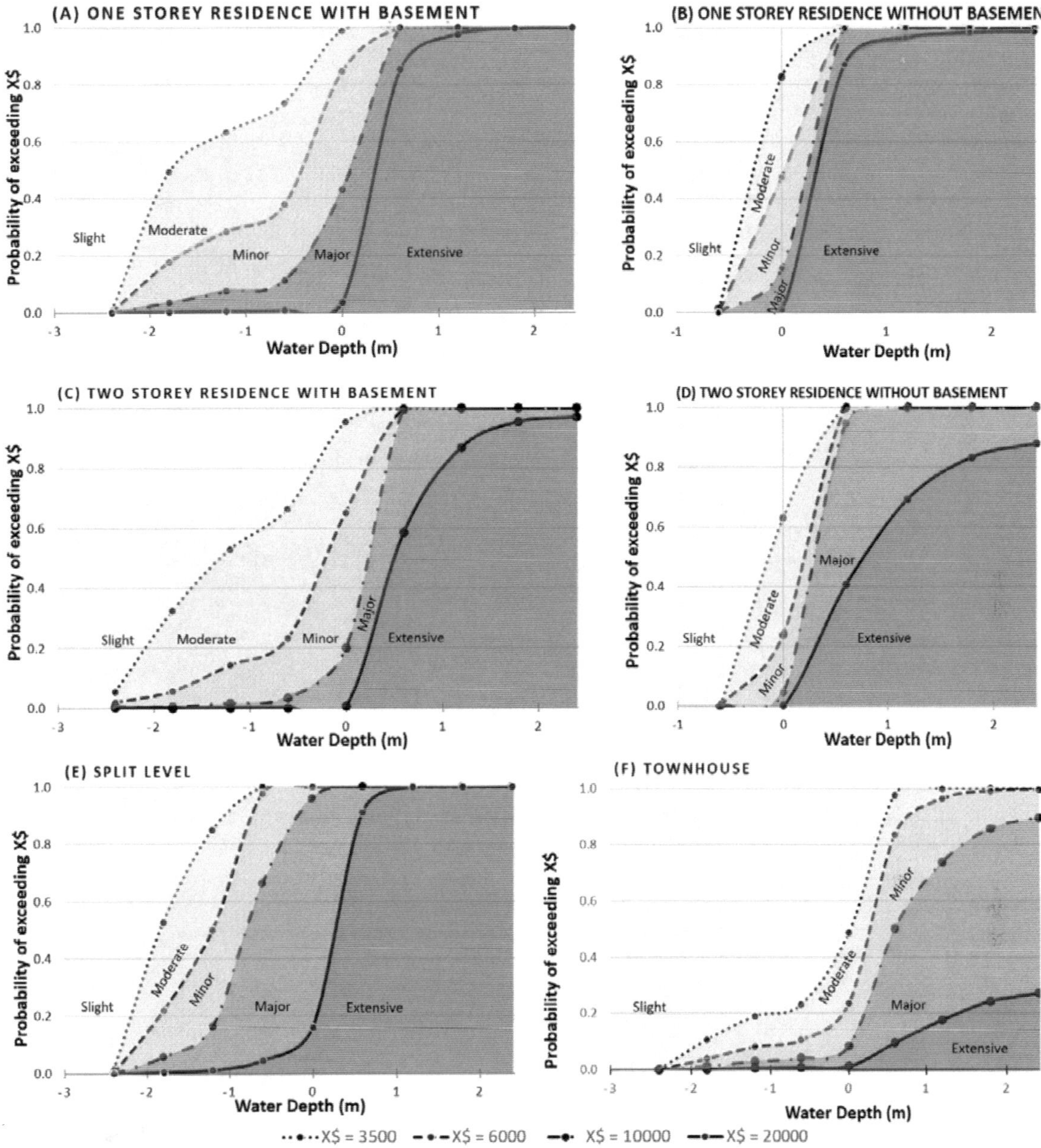

Figure 2: Probabilistic functions of exceeding a given monetary damage level (in 2016 CAD) for 6 structure classes (a – f) based on the 1984 Ontario depth-damage curves.

5 VALIDATION

To test the probabilistic cost functions, a flood risk simulation of residential structures was run. Using data from the city of Fredericton, 116 single family residences, (36 single storey and 80 two-storey buildings) the total estimated flood damages (structural and contents) were computed using ER2 calculator (http://hmcgrat1.ext.unb.ca/ER2_Online/Index.html). Water levels of 0.1 m and 0.6 m were applied as flood hazard input input along with the necessary building details and the original depth-damage curve data (in present day dollars, normalized with total loss value) to estimate damages. The result are shown in Figures 3 and 4. The 36 single story residences with basements have an average building value of $185,500. At

0.6 m water depth the average damage among this sample is $29,558, with a standard deviation of $7,621, Figure 3b. From this sample, the probability of exceedance, using the 5 exceedance values, Table 1, was computed and is plotted in Figure 3a. In this sample, at 0.6 m water depth, 100% of the structures had damage estimates greater than $1, $3,500 and $6,000. 94.44% of the buildings compute damage exceeding $10,000 and 88.89% of these 36 residences exceed $20,000 in losses. These damage estimates line up well with the derived probabilistic cost functions. Damages at 0.1m depth level, in the single story do not align as well with the cost function. At this water level, as much as 97.22% of the structures have damage greater than $1, with 3% reporting no damage at all. Only 22% of the buildings have loss estimates greater than $3,500 and no structures having damage greater than $6,000, Figure 3a, b.

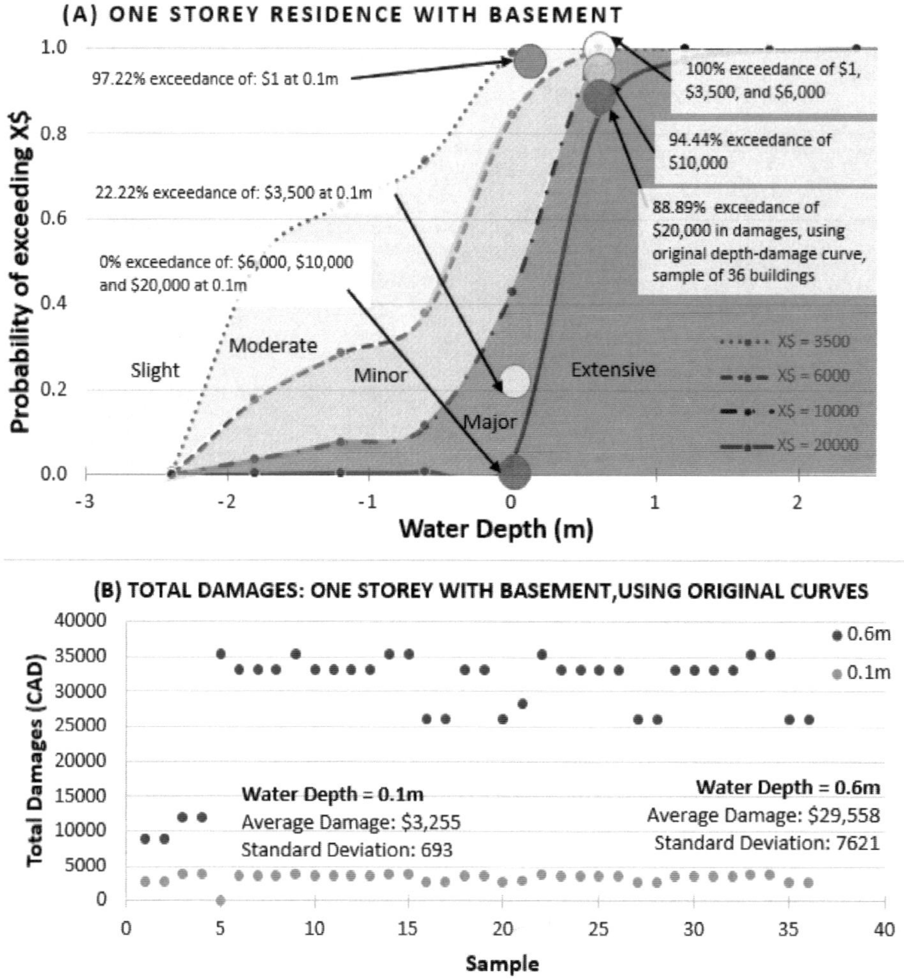

Figure 3 Estimated damage level from a sample of single storey buildings with basements, with simulated flooding of 0.1 m and 0.6 m. (a) sample distribution overlaid on probabilistic cost functions, (b) estimated damage for each buildling in the sample.

These two water levels were also applied to the dataset of two storey buildings, Figure 4. At 0.6 m water depth, it was found that 100% of the properties had total damage estimates greater than $10,000. 58.75% of the structures estimate monetary damage greater than $20,000, which lines up well with the threshold of the probabilistic cost function. At the 0.1m flood depth level, the estimated damages fit well within the probabilistic cost function for all thresholds except the $6,000 threshold. In the sample, 86.25% of the two storey residences has damages exceeding $6,000, however the cost function exceedance range predicts 20 – 70% probability within this damage range.

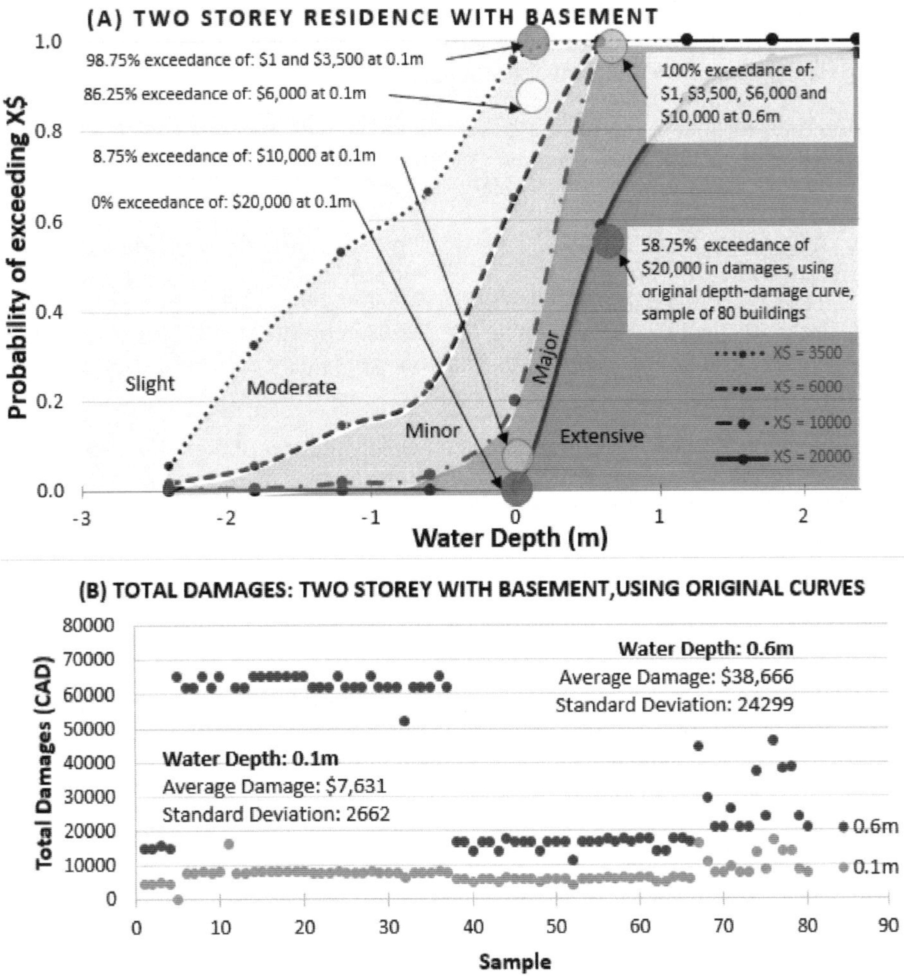

Figure 4. Estimated damage level from a sample of two storey buildings with basements, with simulated flooding of 0.1 m and 0.6 m. (a) sample distribution overlaid on probabilistic cost functions, (b) estimated damage for each building in the sample.

6 CONCLUSIONS

Depth-damage curves represent a standard method to estimate building damage resulting from flood events. These curves are created either synthetically or empirically and represent the average response for a broad class of similar structures in the region for which they were created. Often, the data required to generate and validate these curves are unavailable in all regions, thus, it is common for curves from one community, country or province to be used in other geographic areas. Given that these curves are a representative average, the one-to-one relation of water depth to damage (either in dollars or as a percent of the structure/contents) conveys to the user a fixed damage estimate, with no indication of the uncertainty or variability within the class.

In this research, synthetically derived depth-damage curves from a 1985 report were used as the basis to generate probabilistic cost distribution functions, in an attempt to provide a better idea of the probability of exceeding damages of a specific dollar amount over a range of water depths.

Starting with depth-damage curves from 34 structural classes, the data was converted from 1984 dollars to 2016 dollars. Using the present day currency and the Ministry of Ontario report, which suggests damage curves for 7 basic types of structures are sufficient, the depth-damage curve data were processed and probabilistic cost distribution functions for exceedance of "*$ XX*" were developed. The exceedance classes were based on the Hazus classification for minor and major damage and converted to Canadian dollars.

While the underlying data for this research are based on field surveys conducted in 1984, the described methodology is general and can be applied to any depth-damage data. In addition, the class value ranges as well as the number of classes used can be easily modified.

The probabilistic cost functions were validated against single and two storey residences with basements, using a sample building dataset from Fredericton, NB. Preliminary results indicate that the developed probabilistic cost functions represent the sample datasets well. Further testing of the remaining residential occupancy classes and additional datasets, in other locales, is in progress. Additionally, work is ongoing to access actual claims data to verify the probabilistic cost functions - as the current validation method simply tests one method against another.

The result of this research are probabilistic cost functions which better represent the variability in estimated building damages with respect to a range of flood levels for 7 types of structural classes. The important benefit of these proposed probabilistic cost curves is a more transparent assessment of variability in the expected cost and damages over a range of flood depths.

Acknowledgements

This work is supported by the Public Safety Geoscience Program of Natural Resources Canada – Geological Survey of Canada by the Canadian Safety and Security Program which is led by Defense Research and Development Canada's Centre for Security Science, in partnership with Public Safety Canada. Natural Resources Canada (NRCan) Contribution number 20170367.

References

Dottori, F., Figueiredo, R., Martina, M. L., Molinari, D., & Scorzini, A. R. (2016). INSYDE: a synthetic, probabilistic flood damage model based on explicit cost analysis. Natural Hazards and Earth System Sciences, 16(12), 2577.

FEMA. (2010). HAZUS®MH MR3 Flood Model User Manual., http://www.fema.gov/media-library/

Merz, B., Kreibich, H., Thieken, A., & Schmidtke, R. (2004). Estimation uncertainty of direct monetary flood damage to buildings. Natural Hazards and Earth System Science, 4(1), 153-163.

Natural Resources Canada and Public Safety Canada (2017) Canadian Guidelines and database of Flood Vulnerability Functions, online: http://hazuscanada.ca/sites/all/files/nrc-canadianguidelines-final_2017-03-30_draft.pdf

Paragon Engineering Limited, Ontario Ministry of Paragon Engineering Limited, Ontario Ministry of Natural Resources, "Development of Flood Depth Natural Resources, "Development of Flood DepthDamage Curves for residential Homes in Ontario Damage Curves for residential Homes in Ontario-Vol I - Technical Report" 1985

Plazak, D. (1984). A critical assessment of methodologies for estimating urban flood damages-prevented benefits. Information series (Colorado Water Resources Research Institute); no 52.

Prettenthaler, F., Amrusch, P., & Habsburg-Lothringen, C. (2010). Estimation of an absolute flood damage curve based on an Austrian case study under a dam breach scenario. Natural Hazards and Earth System Sciences, 10(4), 881.

Building Tomorrow's Society
Bâtir la Société de Demain

Fredericton, Canada
June 13 to 16, 2018 / *13 au 16 juin, 2018*

DIRECT MODELLING OF DYNAMIC SOL-STRUCTURE INTERACTION BY FINITE ELEMENTS ANALYSIS

Rioux, Marc-Denis[1,6] / Nollet, Marie-José[2] / Lebœuf, Denis[3] / Koboevic, Sanda[4] / Galy, Bertrand[5]
[1] Ph.D. student, École de Technologie Supérieure (ÉTS), Canada
[2] Professor, École de Technologie Supérieure (ÉTS), Canada
[3] Professor, Laval University, Canada
[4] Professor, Polytechnique Montréal, Canada
[5] Researcher, IRSST, Canada
[6] marc-denis.rioux.1@ens.etsmtl.ca

Abstract: The objective of this paper is to investigate the impact of different types of artificial boundaries on the dynamic behaviour of a free oscillating structure founded at the surface of a soil deposit. The superstructure consists of a moment resisting frame and the soil is considered as a stress dependent medium. 24 DSSI 2D finite elements models are constructed on OpenSees platform. Beam-column elements with fibre sections, including confined and unconfined concrete and steel reinforcement, are employed to represent the frame members. Soil behaviour is modelled using an inelastic multi-yield surface material to a depth of 30 meter in total stress condition. Three different types of lateral boundaries are applied to the soil domain in order to assess their impact on the period elongation and overall damping of the structure under free vibration. Different soil width-to-depth ratios (ϖ=w/H) are considered to evaluate the influence of the distance between the lateral boundaries and the structure. The study showed that the best approach to model the artificial lateral frontier for the problem of the free vibration of a structure located at the surface of an inelastic soil domain (depth 30 meter modelled as a stress dependant medium) is the viscous lateral boundary, using large lateral distance between the structure's footing and the lateral boundary ($\varpi \geq 8$).

1 INTRODUCTION

The current engineering practice rarely considers dynamic soil structure interaction (DSSI) in the design process of new buildings. Even though DSSI is generally perceived to have a beneficial impact on seismic response of the superstructure, a controversy still exists on the subject (Mylonakis and Gazetas 2000). In recent studies, it was proposed to use DSSI in the process of seismic evaluation of existing structures as it might prevent unnecessary retrofits by reducing the seismic force applied to a structure (Gajan and Kutter 2008, Apari et al. 2017).

There are two main approaches to evaluate the impact of DSSI, the direct methods and the substructure approach (Stewart et al. 2012). Both approaches require dynamic modelling of the soil medium to obtain the foundation input motion (Mengi and Tanrikulu 1993). The proper assessment of the extent of the far field is the key aspect in the modelling of the soil. While theoretically the soil medium extend to infinity, in the context of finite element analysis it is necessary to truncated it thereby creating artificial boundaries that do not exist in reality. The infinite soil medium is then modelled as a reduced region called the finite domain (Wolf 1985).

The artificial boundaries must prevent, or at least minimize, wave reflection back into the finite domain for the model to be of any value. This problem has been examined by Lysmer and Kuhlmeyer (1969) who proposed to add dashpot at the numerical lateral boundary to absorb incoming wave and prevent reflection back in the model. This approach can also be used to model the flexibility of the underlying soil deposit at the bottom boundary (Lysmer 1978). Another method to model the artificial boundary is the use of the periodic lateral boundary, better known as the Tied Degree Of Freedom approach (TDOF) (Zienkiewicz et al. 1989). Other methods have been proposed over the years to assess this problem. Comparisons for the efficiency of these methods applied to linear elastic or pseudo elastic soil have been published (Mengi and Tanrikulu 1993, Kontoe et al. 2007).

In the context of DSSI, where large loads are transferred from the structure to the soil, it appears crucial that the soil model be stress dependent and that correct assessment of initial state of stress be reproduced. Several reconnaissance reports show that non-linear behaviour of soil occurred during earthquake events and had a major impact on the dynamic behaviour of the supported structure in the case (Avilès and Pérez-Rocha 1998, Gajan and Kutter 2008, Giorgini et al. 2012, Raychowdhury and Singh 2012). Non-linear DSSI behaviour have been observed even under moderate shaking events (Pecker 2011). Thus, the modelling the soil as a linear elastic homogenous medium does not seem adequate. To the best of the author's knowledge, studies examining the efficiency of artificial boundaries for the problem of inelastic DSSI with a soil model that is stress dependent have not been reported in the literature.

In this paper, the impact of different types of artificial boundaries on the dynamic behaviour of a free oscillating structure founded on the surface of a 30 metres soil deposit is investigated. The soil is considered as a stress dependent medium.24 DSSI 2D finite elements models are constructed on OpenSees platform. Three different types of lateral boundaries are applied to the soil domain. Their impact on the response of the soil-structure system is traced through the period elongation and overall damping. Different width-to-depth ratios of soil medium (ϖ=w/H) are considered to evaluate the influence of the distance between the lateral boundaries and the structure.

2 METHODOLOGY

2.1 Description of the Analysis

To reach the objective, a series of 24 DSSI models are constructed using 2D finite elements. The models are developed in the free software Opensees (OS) (McKenna 1997, Mazzonial et al. 2006). Three different types of artificial lateral boundaries are applied to the soil domain in order to assess their impact on the period elongation and overall damping of the structure under free vibration. In each model, the structure considered is a 3-storey 3-bays reinforced concrete (RC) moment resisting frame 12.2m high, with exterior bay width equal to 8.5m and interior bay width equal to 3m for a total width of the structure of 20m. The frame has 4 distinct footings located at the surface of the soil deposit with dimension L=2m. This type of structure represents a common structural system build between 1960 and 1970 for institutional buildings in the province of Quebec (Lefebvre 2012). The soil considered is a sand deposit in total stress condition with a shear wave velocity (V_s) of 270 m/sec and depth (z) of 30m.

Four different ratios of soil widths to depth (ϖ=w/H) are considered to evaluate the influence of the distance between the lateral boundaries; 0.667, 2, 5.667, 8 (Figure 1 a). The width w is the distance between the structure's footing and the lateral boundary. For each value of ϖ, three different types of artificial boundaries are considered for the soil domain. The selected artificial boundaries were chosen on the basis of simplicity of implementation in a numerical finite element model in the prospect of being used in the practice. The first artificial lateral boundary used is the viscous artificial boundary (Lysmerm and Kuhlmeyer 1969) which consists of discrete elements (dashpots) added at the limits of the finite domain to absorb incoming waves and prevent them from bouncing back into the model. The second artificial boundary used is known as the Tied Degree Of Freedom (TDOF) (Zienkiewicz et al. 1989). In this approach, equal displacements are imposed on nodes on lateral boundaries at equivalent height from the bottom boundary in their respective DOF. In the last artificial boundary used, referred to in the text as the fixed lateral boundary (FLB), all lateral nodes are simply fixed against lateral displacements. A view of the geometric aspect of the model including the soil and the structure is given in Figure 1 (a).

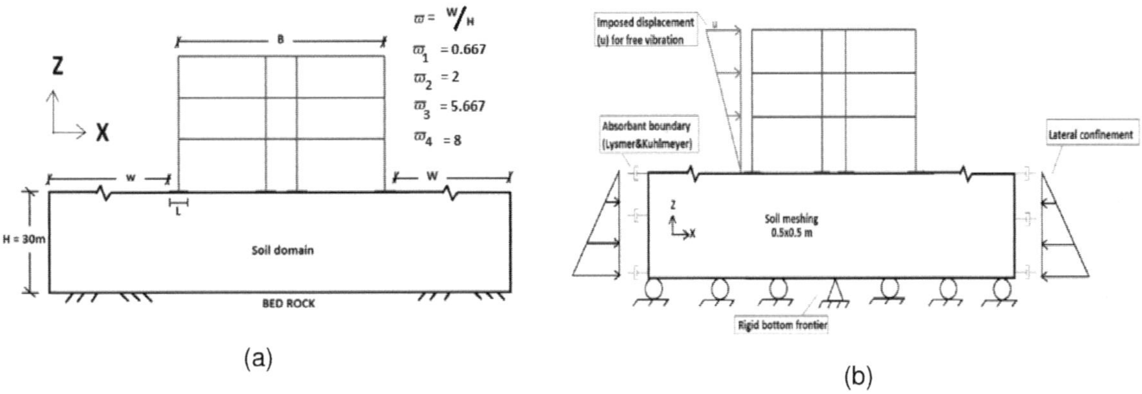

Figure 1 : (a) View of the geometric aspect of the model; (b) Schematic view of the numerical model for the case of viscous artificial lateral boundary and rigid bottom condition

For each ratio ϖ and each type of artificial lateral boundary, the bedrock rigidity is either considered infinite, as depicted in Figure 1 b), or considered finite, and represented using a series of dashpot included to the bottom's frontier nodes to account for the finite rigidity of the underlying bedrock (Lysmer 1978). Properties of bedrock used for the half-space are density (ρ) taken as 2 400 kg/m^3 and shear wave velocity (V_s) taken as 760 m/sec. Table 1 resume the different models tested.

Table 1 : List of tested models and characteristic parameters

Bedrock	$\varpi = {}^{W}/_{H}$	FLB	Viscous	TDOF
Infinite rigidity	0.667	1.A.1	2.A.1	3.A.1
	2	1.A.2	2.A.2	3.A.2
	5.667	1.A.3	2.A.3	3.A.3
	8	1.A.4	2.A.4	3.A.4
Finite rigidity	0.667	1.B.1	2.B.1	3.B.1
	2	1.B.2	2.B.2	3.B.2
	5.667	1.B.3	2.B.3	3.B.3
	8	1.B.4	2.B.4	3.B.4

2.2 Numerical Modelling – Soil

The soil domain is constructed using 0.5x0.5m quadratic elements with a single stabilized gauss point (McGann et al. 2012). The soil mesh is 30 meters deep and has variable ratio ϖ of width to depth, as shown in Figure 1 a) and Table 1. Effect of passive lateral earth force on the soil domain is modelled using loads at each node of the lateral boundaries. The values of the loads to apply are obtained by recording stress in high stiffness spring attached to the artificial boundaries in the first static consolidation analysis steps. The springs are then removed and the forces recorded in each spring are applied to the associated node.

The analysis is conducted in total stress condition, i.e. pore water pressures are not considered. Soil behaviour is modelled using an inelastic multi-yield surface material of the OS library called pressureDependentMultiyield (PDMY) (Yang et al. 2003). Instead of having a single yield limit, the PDMY material defines a nested series of surface which allows representing the small strain behaviour displayed by natural soils. Figure 2 shows the series of yield limits in the principal effective stress space and in the deviatoric plane. As can be seen, the shape of individual yield limit is similar to the Drucker-Prager yield limits (Drucker and Prager 1952).

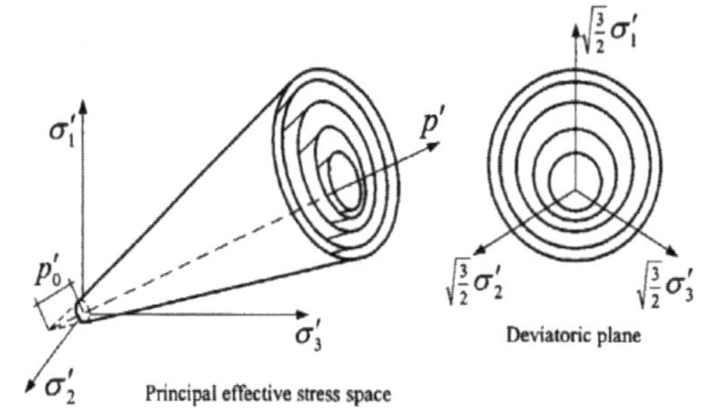

Figure 2 : Conical yield surfaces in principal stress space and deviatoric plane - (Yang et al. 2003)

Values of the soil parameters used in the PDMY model are shown in Table 2, where ρ is soil mass density, φ is the angle of friction of the soil, γ_{max} is the maximum shear strain, P_r is the reference pressure for G_{max} and K_{max}, and ν is Poisson's ratio.

Table 2 : Value of the main parameters used in definition of the PDYM material

ρ (kg/m³)	φ (°)	γ_{max} (%)	P_r (kPa)	ν
1635	37	1	101.3	0.27

The value of the small strain shear modulus G_{max} is calculated from Eq. 1 (Seed and Idriss 1970):

$$[1]\ G_{max} = 21{,}7K_{2-max}P_{atm}\sqrt{\sigma'_m/P_{atm}}$$

where P_{atm} is the atmospheric pressure (≈ 101.3kPa), σ'_m is the mean effective confining pressure, ν Poisson's ratio and K_{2-max} is a parameter establishing the link between the shear modulus value and the confinement pressure, obtained from linear interpolation between value given in the work of Seed and Idriss (1970) or in tabulated format (Kramer 1996). The value of the small strain bulk modulus K_{max} is directly obtained by Eq. 2:

$$[2]\ K_{max} = \frac{2G(1+v)}{3(1-2v)}$$

To ensure a proper choice of G_{max} considering the weight of the structure, an iterative procedure is used. Initial values for G_{max} are chosen as to obtain a V_s of 270 m/sec on the whole soil domain. Static analyses are then conducted and stress state at each element is recorded. Updated value for G_{max} & K_{max} are then computed from equations 1 and 2 and the static analysis is resumed until a stable solution is reached. During the dynamic phase of the analysis, the dependency of the soil to shear strain is directly taken into account in the PDMY material at each time steps based on a hyperbolic relation (Seed and Idriss 1970). Most models used in previous studies do not consider the initialisation of the state of stress and/or the dependency of the shear modulus to the shear strain as the soil is usually modeled as an elastic medium.

2.3 Numerical Modelling – Structure

Fibre sections, including confined and unconfined concrete and reinforcement steel are used to model the structural elements. The inelastic stress-strain relationships considered for structural elements are Concrete02 (Hisham and Yassin 1994) and Steel02 (Filippou et al. 1983) of the OS library. Although the structure remains elastic under free vibration, the non-linearity of the elements was modelled for future analyses. Concrete02 requires input of f'_c, the concrete's compressive strength at 28 days, ε_{c0}, the concrete strain at maximum strength, f'_{cu}, the concrete crushing strength, ε_{cu}, the concrete strain at crushing strength, λ, the ratio between the unloading and initial slope at ε_{cu}, f_t, the tensile strength and E_{ts}, the tension softening stiffness. Steel02 requires input of f_y, the yield strength, E_0, the initial elastic tangent, b, the strain-hardening ratio and parameter to control the transition to plastic branches (R0, CR1, CR2) as well as an optional isotropic hardening parameter. Details about the values used for these parameters and other structural considerations can be found in Apari-Lauzier (2016).

2.4 Free Vibration Analysis

Free vibration of the structure is initiated by a small displacement applied on the left of the frame in accordance with the first modal shape with maximal amplitude of 10^{-6} m. The motion of the top left node of the structure is then recorded and the displacement relative to the base of the footing displacement is calculated. No slipping or uplifting is considered on the interfaces between the soil and the four footings. Numerical period is taken as the mean value obtained between several peaks of the recorded relative displacement. It is then compared with the theoretical value obtained for the classical case of a single degree of freedom (SDOF) elastic oscillator with horizontal, vertical and rotation spring. Theoretical elongation of the SDOF elastic oscillator is given by Eq. 3 (Stewart et al. 2012, Veletsos and Meek 1974):

$$[3]\ \frac{\hat{T}}{T} = \sqrt{1 + \frac{k}{k_x} + \frac{kh^2}{k_{yy}}}$$

where $\hat{T}$ is the natural period of the structure on a flexible base, T is the natural period of the structure on a fixed base, k the rigidity of the structure, k_x the dynamic stiffness of the horizontal spring, k_{yy} the rotational stiffness and h the height of the SDOF model (≈2/3 Height of the structure). The logarithm decrement method, Eq. 4 (Paultre 2005), is used to estimate the damping of the structure (β) during the free vibration:

$$[4]\ \xi \approx \frac{\delta_m}{2\pi} = \frac{1}{2\pi m} \ln\left[\frac{u_n}{u_{n+m}}\right]$$

where u_n is the displacement at cycle n and u_{n+m} is the displacement after m additional cycles. Values of damping obtained are compared with theoretical values expected for the case of a flexible base structure with consideration of foundation damping (β_f) (Stewart et al. 2012):

$$[5]\ \beta = \beta_f + \frac{1}{(\hat{T}/T)^n}\beta_i$$

where β_i is the structural damping for the structure considering fixed base condition and exponent n is taken as 3 for linearly viscous structural damping and 2 otherwise (Stewart et al. 2012). The development of Eq. 3 and Eq. 5 are based on the hypothesis of a SDOF elastic system built over an elastic infinite domain. Since the shape of the imposed displacement respect the first modal shape of the structure and that displacement in both the structure and the soil remains very small, the test conditions for the numerical DSSI finite element model are close to those used in the derivation of the theoretical relation. Hence Eq. 3 and Eq. 5 will serve as benchmark to evaluate the efficiency of the evaluated models presented in Table 1.

3 NUMERICAL RESULTS

In each analysis given in Table 1, the free oscillation is recorded during a 2.5 s interval. Time steps are approximately 0.0005 s with 5000 steps considered in each model. Figure 3 shows an example of the relative lateral displacement of the left top corner of the frame. Damping and difference between numerical and theoretical ratios of period elongation for all models are presented in Figure 4.

Numerical results of models with fixed lateral boundaries and rigid bedrock and increasing value of the ratio ϖ (1.A.1, 1.A.2, 1.A.3 and 1.A.4) are shown in Figure 3 a). They indicate that increasing the value of the ratio ϖ, in other words, increasing the distance between the lateral boundaries, have a significant impact on the recorded displacements. In model 1.A.1 (ϖ = 0.667), the displacement shows a transitory phase followed by a smooth permanent phase with damping ≈1.7%. In model 1.A.2 (ϖ = 2), following the transitory phase, the model fails to develop a smooth permanent phase, as shown by the failed peak at 0.5 s. Model 1.A.3 (ϖ = 5.667), develops a similar trend. Model 1.A.4 (ϖ = 8), develops a permanent phase but the permanent variation in amplitude shows that the motion is perturbed. Estimation of periods or damping from models 1.A.2 and 1.A.3 is irrelevant. Pattern displayed by these models indicates that an incoming wave disrupted the free vibration of the building. Since all movements originate from the structure, the disturbance is an effect of the boundary, lateral and/or bottom.

Results from models with FLB and finite rigidity of the bedrock and increasing value of the ratio ϖ (1.B.1, 1.B.2, 1.B.3 and 1.B.4) are shown in Figure 3 b). After the initial transitory phase, the model develops in each case a smooth permanent shape. However, damping differs greatly. Model 1.B.1 (ϖ = 0.667) displays a damping ≈2,25% whilst models 1.B.2 (ϖ = 2), 1.B.3 (ϖ = 5.667) and 1.B.4 (ϖ = 8) display increasing level of damping with increasing value of ϖ , from 4.5% for model 1.B.2 to 7.4% for model 1.B.4. This trend indicates that in model 1.B.1, the lateral boundary of the model is close enough to limit the impact of the finite rigidity of the bedrock. In model 1.B.2, lateral boundaries are farther from the structure, thus the impact of the finite rigidity of the bedrock increases leading to an increase in the damping value.

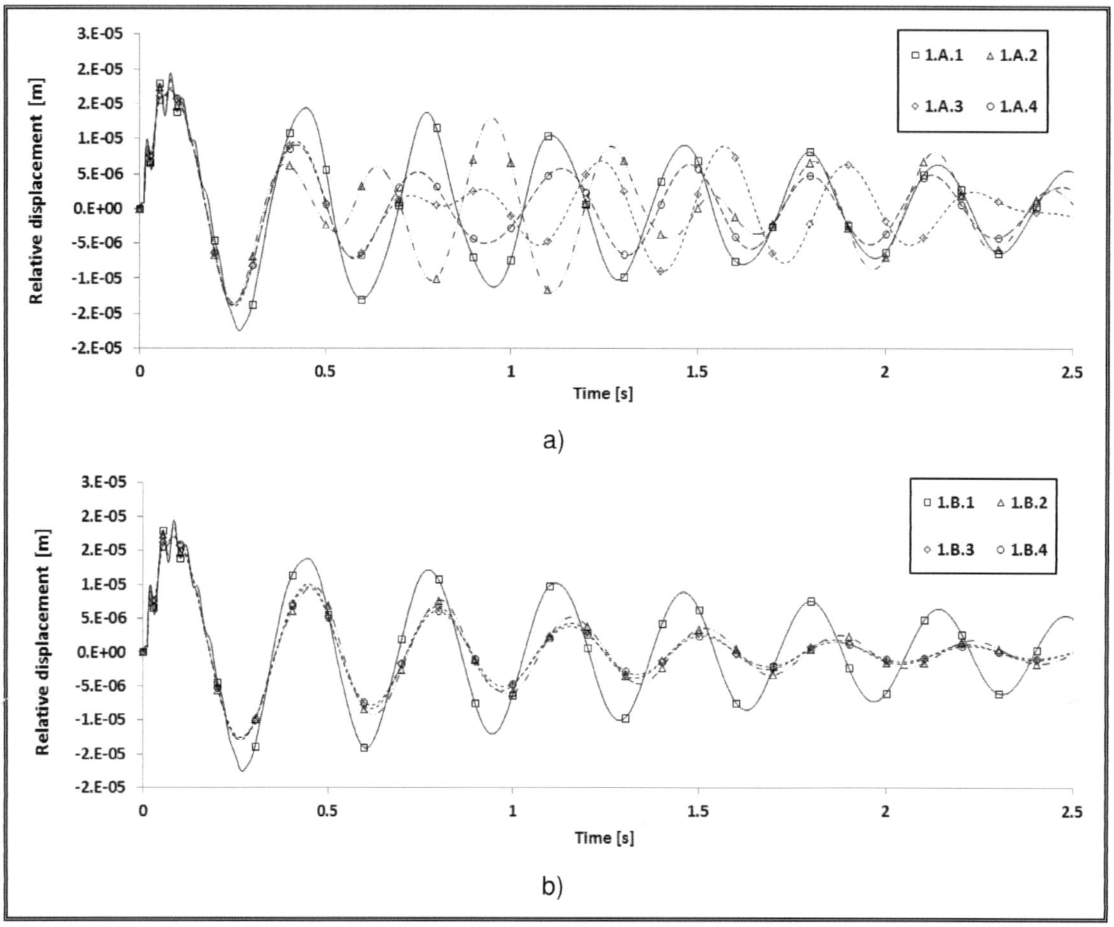

Figure 3 : Relative displacement of the top left node of structure for the case of FLB: a) rigid bedrock b) bedrock with finite rigidity (see Table 1)

Results of models with viscous lateral boundaries and rigid bedrock and increasing value of the ratio ϖ (2.A.1, 2.A.2, 2.A.3 and 2.A.4) show, independently of the value of ϖ, an initial transitory phase followed by a smooth permanent decay of the movement. Differences between the models are noted when examining their respective level of damping and period elongation. Model 2.A.1 (ϖ = 0.667) has higher damping than 2.A.2 (ϖ = 2) which in turn has higher damping than 2.A.3 (ϖ = 5.667). However, damping of model 2.A.4 (ϖ = 8) is higher than damping of model 2.A.3. It was expected that damping would decrease with increasing value of ϖ until a stable value was reached, as distance between the lateral dashpot and the structure is increased. When checking the ratio of period elongation, it is observed that the discrepancy between the theoretical value and the numerical value decreases with increasing ϖ, from 11.3% for model 2.A.1 to approximately 5% for model 2.A.4.

Results of models with viscous lateral boundaries with finite rigidity of the bedrock and increasing value of the ratio ϖ (2.B.1, 2.B.2, 2.B.3 and 2.B.4) are similar to results obtained from models with rigid bedrock. Again, the trend does not seem to be much influenced by ϖ. However, when comparing the elongation ratio, the discrepancy between theoretical and numerical is higher than in the viscous model with rigid bedrock. The difference in elongation ratio increases with increasing value of ϖ in all models except for 2.B.4 (ϖ = 8), remains relatively stable (+0.35%). When considering the damping, the impact of the finite rigidity of the bedrock appears. As value of ϖ increases damping increases and reaches the same value as in model 1.B.4 (ϖ = 8), around 7.4%.

Results of models with TDOF lateral boundaries and rigid bedrock and increasing value of the ratio ϖ (3.A.1, 3.A.2, 3.A.3 and 3.A.4) are similar, in shape, to those obtained for the case of FLB. For model 3.A.1 (ϖ = 0.667), the initial transitory phase is followed by a smooth decay in movement. Models 3.A.2 (ϖ = 2) and 3.A.3 (ϖ = 5.667) however, fail to develop a smooth permanent phase, similar as what was observed with models 1.A.2 and 1.A.3. This suggests, again, that an incoming wave did impact the movement. Model 3.A.4 (ϖ = 8) develops a seemingly permanent phase, although the pattern is less influenced by wave reflection than model 1.A.4. Period elongation ratio and damping for model 3.A.2 and 3.A.3 were not computed as the displacement pattern was irrelevant. When comparing the elongation ratio for model 3.A.1 and 3.A.4, very little variation is observed (+0.55%). The same pattern is seen on the level of damping, which is stable around 2.8%.

Results of models with TDOF and finite rigidity of the bedrock and increasing value of the ratio ϖ (3.B.1, 3.B.2, 3.B.3 and 3.B.4) all show an initial transitory phase followed by a smooth permanent movement. From the result, it is seen that the ratio of period elongation is little influenced by increasing value ϖ. This is not true, however, for the damping, which increases with increasing value of ϖ. For model 3.B.4 (ϖ = 8), the level of damping is roughly equal to that of model 2.B.4 (ϖ = 8), indicating that the impact of the boundary on the response is predominantly coming from the bottom frontier.

Figure 4 summarizes the values of damping for all models obtained from logarithm decrement of the relative displacement curves (Eq. 2). It can be seen that when the finite rigidity of the bedrock is modelled along with a very large soil domain on each side of the building, the damping values for all types of lateral boundaries tend to increase. When the bedrock is considered rigid, only the viscous artificial lateral boundaries does perform well, with larger value of soil domain leading close to the expected value of damping for this flexible base system.

Figure 5 summarizes the values of the difference between theoretical and numerical ratios of period lengthening. As it can be seen, increasing the value of ϖ did impact the values. When the bedrock was considered rigid, the ratio did not change for the TDOF lateral boundaries. The ratio decreases in the case of the viscous lateral boundaries. When the finite rigidity of the bedrock was modelled, all types of models converged with increasing value of width up to a maximum of about 7.6%.

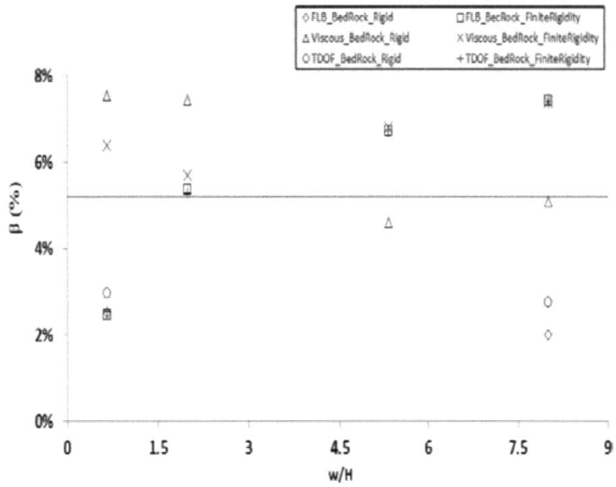

Figure 4 : Damping from logarithm decrement for the different models tested

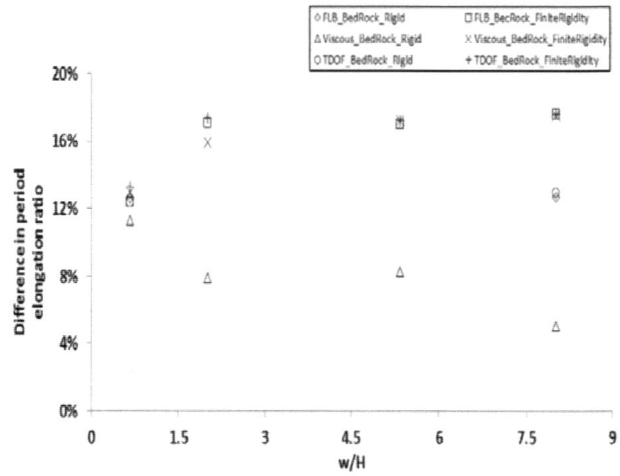

Figure 5 : Difference between theoretical and numerical ratios of period lengthening

When the result of the damping and of the difference between theoretical and numerical ratios of period lengthening are considered together, the following observations can be drawn:

- Fixed lateral boundaries performed poorly with increasing value of ϖ, or the distance between the lateral boundaries, when the bedrock was assumed rigid;
- When the finite rigidity of the bedrock was modelled, FLB period lengthening ratios were between those obtained by the viscous and TDOF methods for all values of ϖ assessed, whilst the damping value was similar for all three methods;
- Viscous lateral boundaries with rigid bedrock over damped the model with low value of ϖ. With larger value of ϖ, the method gave the best damping level of all approaches tested as well as the lowest difference between expected and numerical ratios of period lengthening;
- TDOF lateral boundary with rigid bedrock performed poorly with increasing value of ϖ;
- When the finite rigidity of the bedrock was modelled, the TDOF lateral boundary approach gave the same level of damping and of period lengthening as the FLB.

The viscous lateral boundaries with the ratio $\varpi = 8$ and rigid bedrock assumption did perform better than any other approach tested for the problem considered here. For the largest value ϖ, the difference with the expected ratio of period lengthening is ≈5% whilst the value of damping is within 1% of the expected value for this flexible base system.

4 CONCLUSION

2D finite element simulations were carried out to investigate the impact of different type of artificial boundaries on the dynamic behaviour of a free oscillating structure founded at the surface of a soil deposit modelled as a stress dependent medium. 24 DSSI models were evaluated for different type of artificial lateral boundary and different ratio of soil width to soil height. Original aspect of the work include assessment of the impact of the initial state of stress in the soil deposit on the value of the small strain shear modulus and the consideration of the shear strain on the evolution of the shear modulus during each analysis steps.

The results show the best approach to model the artificial lateral frontier for the problem of the free vibration of structure located at the surface of an inelastic soil domain of depth 30 meter modelled as a stress depend medium is the viscous lateral boundary, using large lateral distance between the structure's footing and the lateral boundary. This method shows good agreement with expected value of damping with larger value of the ratio ϖ while at the same time having good agreement with theoretical ratio of period elongation. The two other approaches valuated showed increasing gap between theoretical and numerical period elongation ratio as the ratio ϖ increased.

Acknowledgements

The financial support of the Fond Québécois Recherche sur la Nature et les Technologies (FRQNT) and the Centre d'Études Interuniversitaire des Structures sous Charges Extrêmes (CEISCE) is gratefully acknowledged.

References

Apari-Lauzier, J.-S. (2016). L'impact de l'interaction sol-structure sur l'évaluation sismique des cadres rigides en béton armée existants situés dans l'est du Canada. École Polytechnique of Montreal, Master thesis.

Apari, J.-S., Koboevic, S., and Nollet, M. J. (2017). Impact of soil-structure interaction effects on seismic assesment in moderate seismic zones. *16th World Conference on Earthquake Engineering*, Santiago, Chile.

Avilès, J. and Pérez-Rocha, L. E. (1998). Site effects and soil-structure interaction in the valley of Mexico. *Soil Dynamics and Earthquake Engineering* 17(1): 29-39.

Drucker, D. C., and Prager, W. (1952). Soil mechanics and plastic analysis for limit design. Quarterly of Applied Mathematics, 10(2), 157-165.

Filippou, F. C., Popov, E. P and Bertero Vitelmo V. (1983). Effects of bond deterioration on hysteretic behavior of reinforced concrete joints. Berkeley, Calif: Earthquake Engineering Research Center, University of California.

Gajan, S. and Kutter, B. L. (2008). Capacity, Settlement, and Energy Dissipation of Shallow Footings Subjected to Rocking. *Journal of geotechnical and geoenvironnemental engineering* 134(8): 1129-1141.

Giorgini, S., Cubrinovski, M., Pampanin, S., Carr, A. J., and Moghaddasi, M. (2012). Integrated foundation-structure modelling of a case study from the Christchurch 2011 earthquake. *15th World Conference on Earthquake Engineering.*

Hisham, M. and Yassin, M. (1994). Nonlinear Analysis of Prestressed Concrete Structures under Monotonic and Cycling Loads, University of California, Berkeley. Ph.D. thesis.

Kontoe, S., Zdravkovic, L., Potts, D. M., and E. Salandy, N. (2007). The use of absorbning boundaries in dynamic analyses of soil-structure interaction problems. *4th International Conference on Earthquake Geotechnical Engineering*, Thessaloniki-Greece.

Kramer, Steven Lawrence. (1996). *Geotechnical earthquake engineering.* Upper Saddle River, N.J.: Prentice Hall.

Lefebvre, K. (2012). Etude du comportement sous charges laterales des ossatures de beton arme avec murs de remplissage de maconnerie, construites avant les annees 1960. (Doctorat), École de technologie supérieur, Montréal, Canada.

Lysmer, J. (1978). *Analytical procedures in soil dynamics.* Berkeley: College of Engineering, University of California.

Lysmerm, J. and Kuhlmeyer, R. L. (1969). Finite Dynamic Model for Infinite Media. *Journal of the Engineering Mechanics Division* 95(4): 859-878.

Mazzoni, S., McKenna, F. T., Scott, M., and Fenves, G. L. (2006). Open System for Earthquake Engineering Simulation (Opensees). Pacific Earthquake Engineering Research Center (PEER), University of California, Berkeley, CA.

McGann, C., Arduino, P. et Mackenzie-Helnwein, P. (2012). Stabilized single-point 4-node quadrilateral element for dynamic analysis of fluid saturated porous media. *Acta Geotechnica*, vol. 7, no 4, p. 297-311.

McKenna, F. T. (1997). Object-Oriented Finite Element Programming: Frameworks for Analysis, Algorithms and Parallel Computing. Civil Engineering, University of California, Berkeley. Ph.D. thesis
Mengi, Y., & Tanrikulu, A. K. (1993). Absorbing Boundary Conditions in Soil-Structure Interaction Analyses. In P. Gülkan & R. W. Clough (Eds.), Developments in Dynamic Soil-Structure Interaction (pp. 111-146). Dordrecht: Springer Netherlands.
Mylonakis, G. and Gazetas, G. (2000). Seismic soil-structure interaction: beneficial or detrimental? *Journal of Earthquake Engineering* 4(3): 277-301.
Paultre, P. (2005). Dynamique des structures - application aux ouvrages de génie civil.
Pecker, A. (2011). Influence of non linear soil structure interaction on the seismis demand in bridges. *International conference on Innovations on Bridges and Soil-Bridge Interaction*. Athens.
Raychowdhury, P. and Singh P. (2012). Effect of nonlinear soil-structure interaction on seismic response of low-rise SMRF buildings. *Earthquake Engineering and engineering vibration* 4(11).
Seed, H. B., Idriss, I. M. (1970). Soil moduli and damping factors for dynamic response analyses. Report No. EERC 70-10, Earthquake Engineering Resource Center, University of California, Berkley, California.
Stewart, J., Crouse, C.B., Hutchinson, T. C., Lizundia, B., Naeim. F and Ostadan F.. 2012. Soil-Structure Interaction for Building Structures. National Earthquake Hazards Reduction Program.
Veletsos, A. S. and Meek, J. W. (1974). Dynamic behaviour of building-foundation systems. *Earthquake Engineering & Structural Dynamics* 3(2): 121-138.
Wolf, J. P. (1985). Dynamic soil-structure interaction: Prentice-Hall.
Yang, Z., Elgamal, A., and Parra, E. (2003). Computational Model for Cyclic Mobility and Associated Shear Deformation. *Geotechnical and Geological Engineering*, 129(12), 1119-1127.
Zienkiewicz, O. C., Bicanic, N., and Shen, F. Q. (1989). Earthquake Input Definition and the Trasmitting Boundary Conditions. In I. S. Doltsinis (Ed.), *Advances in Computational Nonlinear Mechanics* (pp. 109-138). Vienna: Springer Vienna.

Building Tomorrow's Society
Bâtir la Société de Demain

Fredericton, Canada
June 13 to 16, 2018 / *13 au 16 juin, 2018*

FLOOD EARLY WARNING SYSTEMS, MISCONCEPTION AND CHALLENGES - THE CASE OF COLOMBIA

Aguirre, A.[1,2], López, C. [1], Osorio, A. [1], Rivera, L[1], Toro, A. [1], Chang, P. [1]
[1] Universidad Nacional de Colombia Sede Manizales, Colombia
[2] aaguirref@unal.edu.co

Abstract: In Colombia, highly populated urban centers are often located in extremely mountainous terrain. Overall, steep slopes and deep valleys is characteristic of the territory of the Andean mountain range. Furthermore, high climate variability associated with the intertropical zone causes extreme rainfall events throughout the year. In such environment, intense precipitations over steep terrain promptly saturates the watershed and causes overland runoff to rapidly concentrate in natural channels and rivers. The sudden rise of the waterline along the river way threatens vulnerable populations including risks to infrastructure and personal property. Preventive measures to such threat include the implementation of early warning systems in local communities in order to mitigate and limit damages. In order to implement such systems a correct and detailed understanding of the catchment is required including geomorphological parameters and historical rainfall data series of the area allowing one to correctly determine alert thresholds. This study reveals how standard protocols and methodologies for the implementation of channel flood routing early warning systems have shown limited success in Colombia, where limited and/or incomplete data series are available. In addition, common hydraulic modeling softwares are generally ill adapted to the characteristics of the terrain, and watershed response time requires real-time analysis and/or precipitation model prediction as well as up to date monitoring equipment. Difficulties in model calibration and alert threshold determination are also brought into question.

1 INTRODUCTION

Early warning systems (EWS) have been successfully implemented over the years as autonomous or semi-autonomous alert mechanisms meant to respond in a timely manner to various risks associated with flood events (Gestion del riesgo, 2017), thus mitigating the risk to human life and/or damage to personal or public property. Such system allows for the timely warning of any specific pre-established changes occurring over the watershed or channel way. A EWS is comprised of a number of elements. Firstly, a number of monitoring stations are generally set up to measure hydrometeorological variables such as precipitation intensity or channel waterline evolution. The whole system is integrated through a database archiving process with transmitter relays that communicate between the network. Finally, a base station is responsible for data processing, determining if a given alert threshold has been met, according to perceived risk. When deemed necessary, warning stations may generate a public alert or initiate evacuation protocols or trigger additional monitoring protocols (UMAC, 2011) as may be required. Figure 1, shows the general organization of such system with its different components.

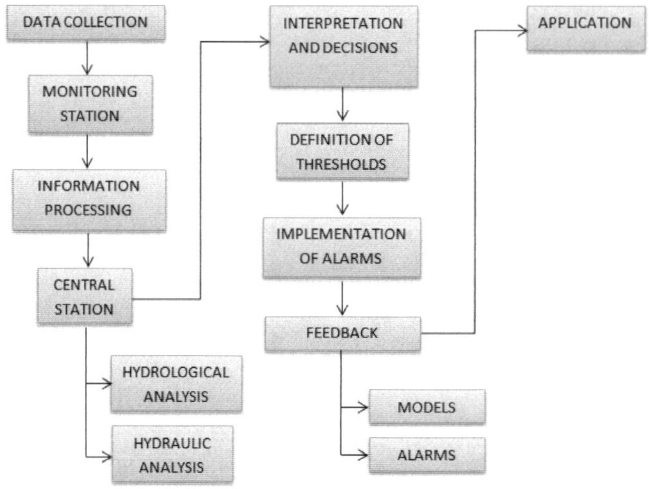

Figure 1. Components of a EWS, Rivera (2018)

Design and implementation of EWS usually require a detailed knowledge of the hydrological conditions of the watershed and the hydraulic properties of the main channel way. Also, a detailed understanding of the geophysical properties of the study area is required in order to allow one to correctly assess data and efficiently support the decision-making process of public authorities or decision makers (UNISDR, 2006).

In addition, monitoring stations are to be calibrated and maintained over time to correctly quantify the hydrometeorological conditions of the watershed. The EWS is often developed and improved over time as the network response is compared to flood events, striving to limit for example, false positives. Eventually appropriate alert thresholds are determined (Diakakis, 2012) considering a given set of risk criteria and thus determining a proportionate response plan considering previously identified vulnerabilities.

In the mountainous region of Colombia, the environment is affected by intense precipitation biannually, that is characteristic of the Inter Tropical Confluence Zone. Torrential precipitation events are typical for the region and combined with the general topography of the area leads to rapidly changing flow conditions with intense run off. This is characteristic for mountainous catchments (Mintegui et al., 2008) and often leads to flash flood events. In the Andes, rural urban centers have been situated along the floodplains. Often geographically isolated and dispersed, such small communities are based on a local agricultural economy. River ways are often the only means to access such communities. Little control is exerted over such areas and limited urban planning has been considered in the development of such settlements aggravating the situation further. Figure 2 shows a typically rural community of the Andeans mountains. On the other hand, major regional centers are also exposed to floods, as they are generally located along the main national river ways. The characteristics of such areas, including population density and exposed public amenities, often lead to greater vulnerabilities when extreme precipitation events occur.

Figure 2. Impoverished dwellings, Rio Blanco Caldas, F. Yara (2017)

On March 31st, 2017, a catastrophic flashflood impacted the small community of Mocoa without forewarning and 300 inhabitants perished and 2000 more were directly affected. Figure 3 illustrates the magnitude of the disaster in Mocoa.

Figure 3. Flashflood Mocoa (El Pais, 2017)

Following the event, the reliance on such local monitoring EWS in support of public safety have been brought into focus. Although various predictive or real time assessment technologies are available in the context of monitoring and alert networks systems, they remain costly and are hardly accessible for small communities. In such context, one understands the importance of correctly establishing emergency protocols and strategically implementing EWS, striving to protect local populations and property (Mendoza et al., 2015). This case study reviews the Colombian context in implementing EWS and highlights current challenges and limitations.

2 NATIONAL CONTEXT

2.1 Data availability and accuracy

In order to correctly implement an early warning monitoring system, valid historical data series that provide for statistically meaningful information is required. Few regions in Colombia are provided with sufficient historical data set for precipitation or river discharge for example to allow for data interpolation, even less so to correctly allow for model calibration. The ability to actually extrapolate from such hydrometeorological data is rarely considered meaningful as was found in various Andes regions (Carrera-Villacrés et al. 2016). Before being operational, a monitoring network and its associated alert systems must be calibrated and validated over a time period including all its components and subcomponents. When available, historical data is often difficult to acquire as field data and monitoring stations are often under the control of privately or semipublic entities or organizations that are not required by law nor are they mandated to disseminate such information.

In Colombia, most rainfall monitoring systems have, in the past, focused on measuring total rainfall over a time window in order to support local agricultural and crop management needs. Monitoring stations used in crop management often lack maintenance and calibration is rarely undertaken, as often limited financial resources are available. Weathering, equipment decay, recklessness, neglect, collapse of infrastructure, robbery, vandalism are common; affecting or altering the data acquisition process. Where in the past, recording errors and the absence of calibration had little or limited consequences; generally local communities would progressively adapt over time to new local environmental conditions or hydrometeorological changes; the lack of data or an erroneous data set, as part of an EWS may in effect have catastrophic consequences (Kundzewicz, 2013).

2.2 Management and Public Authorities

Where in the past, the responsibility for management and maintenance of monitoring networks fell upon local or regional authorities, the integration of alert systems or EWS is generally controlled at the national level. Often such centrally focused organizations are wanting in local knowledge or, lacking resources, are unable to cope with the operation, calibration and maintenance of specifics stations particularly in remote regions. Historically, in Colombia networks and data gathering systems have been neglected and/or purely and simply abandoned over time by local authorities. Furthermore, lack of communication, confusion, mismanagement and miscommunication have generated lost opportunities, have augmented overall implementation costs, and has subjected authorities to liabilities and caused overall delays (Rogers, 2011).

State of the art equipment, that allow for a rapid (often in real time) determination of rainfall intensity given local weather conditions exits (Arattano, 2008). For example, short wave radar technology would allow for a rapid meteorological prognostic and, when required allow for a quick response time at the local level. In Colombia, such systems are rarely found at the regional level and even less so implemented locally. When available, coverage does not extend over the whole area of the countryside which is often more vulnerable. No radar coverage is available in rural areas. In addition, sourcing of such technology is generally done overseas, through foreign agencies or international private interest, as the country is still lacking in the indigenous development of high technology systems. This situation further complicates the management and implementation processes of such technology and equipment.

2.3 Warning systems

Determining alert protocols with limited or inaccurate information remains challenging if not impossible. One has to recall that alarm threshold are meant as a risk mitigation tool for persons and property. Dissemination of alerts can be done through audible alarms such as sirens, local communication (phone, walkie-talkies, HAM shortwave radios) or even through social networks (SMS, phone messaging or texting apps) available locally in order to timely implement for example evacuation protocols. False positive or false negative are a reality. Public trust is rapidly affected by such events and mistrust affects the viability of the system in itself. Indeed confidence in the EWS is key, and one has to consider the social and cultural components of the affected local population. Often training exercises and education is required in order to ensure that indigenous or affected local populations respond correctly, efficiently and in due time to given alert and protocols.

In addition, monitoring stations have in some cases been installed without the required technical expertise or without considering global conditions of the watershed; have been wrongly situated; or have not been appropriately tuned or vetted lacking statistically meaningful historical data set.

Tragic events have occurred where alarms systems were not prompted, were initiated untimely or even have been delayed, being prompted after the tragedy occured.

2.4 Environmental Factors

Geographical information systems (GIS) are often used in the context of alert and warning systems, as they facilitate the geomorphological analysis of the area under concern (Li, 2011). Given the actual size of the watershed, data scaling determines the accuracy of the digital elevation model. Again, often limited topographical data is available in remote areas and map scale are often inadequate and GIS rending scarce. Public authorities have tended lately to supplement the lack of topographical data with aerial photos or through the use of locally purchased aerial drones. Considering the dense canopy that is encountered over the mountainous area of Colombia (Figure 4), such information is of limited help in determining local topography and in many cases is of limited value in even defining the size and the general layout of the catchment basin. Experience with LIDAR technology have proven to be of limited help in determining accurate topographical information in rural areas considering the dense canopy observed in the Andeans and is currently available for urban centers.

Figure 4. Typical canopy of the Andes mountain (Diego M. Garces, 2013)

Alteration of the watershed over time also present additional difficulties. Through urbanization and far-reaching agriculture, zoning and land use management is seldom enforced and soil geomorphology is heavily impacted (Slavoljub, 2011). For example, in a short period of time, significant erosion caused by intensive deforestation increases runoff and may alter flow patterns. In this case, one has to corroborate available topographical information with field data, while trying to establish a baseline study of the area. Such enterprise is not trivial in a developing country with limited technical resources.

2.5 Use of Numerical Modeling Tools

Accurate hydrological and river hydraulic modelling is of outmost importance in the assessment and implementation of flood EWS network, as one strives to correctly and accurately represent water flow processes occurring within the catchment to a sufficiently accurate degree to allow for such prediction tools to be useful. Lack of historical or field data has already been mentioned as a problem in itself, but this situation also limits the possibilities for a robust calibration process. In effect, an erroneous correlation of precipitation intensity vs channel stage could easily be detected and challenged, and in due course, variable correlation could be improved through model calibration. In the absence of accurate and sufficient field data relating observed events to predicted model response, model calibration is in the best case scenario made difficult, or more commonly, rendered impossible. Weather research and forecasting model (WRF) are available to model and predict macroscopic weather events but require downscaling in order to correctly predict or model local or regional events. Currently WRF models are being calibrated and validated but have not been implemented yet as forecasting tools. Research in this area is progressing.

Furthermore, determination of channel characteristics is essential in correctly modelling channel flow. Generally, the modelling of alluvial river flow can readily be undertaken considering spatial approximation of the discharge in 1 or 2 dimension, considering a river's width, to depth, to length ratio, as well as mean uniform flow conditions. Open channel hydraulics in the Andes mountains is characteristically different. In Colombia, channel flow generally does not meet non uniform flow condition showing highly non hydrostatic pressure gradient with a significant vertical dimension component to the flow. The topography is notable for its steep gorges, deep valleys, sharp rises, sudden water falls, adverse slopes and sinuous flow patterns (Anderson et al. 2016). Figure 4a shows the elevation derivatives for Colombia. One can observe distinctively 25% to 50% or greater mean slope in the Andes mountain range. Whereas the watershed hydrograph response or concentration time following a given rainfall is ordinarily understood in terms of hours or even days in North America, that would allow for progressive approximation of the discharge over time. In the Andes, it is not uncommon for the catchment response time to be considered in terms of a few minutes or even to be "immediate", requiring dynamic wave routing that may be more related to dam break theory that to the natural flow of a river flow. Figure 4b show the annual precipitation with 3000-5000mm not being uncommon for the mountain region. Accurately modelling such flow processes is not trivial.

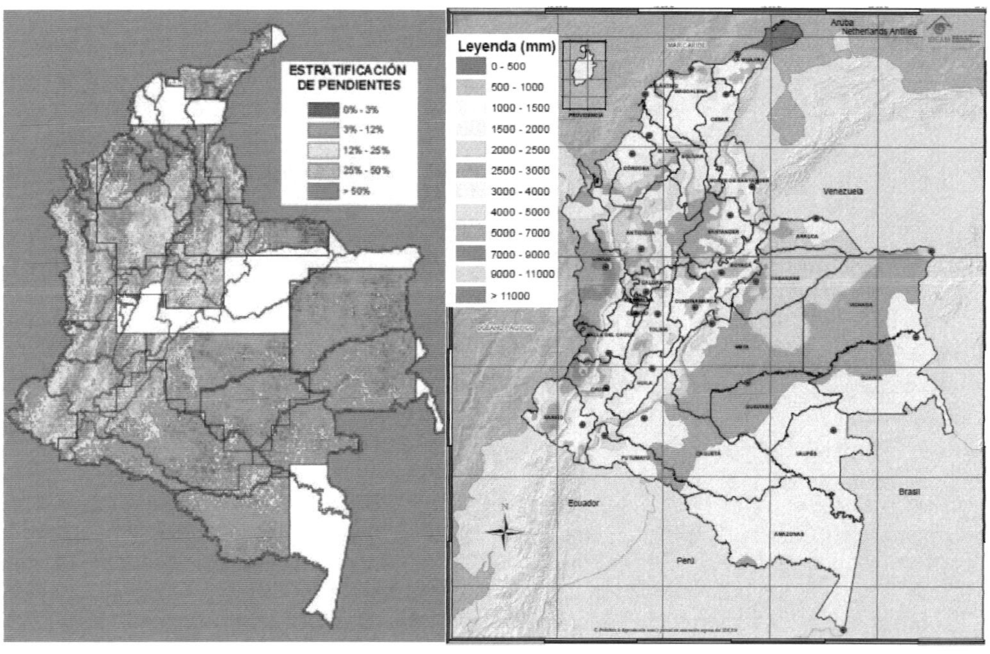

Figure 4a. Average Slope in Colombia (Biesimci, 2018) Figure 4b. Rainfall distribution over Colombia (IDEAM,2014)

2.5.1 Hydrological models

An accurate digital elevation model of the area is often required to allow one to determine the precipitation-runoff correlation of the catchment basin, hence determining its influence on channel flow stages over time.

Hydrological modelling strive to represent the interaction processes of the different components of the hydrological cycle, mass balancing the various intakes as suggested conceptually by equation 1 and presented as a flow diagram in Figure 5.

[1] $R = P - ET - I - Inf$

Where R represents direct runoff, P is associated with precipitation, ET is evapo-transpiration, I is interception and Inf infiltration.

Hydrological models may be lumped, semi distributed or distributed. The former being more common and generally easier to implement requiring less input variables than equivalent distributed models. On the other hand, lumped models require further simplifications and are hence limited in their ability to represent the hydrological balance.

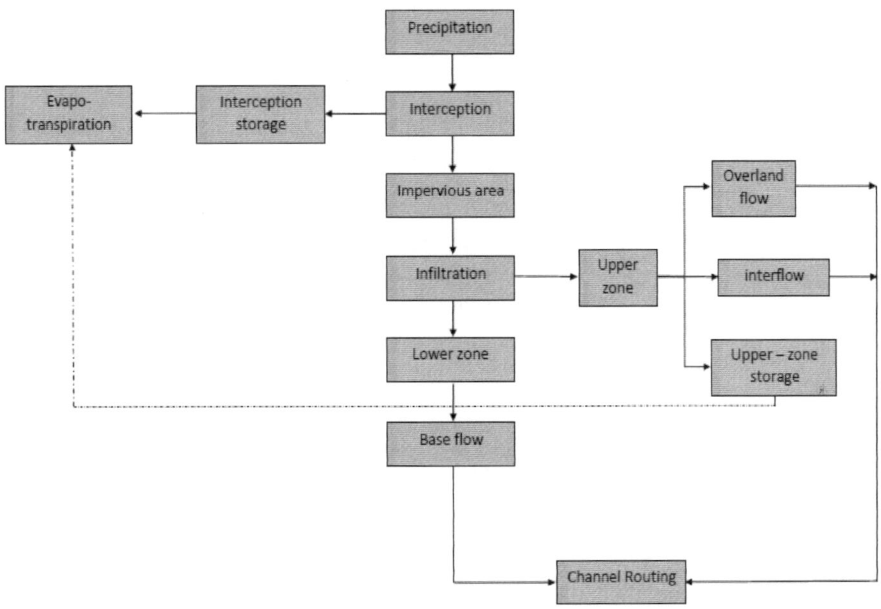

Figure 5. Flow diagram of a model for runoff estimating, Aguirre (2018)

Available integrated GIS models, such as infSWMM and PCSWMM, able to automatically delineate multiple watersheds of varying size.

As the Andean mountains present high climatic and terrain variability, e.j. slopes, soil and sub soil formations, canopy characteristics as well as intense torrential precipitation over the rainy season and periods of extensive dryness, distributed models are generally preferred to model such catchments although they may be more costly and computationally demanding requiring multivariable correlations (Modelos Hidrológicos, Cabrera). Inasmuch as one believes simplified models may be limited in their use as they tend to streamline the phenomena disproportionally, they are generally most often used in practice.

2.5.2 Open Channel Hydraulics models

As hydrological models establish average runoff over time, river hydraulics will determine flood routing characteristics of the event. Over the last 30 years, numerical models have evolved significantly and a broad range are nowadays available. Although more advanced hydraulic flow models like XPSWMM, Mike She, Mike 21, Deltares, HEC-Ras 2D and SWMM incorporate features such as unsteady conditions with a high resolution model of the terrain, readily available one-dimensional uniform or gradually varied flow models based on finite difference approximation and application of the energy equation are more common. Although such models may incorporate modelling of flow regime change and local effects induced by the presence of bridges, piers, jetties, culverts and causeways but are often limited in their use by not allowing for bed condition changes through erosion over time hence affecting changes in the waterline, and requiring the hydrostatic pressure conditions to be met (Cea and Bladé, 2007).

On the other hand, the 2D St-Venant equations may be implemented for shallow flow conditions considering hydrostatic pressure conditions and mild to moderate bed slope.

In all cases, modelling of channel flow processes based on the most common modelling techniques is ill adapted to the Andean mountains were model simplification does not allow one to correctly represent observed flow processes. Indeed, the channel geometry rarely meets the uniform flow conditions, the hydrostatic pressure condition or the one or two dimensional shallow flow requirement.

2.6 Calibration Processes

Flow model calibration requires sufficient data point to validate the channel stage to discharge relationship. In Colombia, few watersheds or channels are provided with gauging stations and when available, monitoring stations are usually restricted to the country's main waterways and navigable channels controlled at the national level. This situation inhibits systematic model calibration and often requires one to calibrate or validate model response based on the limited data that a field monitoring campaign may provide, the personal experience of investigators, or comparison of observed discharge of similar neighboring watersheds establishing correlations when available (Donnelly, 2016). Similarly, hydraulic calibration requires control stations and sufficient knowledge of flow parameters including sediment transport, roughness coefficient, and soil usage amongst other factors. Lacking any of the latter, one is left with field approximation and observation, identifying visually for example, the extent of the flood plain, through witness testimony (oral transmission, hearsay, tradition, local habits, local newspapers and historical photos).

3 CONCLUSION

In Colombia, recent catastrophic events have generated public scrutiny of meteorological monitoring networks and EWS, and a number of networks have been implemented recently at the local level with minimal previous experience in implementing such systems. Shortcomings in technical knowledge, personnel know-how and financial resources, as well as lack of field data and lack of accurate historical data that can be quantified systematically over time have limited the usefulness of such networks.

Hydrological modelling, to be accurate, requires a sufficient number of input variables. Where simplified models may be deemed adequate and less costly they are of limited use in modelling complex interception and run-off dynamics. Hydraulic channel modelling requires one to correctly assess model limitations as it relates to channel flow dynamics.

In such context model calibration and EWS validation and implementation remains an art more than a science at this stage. While recognizing that EWS can make a significant contribution in mitigating risk and improving emergency management processes, the reality is that in Colombia, implementation of such systems remains sketchy at best, often providing a false sense of security. Furthermore, the inability of systems to correctly predict or warn in due time of catastrophic event socially casts doubts on the whole process and on the efficiency and validity of such systems, especially in the case of indigenous populations in remote areas.

Overall, improvement and implementation of EWS in Colombia is hindered by the challenging physical characteristics of watersheds of mountainous regions: topography, concentration time and hydrometeorological conditions to name a few. Nonetheless, such challenges may also present an opportunity, requiring further investigation and improvement in such areas as extrapolation techniques, multiple correlation processes, prediction models, and data optimization. For the time being, the challenge is centered on improving information management including data availability and dissemination as well as extending the overall coverage of monitoring networks. The need for predictive models may very well be required in Colombia, forecasting or anticipating meteorological events before they even occur with real time assessment and analysis of data the implementation of such systems is to be explored further, for example validating downscaled WRF models at the regional level.

Acknowledgements

The authors would like to recognize the financial support of Universidad Nacional de Colombia Sede Manizales through its Grupo de Trabajo Academico en Ingenieria Hidraulica y Ambiental as sponsor of this research study.

References

Alejandra Mendoza, Humberto González, Jorge Buelvas, Sandra Liliana Martínez Rueda. 2015. *Guía Para La Implementación de Sistemas de Alerta Temprana.*

Anderson, Veronica J., Brian K. Horton, Joel E. Saylor, Andrés Mora, Eliseo Tesón, Daniel O. Breecker, and Richard A. Ketcham. 2016. "Andean Topographic Growth and Basement Uplift in Southern Colombia: Implications for the Evolution of the Magdalena, Orinoco, and Amazon River Systems." *Geosphere* 12 (4): 1235–56. https://doi.org/10.1130/GES01294.1.

Arattano, Massimo, and Lorenzo Marchi. 2008. "Systems and Sensors for Debris-Flow Monitoring and Warning." *Sensors* 8 (4): 2436–52. https://doi.org/10.3390/s8042436.

Biesimci. 2018. "MAPAS DE ESTRATIFICACION DE PENDIENTES." 2018. http://www.biesimci.org/altimetria/Estratificacion/indices/Estratificacion.html.

Cabrera, Juan, and Civ Eng. n.d. "Modelos Hidrológicos." consulted http://studylib.es/doc/6826820/calibraci%C3%B3n-de-modelos-hidrol%C3%B3gicos---instituto-para-la-m... March 8 2018

Cardona Arboleda, Omar Darío. 2005. "LA GESTIÓN DEL RIESGO COLECTIVO Un Marco Conceptual Que Encuentra Sustento En Una Cuidad Laboratorio."

Carrera-Villacrés, David Vinicio, Paulina Valeria Guevara-García, Lizbeth Carolina Tamayo-Bacacela, Ana Lucía Balarezo-Aguilar, César Alfonso Narváez-Rivera, and Diana Rosa Morocho-López. 2016. "Relleno de Series Anuales de Datos Meteorológicos Mediante Métodos Estadísticos En La Zona Costera E Interandina Del Ecuador, Y Cálculo de La Precipitación Media." *Idesia (Arica)* 34 (3): 81–90. https://doi.org/10.4067/S0718-34292016000300010.

Cea, Luis, and Ernest Bladé. 2007. "Modelización Matemática En Lecho Fijo Del Flujo En Ríos. Modelos 1D Y 2D En Régimen Permanente Y Variable." *Grupo GIAMA*, 1–30.

Diakakis, Michalis. 2012. "Rainfall Thresholds for Flood Triggering. The Case of Marathonas in Greece." *Natural Hazards* 60 (3): 789–800. https://doi.org/10.1007/s11069-011-9904-7.

Diego M. Garces. 2013. " Cascada en la vía al puerto de Tumaco, Reserva Natural La Planada, Chocó biogeográfico." http://www.wwf.org.mx/?207698/Convocatoria-Becas-Fondo-Colombia-Biodiversa#

Donnelly, Chantal, Jafet C.M. Andersson, and Berit Arheimer. 2016. "Using Flow Signatures and Catchment Similarities to Evaluate the E-HYPE Multi-Basin Model across Europe." *Hydrological Sciences Journal* 61 (2). Taylor & Francis: 255–73. https://doi.org/10.1080/02626667.2015.1027710.

El País. 2017. "Así Es El Plan Con El Que Buscan Devolver La Esperanza a Mocoa." 2017. http://www.elpais.com.co/colombia/asi-es-el-plan-con-el-que-buscan-devolver-la-esperanza-a-mocoa.html.

Gestion del riesgo. 2017. "Sistemas de Alerta Temprana (SAT)." 2017. portal.gestiondelriesgo.gov.co/Paginas/SAT.

IDEAM. 2014. ". Atlas climatológico de Colombia." http://atlas.ideam.gov.co/visorAtlasClimatologico.html.

Kundzewicz, Zbigniew W. 2013. "Floods : Lessons about Early Warning Systems." *Emerging Lessons from Ecosystems*, 379–400.
Li, Xue Ping, and Yun An Li. 2011. "Design of GIS-Based Monitoring and Early-Warning System of Landslide Hazard in Diao Zhongba." *Energy Procedia* 16 (PART B): 1174–79. https://doi.org/10.1016/j.egypro.2012.01.187.
Mintegui Aguirre, JA, and JC Robredo Sánchez. 2008. "Estrategias Para El Control de Los Fenómenos Torrenciales Y La Ordenación Sustentable de Las Aguas, Suelos Y Bosques En Cuencas de montaña.[Strategies for the Control of Torrent Phenomena and the Sustainable Management of the Water, Soils and Forests of M."
Rogers, D, and V Tsirkunov. 2011. "Costs and Benefits of Early Warning Systems." *Global Assessment Report on Disaster ...*, 16. https://doi.org/10.1017/CBO9781107415324.004.
Slavoljub Dragicevic, Nenad Zivkovic Mirjana Roksandic Stanimir Kostadinov Ivan Novkovic Radislav Tosic Milomir Stepic Marija Dragicevic and Borislava Blagojevic. 2011. "Land Use Changes and Environmental Problems Caused by Bank Erosion: A Case Study of the Kolubara River Basin in Serbia," 6604–8. https://doi.org/10.5772/2728.
UNISDR, Estrategia Internacional para la Reducción de Desastres de las Naciones Unidas. 2006. "Desarrollo de Sistemas de Alerta Temprana : Lista de Comprobación," 10.
V.M Aristizabal, B. Botero, J.J. Velez. 2015. *Manual de Hidrología Para Obras Viales Basado En El Uso de Sistemas de Información Geográfica.*
Wiki, UMAIC. 2011. "Sistema de Alerta Temprana." 2011.

Building Tomorrow's Society
Bâtir la Société de Demain

Fredericton, Canada
June 13 to 16, 2018 / *13 au 16 juin, 2018*

FLOOD RISK ASSESSMENT IN CANADA: ISSUES OF PURPOSE, SCALE, AND TOPOGRAPHY

Elshorbagy, Amin[1,2,3], Bharath, Raja[1] and Ahmed, Mohamed[1]
[1] Department of Civil, Geological, and Environmental Engineering, University of Saskatchewan, Saskatoon, Canada
[2] Global Institute for Water Security, University of Saskatchewan, Saskatoon, Canada
[3] amin.elshorbagy@usask.ca

Abstract: Flood risk assessment in Canada is a top priority in both science and policy agendas. Fluvial and pluvial flood assessment, which is the focus of this paper, requires reliable flood mapping in Canada, which in turn raises issues regarding purpose, scale, and topography. Large scale planning of developments, landuse change and zoning, or specific engineering design are examples of different purposes that require different scales for flood mapping. The different topographies across Canada also dictates different approaches for flood modeling and mapping. A national flood risk assessment in Canada, using fine resolution global and national datasets, is presented. A national flood hazard map is prepared using a 20m resolution DEM to identify, for each pixel, the distance from and the height above the nearest river. An exposure map is prepared by using landuse and the satellite-based nightlight data to determine the value of each pixel. A national economic flood risk map is then produced, and subsequently overlaid with population density information to produce a socioeconomic flood risk map for Canada. At local scale, where more detailed flood hazard information is needed, hydraulic models are developed to better map the flood extent that corresponds to specific flood quantiles. A probabilistic flood hazard map (PFHM) for the Qu'Appelle River reach is produced by perturbing input and model parameters within expected ranges of uncertainties using a combined 1D/2D HEC-RAS as a hydrodynamic model.

1 INTRODUCTION

Floods in urban centres and major cities attract attention because of the tangible damage (to life and property) that they cause to large number of people. In Canada, floods claimed the lives of nine people and caused around $9.0 billion in damages since 2000 (Jacob and Church, 2011). Nonetheless, there are several aspects of flood damages that afflict infrastructures, such as highways, vulnerable residents in remote areas, critical powerlines, and environmental services and ecosystems that do not receive equally fair attention. The traditional approach of producing local flood maps in each municipality is insufficient because it lacks consistency in the way flood risks are assessed in different areas in Canada, and it does not allow policy makers to make proper investment decisions with regard to flood damage mitigation. There is enough evidence that societies' flood exposure and vulnerability are increasing both globally (Ceola, 2014) and in Canada (Elshorbagy et al., 2017). For example, Elshorbagy et al. (2017) show that the city of Fort McMurray, Alberta, has expanded significantly between 1992-2013 on the highest flood hazard areas.

The lack of large extent, Canada wide, flood hazard and risk maps can lead to planning, development, and investment decisions that are not flood-smart. FloodNet, a Canada wide strategic research network, started in 2014, and is funded by NSERC to establish flood forecast and flood risk assessment approaches across

Canada (www.nsercfloodnet.ca). The work presented in this paper is part of Theme 4 of FloodNet, which is centered around physical and socioeconomic flood risk assessment in Canada. The objective of this paper is to provide a brief but comprehensive framework of flood mapping and flood risk assessment needs in Canada.

The terms of flood hazard exposure, vulnerability, and risk have some different definitions by different researchers and practitioners, however, we follow the most common definitions provided by UNISDR (2009), IPCC (2012), and Colleantuer et al. (2015). Hazard refers to the disaster itself or its potential occurrence, and usually quantified by engineers as the probability of occurrence of a flood event (de Moel et al. 2009). Intuitively, a low-lying area that is close to a river has a higher level of flood hazard than an area of a higher elevation that is far removed from the river. Elshorbagy et al. (2017) used the distance from and elevation above the river as two indicators of the flood hazard level, which can be better described as flood prone level, of any land pixel. Exposure is given by the economic and intrinsic values that are present at the location involved (IPCC, 2012). Population density, capital investment, and land or property value are indicators of flood exposure. Vulnerability is the capacity of the society to deal with the flood event, which is the state of susceptibility to harm from exposure to floods (Adger, 2006). Risk is the product of hazard, exposure, and vulnerability.

2 LARGE EXTENT FLOOD RISK ASSESSMENT

Parameters representing the concepts of hazard and exposure were identified and, subsequently, a flood risk index was developed. Flood hazard was estimated using elevation above the nearest drainage (EAND), which is similar to height above nearest drainage (Renno et al., 2008) and distance from the nearest from the nearest drainage (DFND). Both parameters were derived from a Canadian digital elevation model (DEM) obtained from natural Resources Canada (http://geogratis.gc.ca/site/eng/extraction). The 20 m DEM resolution was used in this study. Hazard values, calculated as the product of EAND and DFND, were reclassified into five different hazard classes as shown in Table 1. Eventually, a topography-based classified flood hazard map (CFHM) was created for Canada using ArcGIS, and more details of the methodology can be found in Elshorbagy et al. (2017).

Table 1: Classes of EAND, DFND, and the resultant classified flood hazard for Canada

EAND (m)	Class	DFND (m)	Class	Hazard	Class	Hazard level
≤ 2.0	5	≤ 1000	5	21 – 25	5	Severe
2.1 – 4.0	4	1001-2500	4	16 – 20	4	High
4.1 – 6.0	3	2501-5000	3	11 – 15	3	Medium
6.1 – 8.0	2	5001-10000	2	6 – 10	2	Low
> 8.0	1	> 10000	1	1 – 5	1	Very low

The CFHM was validated quantitatively and qualitatively against a flood inundation map developed using hydraulic modeling by the city of Calgary (Government of Alberta, 2013). Validation is meant to assess that the product provides useful information for locating the areas at higher flood hazard (Biondi et al., 2012).

Landuse map for Canada (Latifovic et al., 2012), available at a spatial resolution of 250 m (www.cec.org/tools-and-resources/map-files/land-cover-2005), was used a flood exposure index that reflects the land value. The second exposure parameter considered in this study is nightlights. The nightlight values, based on satellite imagery and defined by a digital number (DN) ranging from 0 to 63 to reflect the degree of luminosity, were classified for Canada into five different nightlight classes. More details on the nightlight classes for Canada can be found in Elshorbagy et al. (2017). An integrated flood exposure index was calculated as the product of landuse value and DN. Finally, flood risk across Canada was calculated as the product of hazard and exposure, as local vulnerability information was not available. Similar to hazard classes, flood risk values were reclassified into five classes, ranging from Severe to Very low.

3 LOCAL SCALE FLOODPLAIN MAPPING

The Qu'Appelle River basin (QRB) in Saskatchewan serves as a significant tributary of the Assiniboine River and flows for a total of 460 km. The case study considered here is located within the QRB and consists of reaches from two rivers: Moose Jaw and Qu'Appelle as shown in Figure 1. The reach under study is 113 km in length and begins at a streamflow gauging station, called Moose Jaw above Thunder Creek (05JE001), and ends just below the city of Lumsden. The Moose Jaw River up to the confluence flows within a defined valley with a limited floodplain and hence, can be modeled with a 1D hydraulic model. The Qu'Appelle River reach from the confluence up to Lumsden has an average channel width of 30 m, whereas the floodplain extends beyond 1 km, and has wetland-dominated terrain, which is suitable for 2D hydraulic modeling. The gauging station 05JE001 was considered as the upstream boundary condition for the Moose Jaw River.

HEC-RAS 5.0.3 was used in this study to perform both 1D and 2D hydraulic modeling. Data requirements for HECR-RAS include cross sections that provide topographic information, friction parameters in the form of manning's roughness coefficient n, and flow data.

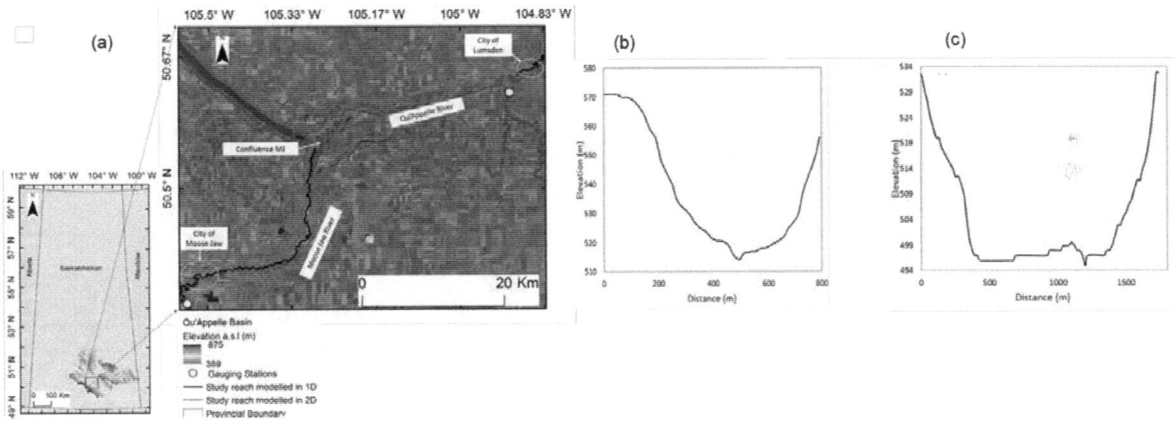

Figure 1: (a) Location of the Moose Jaw and the Qu'Appelle Rivers, (b) typical channel cross section along the Moose Jaw River, and (c) cross section of the Qu'Appelle River.

The hydraulic model was calibrated for the 2011 flood in the Qu'Appelle valley. Daily streamflow measurements were used for a period of 35 days that contained the flood event (April 1 – May 4, 2011). The observed peak flows at Moose Jaw River, Wascana Creek, and at Lumsden are 197 m3/s, 95 m3/s, and 300 m3/s, respectively. Roughness values were lumped into two values, one for the channel and one for the floodplain. The calibrated roughness values were found to be 0.032 and 0.041 for the channel and the floodplain, respectively. Both the roughness parameters (channel and floodplain) and the flood hydrograph were considered to be uncertain, and were varied within reasonable ranges to assess their effect on the flooding extent and depth. Roughness values were sampled in the range of 0.01 to 0.05. Similarly, the flow hydrographs were perturbed to account for 40% uncertainty level (Di Baldassarre and Montanari, 2009). We used the method proposed by Savage et al. (2016) for perturbing the flood hydrograph. A total of 5000 samples for the roughness parameters and the hydrographs were generated to produce probabilistic flood maps in the study area. Eventually, the probability of inundation (PoI), depth of inundation (DoI), and the coefficient of variation with regard to DoI were obtained for each pixel in the study area.

4 RESULTS AND ANALYSIS

4.1 Large Extent Topography- and Nightlight-based Flood Risk Assessment

The topography-based classified flood hazard map (CFHM) of Canada, developed based on the method explained in section 2, is shown in Figure 2. Large areas of the country are classified under severe and high levels of flood hazard due to their elevation and proximity to rivers. However, most of these areas have negligible human presence. The CFHM can be useful for large-scale planning and development, where avoiding encroachment into flood hazardous area is recommended. It can also help identifying critical structures (e.g. highways) and small remote communities that are at stake because of being located within, or surrounded by, high and severe flood hazard areas. A comparison between the CFHM developed in this study and the hydraulic-modeling-based 100-year inundation map is shown in Figure 3. There is good agreement between the 100-year flood inundation and the hazard level classified in this study as severe. A Canada wide flood risk map was produced for Canada according to the methods explained in section 2. To identify flood impact on people (social impact) and separate it from economic impact, we overlaid the flood risk map with the population density layer. Figure 4 shows an example of such flood risk map with and without population for the city of Calgary. The central part of the city with high-rise buildings and high population density remains within the highest level of flood risk. The northern and southern parts, which are mainly commercial areas with lower population density, and thus lower social risk, assume reduced levels of overall flood risk (Fig. 4b) in spite of having severe economic flood risk (Fig. 4a).

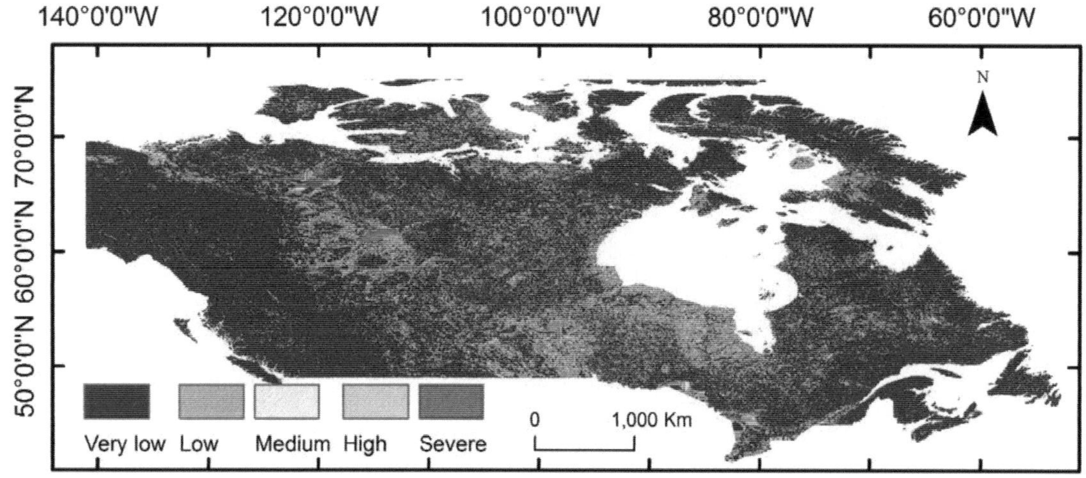

Figure 2: Flood hazard map for Canada obtained using the 20 m DEM.

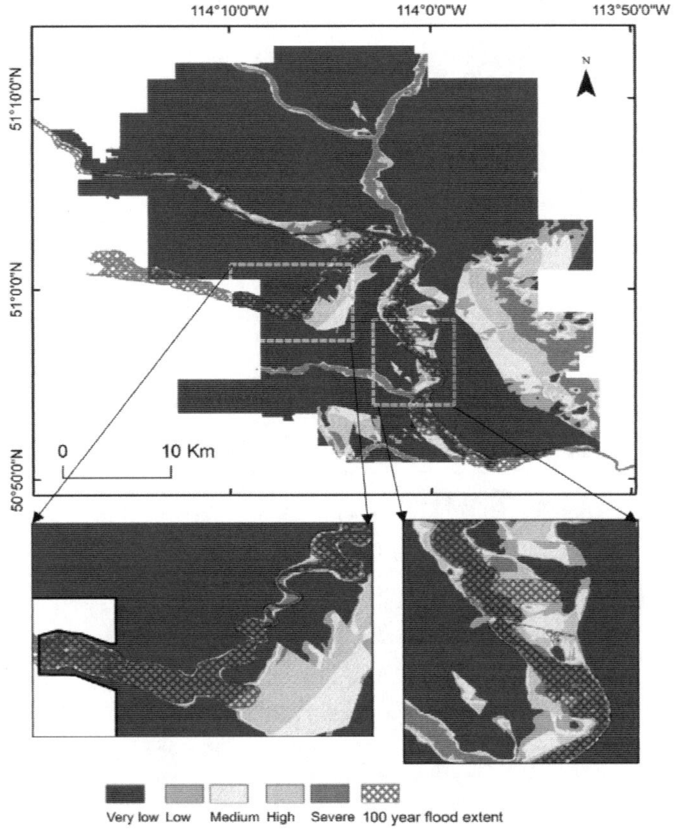

Figure 3: Comparison of the CFHM obtained from the present study and a 100-year flooding map provided by the city of Calgary.

It is important to note that local information is important for estimation of flood hazard and vulnerability. Information on existing flood protection is necessary to provide useful guidelines to decision makers. This information can be used along with the maps provided here to help water resources managers at local scale.

4.2 Probabilistic Flood Mapping

Figure 5 shows the probabilistic flood hazard map (PFHM) with the PoI of each flooded pixel in the floodplain. The PoI values are high at most locations along the reach, and high values of POI (> 0.9) indicate that these locations can be expected to be inundated even when the peak flow is 40% lower than that of the 2011 flood. These high values of PoI can be attributed to the topography of the floodplain along the study reach. The Moose Jaw River is characterized by narrow valley along the river until the confluence with the Qu'Appelle River (Fig 1b). In the Qu'Appele River reach, the floodplain extends over a kilometer and is very flat and thus, water leaving the channel and flowing onto the floodplain is most likely to inundate the entire valley (Fig. 1c). However, even in locations with high PoI, the depth of inundation varies and the variations are shown in Figure 6. In the stretches shown in the figure, the low PoI is associated with those locations where the average DoI is small and vice versa. High DoI is restricted to the channel in the Moose Jaw River, whereas there are multiple locations in the floodplain that have high DoI values in the Qu'Appelle reach. This is due to the presence of potholes in the floodplains that retain water during high flows.

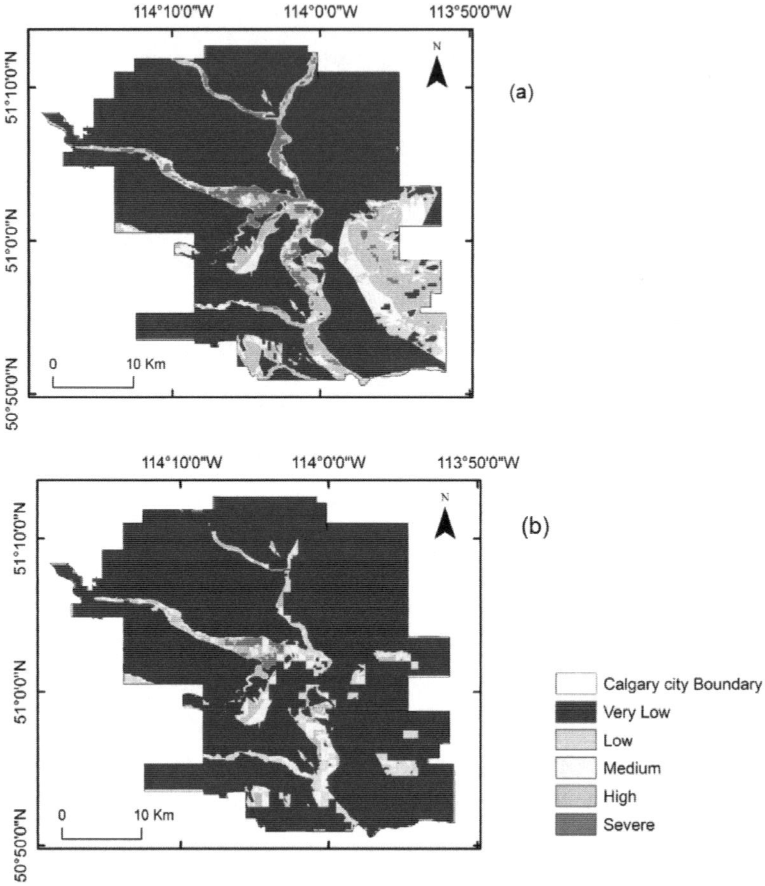

Figure 4: Flood risk map of Calgary, (a) without population, and (b) with population.

To evaluate the large extent flood hazard maps against local area hazard maps obtained using detailed hydraulic modeling, a qualitative comparison was carried out between the CFHM and the PFHM. The CFHM that contained the study reach was extracted from the larger map that contained all of Canada, and the PFHM and the CFHM for a small reach along the Qu'Appelle River is presented in Figure 7. A visual comparison between the two maps indicate that the hazard level "severe" has a slightly wider extent in comparison with the PFHM. The PFHM shows extent of flood on the Qu'Appelle River alone, whereas the CFHM takes into account other streams that flow into the Qu'Appelle River.

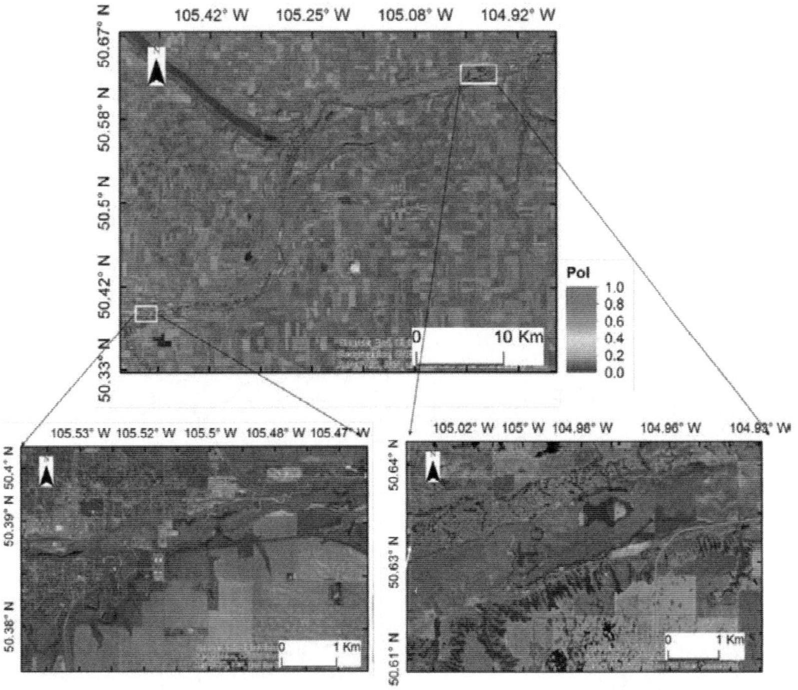

Figure 5: Probabilistic flood hazard map for the entire reach (top) and enlarged portions at a location along Moose Jaw River (bottom left) and along the Qu'Appelle River (bottom right).

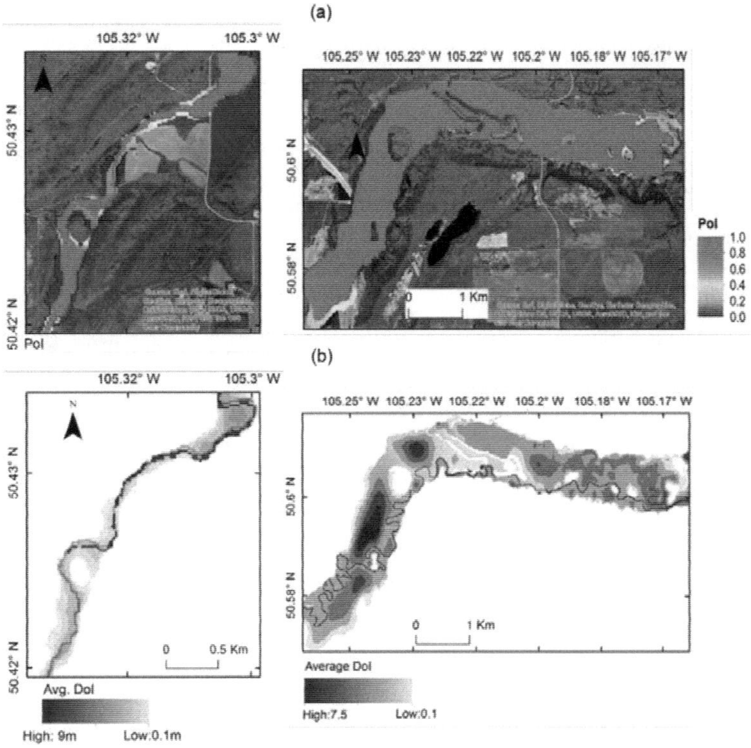

Figure 6: (a) Probability of inundation along the Moose Jaw River (left panel) and the Qu'Appelle River (right panel), and (b) average depth of inundation (DoI) along the same reaches.

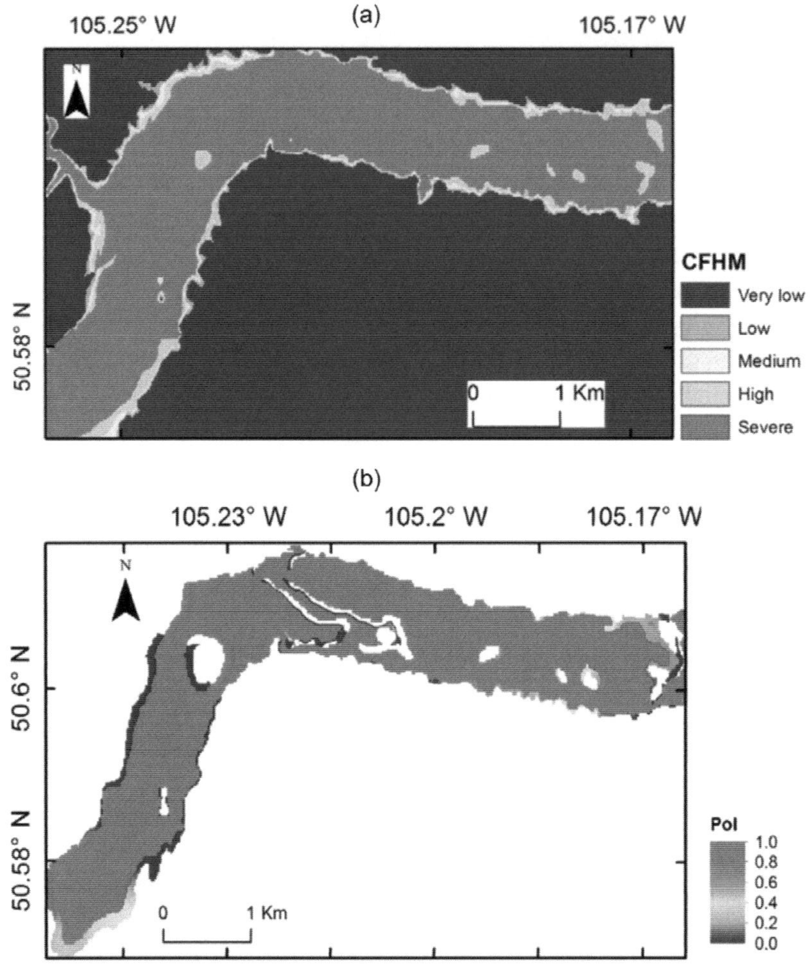

Figure 7: (a) The classified flood hazard map along the Qu'Appelle River, and (b) the probabilistic flood hazard map showing the probability of inundation (PoI) for the same stretch.

It is important to note here that land depressions in the prairies retain a significant amount of surface runoff and significantly affect the streamflow. During snowmelt or heavy rainfall events, these land depressions are interconnected, resulting in landscape flooding. Traditional hydraulic models have limited or no capabilities to simulate such a pluvial flooding pattern over the landscape dominated by land depressions. A new modelling framework is currently underway to simulate the landscape flooding by coupling a rule-based model that utilizes topography data with a simplified hydrological bucket model. The rule-based model is a grid-based simulation approach that can simulate flooding pattern over a certain area by minimizing the differences of water elevation in the surrounding grid cells. The rule-based approach handles the overland flow movement using Manning`s equation, while the bucket model handles the hydrological processes for surface runoff generation. This framework enables determining the quantity and the overland path of surface runoff and shows potential in predicting flooding patterns and the actual connectivity between land depressions using terrain data.

The comparison provided in this paper between the CFHM and the PFHM is a starting point to convert the CFHM of Canada into actual hazard map, where flood zones are related to flood frequency; e.g. 100-year flood. Flood frequency curves at various gauging stations provide stremflow values that correspond to a selected frequency, and consequently, the corresponding water level can be estimated using local rating curves. Spatial interpolation techniques can be used to create water elevation as a continuous surface, which can be used to delineate the inundated area using ArcGIS. Such a product can be used, along with local information on property value, to estimate flood dame and suggested flood insurance premium, which can be a useful flood risk scoring tool for the public and policy makers. As part of the FloodNet, it is planned to produce a web-based tool for flood risk scoring across Canada.

5 CONCLUSION

The topography- and nightlight-based approach adopted in this study for flood risk assessment on a national scale is both useful and practical. Without detailed hydraulic modeling, the flood hazard map of Canada can provide a reliable preliminary assessment of the flood hazard level anywhere in the country. The flood risk map, which integrates both hazard and exposure, including nightlights, is the most useful product as it allows for evaluating the spatial distribution of the expected flood damage, and thus can help in prioritizing government intervention and strategic resource allocation. The severe and high flood hazard areas in Canada are spread over all regions of the country; however, the severe and high flood exposure and risk are concentrated in the southern part of the country around urban centres.

A probabilistic framework to flood mapping, by accounting for uncertainties in inflow hydrographs and roughness parameters, was applied in the present study. The Qu'Appelle River that is prone to frequent floods in the Canadian prairies was selected and a hydrodynamic model was developed for a reach of the river. The variability in the depth of inundation (DoI) was found to be low at locations with high DoI and vice versa, indicating less uncertainty at locations where the DoI is high. The sensitivity analysis results indicated that the influence of channel roughness on flooding extents is more at locations characterized by steep slopes and unidirectional flow, whereas the flood extents were insensitive to roughness parameters and boundary conditions at locations where the channel had a mild slope and flat floodplain characterized by small depressions that retain water during high flows.

Acknowledgements

The authors acknowledges the funds through the NSERC-funded FloodNet and the financial support through the devolved scholarship provided by the University of Saskatchewan to the third author.

References

Adger, W. 2006. Social vulnerability to climate change and extremes in coastal Vietnam, World Dev., 27: 249-269.

Biondi, D., Freni, G., Iacobellis, V., Mascaro, G., and Montanari, A. 2012. Validation of hydrological models: Conceptual basis, methodological approaches and a proposal for a code of practice, Phys. Chem. Earth, 42-44: 70-76.

Ceola, S., Laio, F., and Montanari, A. 2014. Satellite nighttime lights revealing increased human exposure to floods worldwide, Geophys. Res. Lett., 41: 7184-7190.

Collenteur, R., de Moel, H. Jongman, B., and Di Baldassarre, G. 2015. The failed-levee effect: Do societies learn from flood disasters?, Nat. Hazards, 76: 373-388.

de Moel, H., van Alphen, J., and Aerts, J. 2009. Flood maps in Europe – methods, availability and use, Nat. Hazards Earth Syst. Sci., 9: 289-301.

Elshorbagy, A., Bharath, R., Lakhanpal, A., Ceola, S., Montanari, A., and Lindenschmidt, K-E. 2017. Topography- and nightlight-based national flood risk assessment in Canada, Hydrol. Earth Syst. Sci., 21: 2219-2232.

Government of Alberta. Flood hazard studies, available at: http://aep.alberta.ca/water/programs-and-services/flood-hazard-identification-program/flood-hazard-studies/default.aspx (last access: March 2018).

Jacob, M., and Church, M. 2011. The trouble with floods, Can Water Resour, J., 36(4): 287-292.

IPCC. 2012. Glossary of terms, in: Managing the Risks of Extreme Events and Disasters to Advance Climate Change Adaptation, edited by: Field, C. et al., A Special Report of Working Groups I and II of the Intergovernmental Panel on Climate Change (IPCC), Cambridge University Press, Cambrifge, UK, and New York, NY, UA, 555-564.

Latifovic, R., Homer, C., Ressl, R., Pouliot, D., Hossain, S., and Colditz, R. North American land-change monitoring system, edited by: Giri, C., Remote sensing of land use and land cover: Principles and applications, CRC/Taylor & Francis, Boca Raton, Fl, 303-324.

Renno, C., Nobre, A., Cuartas, L., Soares, J., Hodnett, M., Tomasella, J., and Waterloo, M. 2008. HAND, a new terrain descriptor using STRM-DEM: Mapping terra-firme rainforest environments in Amazonia, Remote Sens. Environ., 112: 3469-3481.

Savage, J., Bates, P., Freer, J., Neal, J., and Aronica, G. 2016. When does spatial resolution become spurious in probabilistic flood inundation predictions?, Hydrol. Process., 30(13): 2014-2032.

UNISDR (United Nations International Strategy for Disaster Reduction). 2009. Terminology on disaster risk reduction, Geneva, Switzerland.

Building Tomorrow's Society
Bâtir la Société de Demain

Fredericton, Canada
June 13 to 16, 2018 / *13 au 16 juin, 2018*

FRAMEWORK FOR SEISMIC VULNERABILITY OF HIGHWAY BRIDGE NETWORKS

Abo-El-Ezz, Ahmad[1,3], Nollet, Marie-José[2], McGrath, Heather[1] and Nastev, Miroslav[1]
[1] Natural Resources Canada, Geological Survey of Canada, Québec, QC, Canada
[2] École de technologie supérieure, Montréal, QC, Canada
[3] ahmad.aboelezz@canada.ca

Abstract: In earthquake prone regions, the evaluation of seismic impacts on bridges is crucial to mitigation, emergency and recovery planning for highway networks. The degree of bridge damage determines the cost and time required for repairs and the level of post-earthquake functionality of the bridge determined by its capacity to carry traffic flow. The various losses of bridge functionality induce reduction or disruption of the transportation network, increase costs due to detour or reduced traffic flow and, what is most important, restrict access to emergency routes. This paper presents a framework for development and implementation of seismic vulnerability of highway networks. The proposed framework consists of the following successive models: hazard, exposure, damage and impact. The seismic hazard model generates spatial distribution of the shaking intensity for earthquake scenarios in terms of ground motion intensity measure (IM); the exposure model provides a database of bridge classes broadly defined with respect to their static and dynamic properties; the damage model assesses seismic performance of bridge classes in the network applying respective fragility functions represented as probabilistic relationships between the IMs and the simulated degree of expected damage; whereas the impact model evaluates the post-earthquake traffic-carrying capacity of the highway network based on the predicted damage including repair costs of bridges, road-closures and inspection priority. A case study of application of the proposed framework is presented for damage assessment of a hypothetical bridge network in Quebec City subjected to a magnitude M6 seismic scenario.

1 INTRODUCTION

The evaluation of seismic impacts on bridges is crucial to mitigation, emergency and recovery planning for highway networks. The degree of bridge damage determines the cost and time required for repairs and the level of post-earthquake functionality of the bridge determined by its capacity to carry traffic flow (Werner et al. 2006, Padgett and DesRoches 2007). The various levels of bridge functionality losses typically induce reduction or disruption to the transportation network, increase costs due to detour or reduced traffic flow and, what is most important, potentially restrict access to emergency routes. Following a strong earthquake event, the transportation emergency managers have insufficient time to carry on detailed bridge-by-bridge inspections; yet decisions to keep the traffic flowing or close a given bridge have to be made. The rapid assessment of bridge condition, therefore, is essential for informed decision on the post-earthquake functionality (Lin et al. 2004, Wald et al. 2006). Furthermore, pre-earthquake mitigation planning relies heavily on generation of potential damage scenarios to identify the most vulnerable sections of the transportation system where resources should be put first to achieve maximum cost-benefit of the seismic retrofit (Werner et al. 2006).

This paper describes the methodological development and implementation of seismic vulnerability of highway networks. The successive models for running damage scenarios are described including hazard, exposure, damage and impact models. To demonstrate the capacities of the proposed framework, a case-example is presented for damage assessment of a hypothetical bridge network in Quebec City subjected to a M6 earthquake scenario.

2 FRAMEWORK

The proposed framework for seismic vulnerability of highway bridge networks is presented in Figure 1. The seismic hazard model generates spatial distribution of the shaking intensity for earthquake scenarios in terms of ground motion intensity measure (IM); the exposure model provides a database of bridge classes defined according to construction material, structural system and seismic design level; the damage model assesses seismic performance of the bridge classes in the network applying respective fragility functions represented as probabilistic relationships between the IMs and the simulated degree of expected damage; and the impact model evaluates the post-earthquake traffic-carrying capacity of the highway network based on the predicted damage including repair costs of bridges, road-closures and post-earthquake inspection priority.

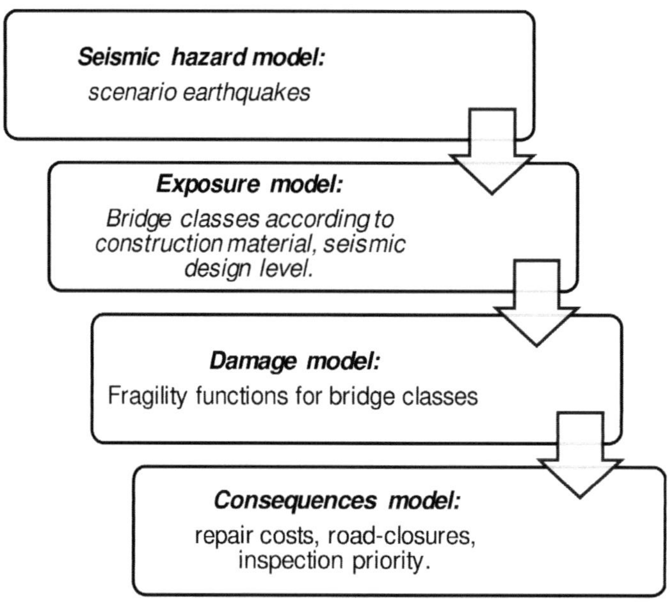

Figure 1: Framework for seismic vulnerability of highway bridges network.

2.1 Seismic Hazard Model

A simple algorithm has been developed with a shake map generation capacity for earthquake events with specified magnitude, distance and simple fault geometry (Nastev et al. 2015). It applies the new generation of ground-motion prediction models for reference response spectral accelerations on rock including PGA and PGV: AA13 for Eastern Canada (Atkinson and Adams 2013) applied in the National Building Code of Canada NBCC 2015 (NRC 2016). The epistemic and aleatory uncertainty can be captured with the provided upper and lower confidence levels. The ground motion intensity is then corrected for the local soil conditions with the amplitude and frequency dependent site amplification factors as functions of the average VS_{30} at each site as defined by NBCC 2015. In Eastern Canada, microzonation have been completed on a regional scale for three cities: Quebec City (Leboeuf et al., 2013), Montreal (Rosset et al. 2015) and Ottawa-Gatineau (Motazedian et al., 2011). It is also available as a 3D-Model for the Lowlands of the Saint-Lawrence Valley (Nastev et al. 2016).

2.2 Exposure Model

The inventory of bridges potentially exposed to ground shaking is the second major input parameter. It can be conducted at a local (bridge) scale by sidewalk and virtual desktop surveys, or at urban or regional scale by interpreting data from municipal bridges or databases from the department of transportation. To simplify the structural analyses, the individual bridge structures are grouped into relative broad classes according to their expected behaviour under seismic loading (Basoz and Mander 1997, FEMA 2012). The following structural parameters are inventoried: year of construction, number of spans (single span or multiple span), super-structure type (reinforced concrete, steel, or wood), pier type (single column bent, multiple column bents, or pier wall), abutment type (monolithic or non-monolithic), bearing type (high rocker bearings, low steel bearings or elastomeric bearings), isolation bearings (with or without), span continuity (continuous, discontinuous, in-span hinges or simply supported). Figure 2 shows an example of bridge categorisation according to the classification scheme proposed in Hazus (FEMA 2012). This classification scheme is specifically intended for seismic assessment.

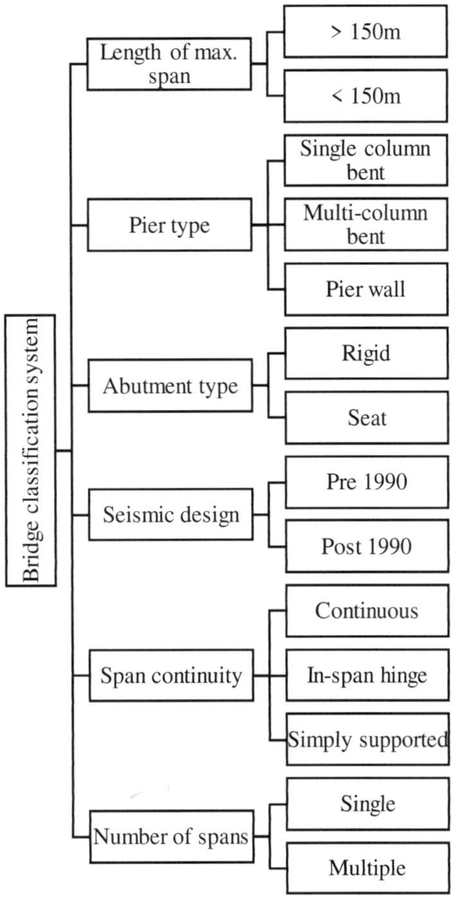

Figure 2: Bridge classification scheme according to Hazus (FEMA 2012).

2.3 Damage Model

The damage model consists of a dataset of fragility functions related to the bridge classes in the exposure model. A set of fragility functions quantifies the conditional probability representing the likelihood that a given bridge structure will meet or exceed specified level of damage for a given intensity measure (IM) of the seismic hazard. There are three main approaches for creating seismic fragility functions: (1) experts' opinion methods estimate the probable damage distribution of bridges when subjected to different earthquake intensities based on a standardised questionnaire completed by experts (ATC 1985); (2)

empirical methods using damage data from post-earthquake field observations (Basoz and Kiremidjian 1999, Shinozuka et al. 2000, Yamazaki et al. 2000); and (3) analytical methods that rely on mechanical or numerical structural models to simulate the seismic response of bridges (FEMA 2012, Tavares et al. 2012). Figure 3 shows an example of a set of fragility function for pre-1990 multi-span continuous concrete bridge classes in Hazus (FEMA 2012) representing the thresholds of attaining or exceeding four limit states (slight, moderate, extensive and complete). The damage state probabilities corresponding to Sa(1.0s)=0.2g is shown as an example on the left side of the figure. On the right, the distribution of five post-earthquake damage level probabilities is shown, and is calculated by the difference between two consecutive limit state curves for Sa(1.0s)=0.2g.

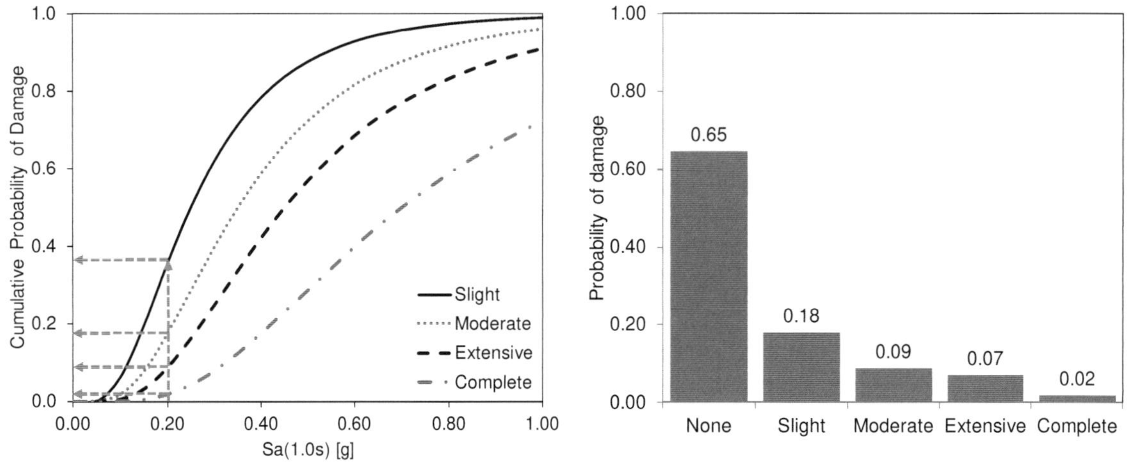

Figure 3: Example fragility functions for pre-1990 multi-span continuous concrete bridge class (left) and the five damage state probability corresponding to Sa(1.0s)=0.2g (right).

2.4 Impact model

Based on the damage assessment results, the negative effects of the generated damage to bridges are quantified using the impact model. It includes: inspection priority, likely immediate post traffic state the bridges can be assigned in terms of inspection priority, likely immediate post event traffic state (Lin et al. 2003) and repair cost ratios (Werner et al. 2006) (Table 1).

Table 1: Bridge damage states and the corresponding, inspection priority and likely post-event traffic states and repair cost ratios.

Damage state	**None**	**Slight**	**Moderate**	**Extensive**	**Complete**
Range of repair cost ratio	0	1%-5%	5%-50%	50%-80%	80%-100%
Average damage ratio	0	3%	25%	75%	100%
Inspection priority	None	Low	Medium	Medium-high	High
Likely post-event traffic state	Open to normal traffic- no restrictions	Open to normal traffic- no restrictions	Open to limited traffic-speed/weight/lane restrictions	Emergency vehicles only-speed/weight/lane restrictions	Closed until shored/braced-potential for collapse

In order to estimate the incurred economic losses, the mean damage ratio (MDR) is computed as the weighted sum of the average damage ratios (Di) multiplied by the probability of being in each damage state P(DSi) (Equation 1). The MDR can then be used to identify the priority rank for inspection.

[1] $$MDR = \sum_{i=1}^{4} \mathrm{Di}\,.\,\mathrm{P(DSi)}$$

3 CASE STUDY

To demonstrate the capacities of the proposed framework, it is applied to assess the inspection priority of for a group of bridges in Quebec City bridges affected by a magnitude 6 hypothetical earthquake event at a distance of about 10km from downtown. A total of 39 bridges were considered for this investigation which represent two classes of simply supported reinforced concrete girder bridges: 20 single span (SSCG) bridges, and 19 multi-span (MSCG) bridges (Figure 4). It should be emphasized that this group of bridges do not represent the complete network in the vicinity of Quebec City. Specific details on the structural systems of the bridges level were not available, and basic information was collected from multiple sources (e.g., Hida 2009, Google Street View, CanVec database https://open.canada.ca) to provide first-order classification of the bridge types for this demonstration study.

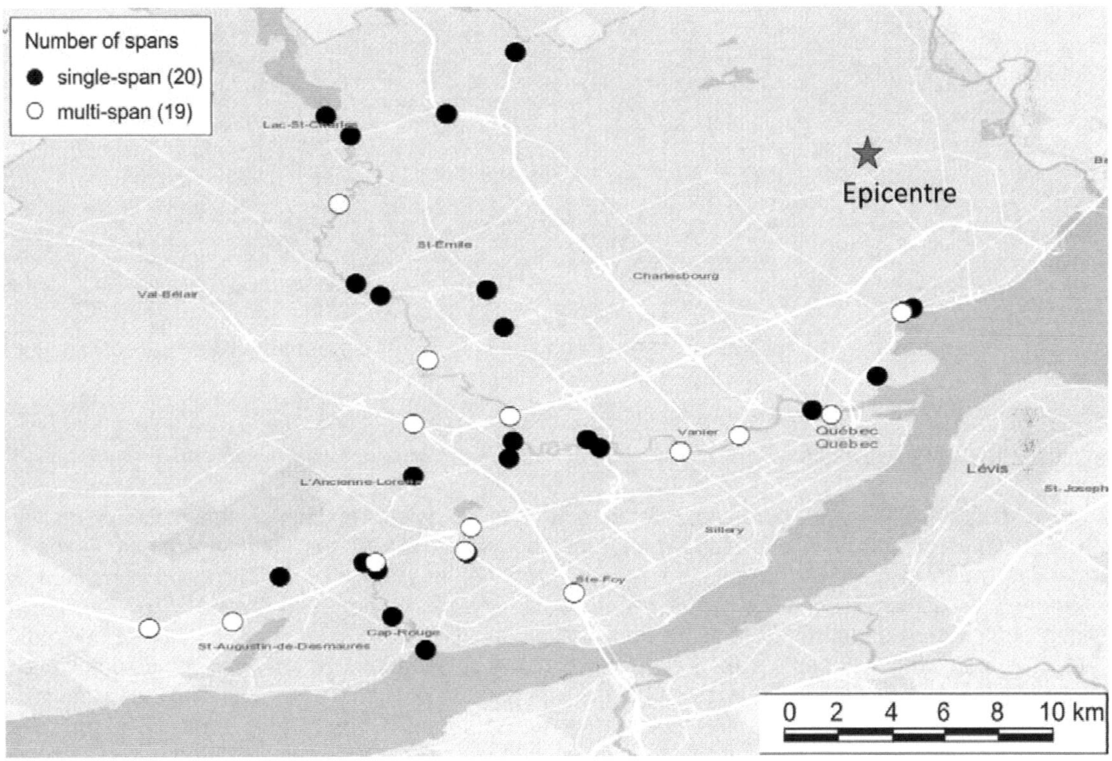

Figure 4: Map showing the location of the 39 simply supported reinforced concrete girder bridges (figure generated by the authors).

The ground motion prediction equation by Atkinson and Adams (2013) was used to estimate the shaking intensity at each bridge location with local site effects computed based on the site classification from the microzonation study for Quebec City (Leboeuf et al., 2013). Currently, there are no available fragility functions developed explicitly for the considered bridges in Quebec City. As a result, Neilson (2005) fragility functions for simply supported SSCG bridges and Tavares et al. (2012) fragility functions for simply supported MSCG bridges were employed.

Figure 5 shows the results of the expected damage states of the bridges and the corresponding inspection priority rank for the M6 scenario, according the computed MDF of each bridge. For example, the values of the MDFs for the 5 bridges with medium inspection priorities are: 16%, 14%, 15%, 8%, and 6%.

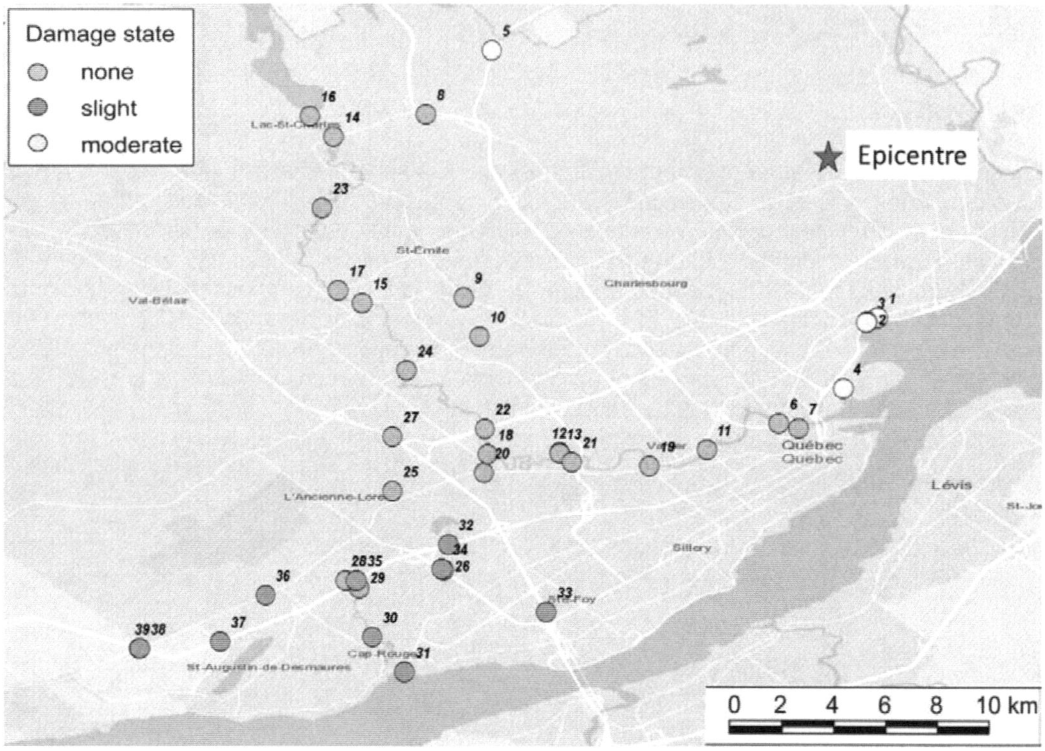

Figure 5: Damage assessment for the M6 earthquake scenario. The numbers indicate inspection priority (figure generated by the authors).

4 CONCLUSIONS

This paper described the methodological development and implementation of seismic vulnerability of highway networks. The successive models for running damage scenarios including hazard, exposure, damage and impact models were discussed. A case study demonstration of the proposed framework was presented for the damage assessment of a hypothetical network of bridges in Quebec City. The network consisted of 39 simply supported reinforced concrete girder bridges with single-span (20) and with multi-span (19). A magnitude M6 earthquake scenario was considered about 10 km from the downtown area. The estimated degree of damage to a given bridge was used to determine the cost and time required for repairs and the level of post-earthquake functionality of the bridge defined by its capacity to carry traffic flow. The results indicate that 5 of the bridges are expected to be moderately damaged. The inspection priority ranking of the bridges based on the MDF could provide highway officials with a risk-informed knowledge for more efficient planning of inspection and repairs activities.

Such rapid assessment of bridge conditions is essential for informed decision making on the post-earthquake functionality. It can also be used for pre-earthquake mitigation planning purposes based on potential damage scenarios. The spatial distribution of damage helps will help identify the most vulnerable sections of the transportation system which will require rapid intervention.

References

Atkinson, G. M. and Adams, J. (2013). Ground motion prediction equations for application to the 2015 Canadian national seismic hazard maps. *Canadian Journal of Civil Engineering*, **40**(10), 988-998.

Basoz, N., and Mander, J. (1999). Enhancement on the Highway Transportation Lifeline Module in Hazus. 1999. *National Institute of Building Sciences*, Washington, DC, USA.

FEMA (2012). Multi-Hazard Loss Estimation Methodology, Earthquake Model, HAZUS-MH MR1 Technical Manual. 2012. Federal Emergency Management Agency, Washington, D.C, USA.

Leboeuf, D., Perret D., Nollet M-J., Lamarche L., Nastev M., and Parent M. (2013). Microzonage sismique des Villes de Québec-Ancienne-Lorette et Réserve indienne de Wendake (Catégories d'emplacement). *Commission géologique du Canada*, Dossier Public 6875, Échelle 1: 50,000, doi: 10.4095/292640.

Lin KW., Turner LL., Wald D. (2014). ShakeCast V3: An Enhanced Tool for Post-Earthquake Response. *Caltrans Division of Research, Innovation & System Information*, Sacramento, CA, USA.

Motazedian, D., Hunter, J. A., Pugin, A., and Crow, H. (2011). Development of a VS30 (NEHRP) map for the city of Ottawa, Ontario, Canada. *Canadian Geotechnical Journal*, **48**(3), 458-472.

Nastev, M., Nollet, M.J., Abo El Ezz, A., Smirnoff, A., Ploeger, S.K., McGrath, H., Sawada, M., Stefanakis, E. and Parent, M. (2015). Methods and tools for natural hazard risk analysis in eastern Canada: Using knowledge to understand vulnerability and implement mitigation measures. *Natural Hazards Review*, ***18***(1), p.B4015002.

Nastev, M., Parent, M., Ross, M., Howlett, D. and Benoit, N., (2016). Geospatial modelling of shear-wave velocity and fundamental site period of Quaternary marine and glacial sediments in the Ottawa and St. Lawrence Valleys, Canada. *Soil Dynamics and Earthquake Engineering*, **85**, pp.103-116.

Nielson, B.G., 2005. Analytical fragility curves for highway bridges in moderate seismic zones. *Doctoral dissertation*, Georgia Institute of Technology, USA.

NRC (2016). National Building Code of Canada, NBCC 2015. *National Research Council of Canada.* Ottawa, Canada.

Hida, H. (2009). Évaluation de la vulnérabilité sismique des ponts municipaux », mémoire de maîtrise, ÉTS, Canada, 175 p.

Padgett JE., DesRoches R. (2007). Bridge functionality relationships for improved seismic risk assessment of transportation networks. *Earthquake Spectra*, **23**(1):115-30.

Rosset, P., Bour-Belvaux M., and Chouinard L. (2015). Microzonation models for Montreal with respect to VS30. *Bulletin of Earthquake Engineering*, **13**(8), p. 2225-2239.

Tavares DH., Padgett JE., Paultre P. (2012). Fragility curves of typical as-built highway bridges in eastern Canada. *Engineering Structures*, **31**(40):107-18.

Wald DJ., Lin KW., Worden B., Turner L. (2006). ShakeCast: Facilitating the use of ShakeMap for post-earthquake decision-making and response within Caltrans and other critical lifeline communities. *Fifth National Seismic Conference on Bridges & Highways*; San Francisco, USA, 2006.

Werner S., Taylor C., Cho S., Lavoie JP., Huyck C., Eitzel C., Chung H., Eguchi R. (2006). REDARS 2 methodology and software for seismic risk analysis of highway systems (technical manual). Multidisciplinary Center for Earthquake Engineering Research (MCEER): Oakland, CA. 2006.

Building Tomorrow's Society
Bâtir la Société de Demain

Fredericton, Canada
June 13 to 16, 2018 / *13 au 16 juin, 2018*

FROM TOP-DOWN TO BOTTOM-UP APPROACHES TO RISK DISCOVERY: A PARADIGM SHIFT IN CLIMATE CHANGE IMPACTS AND ADAPTATION STUDIES RELATED TO THE WATER SECTOR

Seidou, Ousmane[1,3] and Alodah, Abdullah[1,2]
[1] University of Ottawa, Canada
[2] Qassim University, Saudi Arabia
[3] oseidou@uottawa.ca

Abstract: Healthy water resources are key to every nation's wealth and well-being. Unfortunately, stressors such as climate change, land-use shifts, and increased water consumption are threatening water availability and access worldwide. The pressure on water resources is projected to increase dramatically in the future and turn into a global crisis unless bold actions are taken. Researchers and practitioners are therefore under great pressure to develop methodologies and tools that can streamline projected changes into adaptation decisions. The vast majority of climate change adaptation studies use a top-down approach, which essentially consists of using of a limited set of climate change scenarios to discover future risks. However, recent research has identified critical limitations to that approach; for instance, even when multi-model multi-scenario projections are used, not all possible future conditions are covered and therefore plausible risks may be overlooked. There is also no established way to evaluate the credibility of a given projection scenario, making it a challenge to include them in a decision or design framework. The bottom-up approach (which consists of identifying risky situations using stochastically generated climate states then use projections to evaluate their likelihood in the future) was recently put forward to address the limitations of the top-down approach, but several issues remain unaddressed. This paper proposes a methodology that allows the generation of a large number of climate change projections by combining the outputs of weather generators to the outputs of regional climate models, and associate a likelihood to each projection. The new set of projections provide a better coverage of the risk space, and can therefore facilitate the implementation of the bottom-up approach. Applications on the Upper Niger and Bani River Basins in West Africa, implementing 20 regional climate models (RCMs) with two emission scenarios (RCPs), is presented.

1 INTRODUCTION

The scientific community recognizes climate change as one of the most pressing global issues and there is a pressure on researchers to develop methodologies and tools that can streamline projected changes into adaptation decisions. The problem is not new and thousands of researchers have already tried to tackle the problem using a 'top-down' approach, which essentially consists of using of a limited set of climate change scenarios to discover future risks. However, some critical limitations of that approach have been delineated recently; for instance, it has emerged that even when multi-model multi-scenarios projections are used, not all possible future conditions are covered and therefore plausible risks may be overlooked; yet there is also no established way to evaluate the credibility of a given projection scenario, making it a challenge to include them in a decision or design framework. The 'bottom-up' approach (which consist of identifying risky situations using stochastically generated climate states then use projections to evaluate their likelihood in the future) the was recently put forward to address the limitations of the top-down approach but several issues remain unaddressed. A typical (and hypothetical) output of the bottom-up approach is presented on Figure 1. The risk on a watershed is measured using a water stress index, which is calculated as function of a range of future temperature changes (ΔT) and precipitation changes (ΔP). Climate change projections of ΔT and ΔP from a limited number of Regional Climate Models (RCMs) are plotted on the map to give an indication of the likelihood of a particular combination of ΔT and ΔP, and therefore the associated value of water stress.

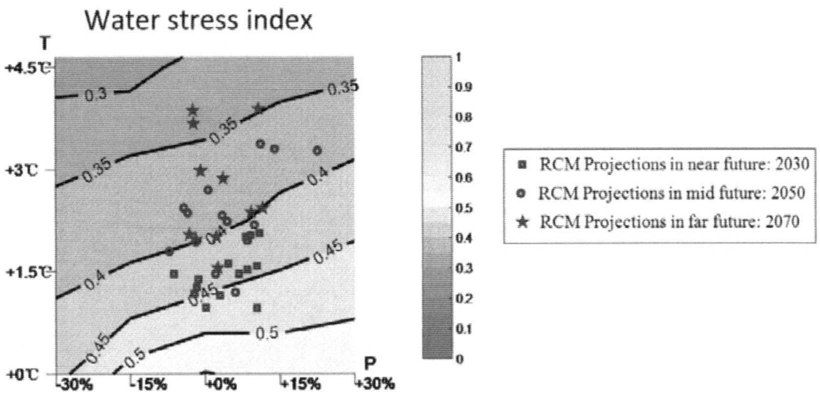

Figure 1: Typical output of the bottom-up approach to risk discovery

There are some obvious limitations to the usefulness of the above figure. The number of points representing the RCMs do not cover the entire space, meaning that the likelihood of certain combinations of ΔT and ΔP cannot be evaluated. All RCMs are treated equally, despite the fact that their performances are unequal in the historical period. A better approach would have been to provide each RCM output with a likelihood derived from how it fares on the historical period. This paper is an attempt to address the above issues using a novel stochastic method that would clone and perturb an initial set of climate change scenarios derived from climate models and generate new ones that would cover the range of variability observed on the historical period. The case study was performed on the Upper Niger and Bani River Basins (UNBB) in west Africa. At the outlet of the UNBB with lies the Inner Niger Delta (IND), which is the largest wetland in West Africa and source of income and food for at least 3 million people. A major concern for people in the area is the prospect that the delta would shrink to catastrophic states under climate change. First, a SWAT rainfall-runoff model is developed and calibrated on the UNBB. The outputs of an ensemble of 20 Regional Climate Models under scenarios RCP4.5 and RCP8.5 were downscaled and used to force the hydrological model to obtain projections of the hydrological cycle in the future. The KNN weather generator is then used to generate 250 random realizations of the historical climate. The 40 downscaled climate time series (from 20 RCMs using two RCP scenarios) were then rescaled to match the variance and/or mean of each realization of the weather generator, leading to 10200 climate projections in the future. A climate response function linking the average volume of water in the Inner Niger Delta to the statistical characteristics of precipitation in the area was calibrated using the SWAT model inputs and outputs on the historical period, and used to generate a high-resolution projection of the water volume in the Inner Niger Delta in the future. The projections are used to calculate the likelihood of particular values of the average volume of water in the IND.

2 MATERIAL AND METHODS

2.1 Study Area

The Niger River is the third longest African river and flows through 5 countries (Guinea, Mali, Niger, Benin, Nigeria), while draining runoff from four others (Burkina Faso, Ivory coast, Chad and Cameroun (Figure 1). The Inner Niger Delta (IND) in central Mali is a large wetland in an otherwise dry area that provides a multitude of ecosystem services to 3 million people. Two rivers flow in the IND: the Niger River at the North, and the Bani River at the south. The area under investigation is the watershed of the Niger river at Diré station, just at the exit of the IND.

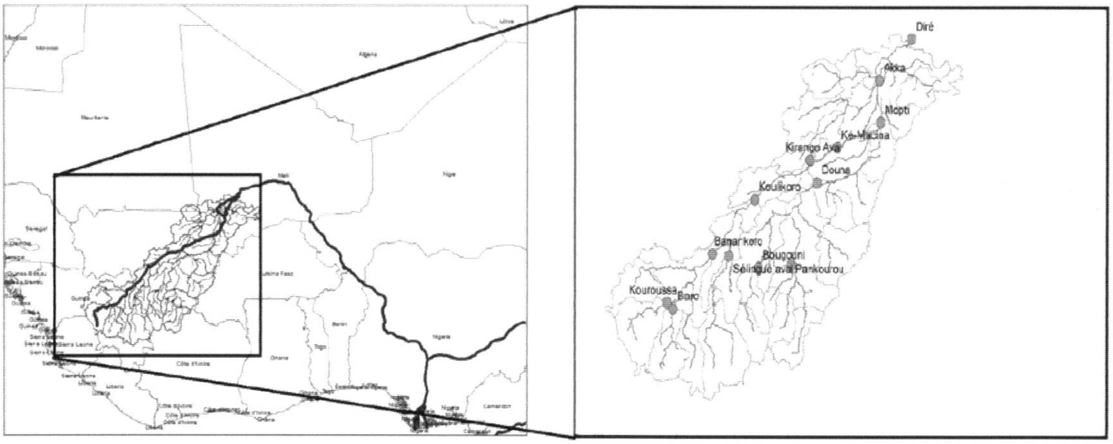

Figure 2: Geographic location of the UNBB

2.2 Hydrological Modeling

SWAT (Soil and Water Assessment Tool: Arnold et al., 1998) was used to develop a rainfall-runoff model of the UNBB. The model has 32 subwatersheds, 188 Hydrologic Response Units, and 4 reservoirs. The Delta is represented in the model by three reservoirs (at Mopti, Akka and Dré) in cascade since it is too large to be represented by a single reservoir. The SWAT reservoir routine was modified to account for the looped rating curves at the three stations in the IND. The other parameters of the model were derived from the following data sets:

1. Digital elevation model: Shuttle Radar Topography Mission (Jarvis et al., 2008),
2. Land use maps: Global Land Cover Facility (Bartholom and Belward, 2005),
3. Soil maps: Food and Agricultural Organization of the United Nations Organization for Education, Science and Culture (FAO/UNESCO, 2003),
4. Climate data from 1979 to 2012: WATCH-Forcing-Data-ERA-Interim (Weedon et al., 2017), and
5. Streamflow data: recorded streamflow at 14 hydrometric (Kouroussa, Baro, Selingué, Pankourou, Banakoro, Koulikoro, Douna, Kirango, Ke-Macina, Mopti, Akka and Diré) gauges throughout the watershed, obtained from the Malian and Guinean ministries of water resources.

The model was calibrated on the 1979-1997 period using the sequential uncertainty fitting (SUFI2) implemented in the SWAT-CUP program (Abbaspour, 2012), and validated on the 1998-2012 period. The Nash–Sutcliffe efficiency coefficient (NS) was used as an objective function in the calibration and validation phase:

$$[1]\ NS=1-\left[\frac{\sum_{i=1}^{n}(O_i-P_i)^2}{\sum_{i=1}^{n}(O_i-\bar{O})^2}\right]$$

where O stands for observed and P for predicted values, and $\bar{O}$ is the mean of the observed values. The calibration was done in a sequential manner starting from most upstream stations and moving towards the outlets.

2.3 Climate Statistics & Performance Indicator Spaces (R_T)

We are going to adopt the same framework as (Brown et al. 2012) where a climate state represented by a time series X_t is summarized by a subset of climate statistics v_T calculated over period T that are relevant to the problem under investigation (e.g. the mean, standard variation of precipitation and temperature). v_T belongs to a multidimensional space that will be called from now on the Climate Statistics Space (CSS). Elements of the CSS can be calculated any time series among observations, downscaled climate models outputs or stochastically generated time series. A climate state from the CSS will yield a level of risk and performance that is measured by a set of risk/performance indicators $\boldsymbol{R}_T$. $\boldsymbol{R}_T$ is obtained by feeding an impact model with time series, and belongs to another multidimensional space called the Risk and Performance Indicator Space (RPIS). A response function $\boldsymbol{f}$ (called the climate response function) can developed to map the climate statistics space to the performance state (i.e. $X_t = \boldsymbol{f}(v_T)))$. We also assume that following Brown et al. (2012), a weather generator was used to generate a large number of stochastic time series. Figure 1 illustrates the relationships between climate time series, climate statistics and risk/performance indicators in the bottom-up framework.

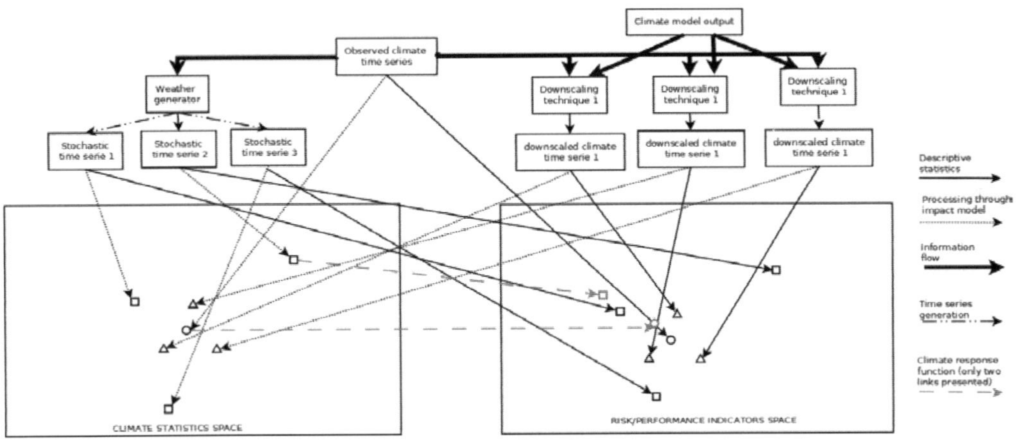

Figure 3: relationships between climate time series, climate statistics and risk/performance indicators in the bottom-up framework.

For the sake of simplicity, the RPIS is restricted to one dimensions corresponding to the average volume of water stored in the IND over a 30-years period.

2.4 Development of The Climate Response Function

While a lot of climate statistics can affect the volume of water in the IND, the dimensions of the CSS were restricted to two. The climate statistics to be included in the CSS were selected as follow:

a) The UNBB was divided into 7 regions shown on Figure 3

b) For each of the regions, and each month of the year M, 12 time series representing the cumulative precipitation on the region between month 1 and month M were generated from observed climate time series and downscaled climate time series. Each time series cover one of the following periods: 1986-2005, 2041-2070 and 2071-2100; These time series are denoted $CumPCP_t^{R,M,P}$ where R is the region (1 to 7), M is the month (1 to 12), P is the period (1986-2005, 2041-2070 or 2071-2100). t varies from 0 to the number of days in P.

c) For each of the regions, and each month of the year M, 12 time series representing the average temperature over the region between month 1 and month M were generated from observed climate time series and downscaled climate time series; Each time series cover one of the following periods:

1986-2005, 2041-2070 and 2071-20100; These time series are denoted $AvTMP_t^{R,M,P}$ where R is the region (1 to 7), M is the month (1 to 12) and P is the period (1986-2005, 2041-2070 or 2071-2100).

d) For each of the above time series, a time series $vIND_t^{R,M,P}$ of water stored in the IND was also extracted from corresponding SWAT simulations and used to calculate the average volume of water stored in the IND

e) The mean and standard deviation in time of $CumPCP_t^{R,M,P}$ and $AvTMP_t^{R,M,P}$ were included in a pool of potential predictors for the average in time of $vIND_t^{R,M,P}$.

f) The selection of the variables to include in the CSS was done in two steps. First, stepwise regression was used to screen the pool of potential predictors and retain only the ones which have a meaningful relation with the predictand. The two predictors that passed the screening and have the highest correlation with the predictand are finally retained as dimensions of the CSS.

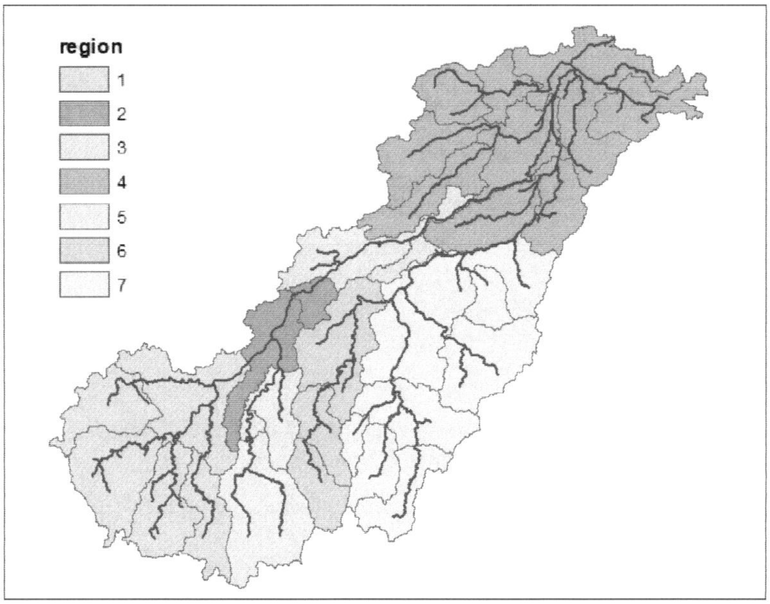

Figure 4: Regions in the UNBB

2.5 Climate Change Projections and Downscaling

The CORDEX (COordinated Regional Downscaling Experiment: Giorgi et al., 2009) aims to simultaneously evaluate global climate models and produce climate projections for use in impact and adaptation studies, mainly through dynamical downscaling. A set of 40 regional climate models outputs (20 different climate models, two RCP scenarios: RCP4.5 and RCP8.5) were obtained from the CORDEX team. The list of the regional climate models, the institution in which they were developed along with the driving Global Circulation Model (GCM) are provided in Table 1. The outputs of the RCMs were downscaled using quantile-quantile mapping following Amadou et al. (2014)

Table 1: list of Regional Climate Models used in the study

RCM	CANRCM4	CCLM-4-8-17	CCLM-4-8-17	CCLM-4-8-17	HIRHAM5-v1
Institution	CCCma (Canadian Centre for Climate Modelling and Analysis, Victoria, BC, Canada)	Climate Limited-area Modelling Community (CLM-Community)	Climate Limited-area Modelling Community (CLM-Community)	Climate Limited-area Modelling Community (CLM-Community)	Danish Meteorological Institute
Driving GCM	CanESM2	ICHEC-EC-EARTH	MOHC-HadGEM2-ES	MPI-M-MPI-ESM-LR	NCC-NorESM1-M
RCM	**RACMO22T**	**RCA4-v1**	**RCA4-v1**	**RCA4-v1**	**RCA4-v1**
Institution	Royal Netherlands Meteorological Institute	Swedish Meteorological and Hydrological Institute, Rossby Centre	Swedish Meteorological and Hydrological Institute, Rossby Centre	Swedish Meteorological and Hydrological Institute, Rossby Centre	Swedish Meteorological and Hydrological Institute, Rossby Centre
Driving GCM	ICHEC-EC-EARTH	CCCma-CanESM2	CNRM-CERFACS-CNRM-CM5	CSIRO-QCCCE-CSIRO-Mk3-6-0	ICHEC-EC-EARTH
RCM	**RCA4-v1**	**RCA4-v1**	**RCA4-v1**	**RCA4-v1**	**RCA4-v1**
Institution	Swedish Meteorological and Hydrological Institute, Rossby Centre	Swedish Meteorological and Hydrological Institute, Rossby Centre	Swedish Meteorological and Hydrological Institute, Rossby Centre	Swedish Meteorological and Hydrological Institute, Rossby Centre	Swedish Meteorological and Hydrological Institute, Rossby Centre
Driving GCM	IPSL-IPSL-CM5A-MR	MIROC-MIROC5	MOHC-HadGEM2-ES	MPI-M-MPI-ESM-LR	NCC-NorESM1-M
RCM	**RCA4-v1**	**REMO2009-v1**	**REMO2009-v1**	**REMO2009-v1**	**WRF331-v1**
Institution	Swedish Meteorological and Hydrological Institute, Rossby Centre	Helmholtz-Zentrum Geesthacht, Climate Service Center, Max Planck Institute for Meteorology	Helmholtz-Zentrum Geesthacht, Climate Service Center Germany	Helmholtz-Zentrum Geesthacht, Climate Service Center Germany	Uni Research and the Bjerknes Centre for Climate Research
Driving GCM	NOAA-GFDL-GFDL-ESM2M	ICHEC-EC-EARTH	IPSL-IPSL-CM5A-LR	MOHC-HadGEM2-ES	NCC-NorESM1-M

2.6 Weather Generator and Climate Time Series Likelihood Estimation

The weather generator proposed by Sharif and Burn (2007) was used to generate 250 realizations of the historical climate. This approach is based on a traditional resampling of the historical data but a random component is added to the individual resampled data points in order to reproduce values that are not in the historical records. For temperature variables, the temporal window and the number of nearest neighbor's k are arbitrarily chosen to be 14 days and 7 neighbors. It is assumed that the representation of the synthetically generated time series in the CSS will represent the natural climate variability. The density of the points can be converted into a two-dimensional probability distribution as illustrated in Figure 5, and used to assess the likelihood of other climate time series.

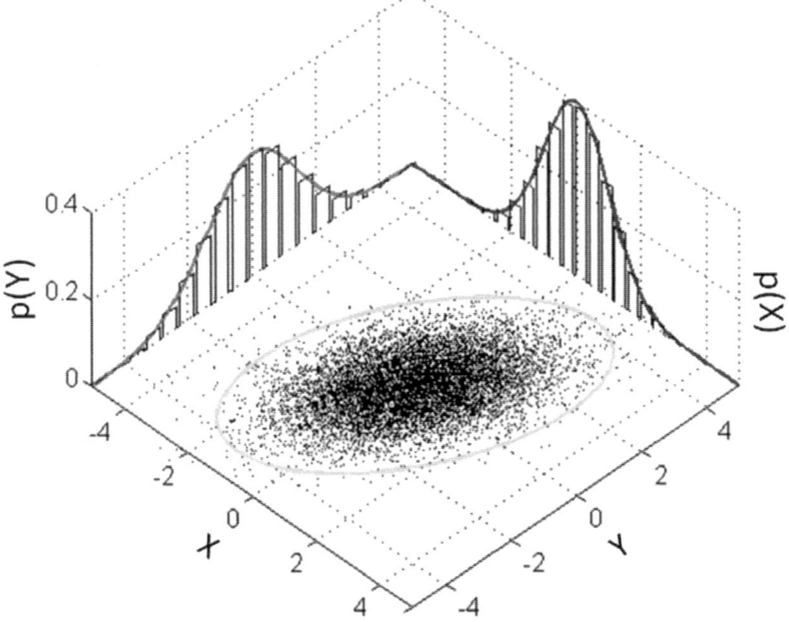

Figure 5: Illustration of the conversion of a cloud of points in a two dimensional space into a bidirectional probability density function.

2.7 Combining Weather Generator Outputs and Downscaled Climate Models Outputs To Generate High Resolution Projections

One of the major limitations of the 'top-down' approach is that the number of scenarios is too low to fully explore the risk space. The problem may be mitigated by cloning projections generated by climate models and slightly 'perturbing' them so that the new set of scenario will have the same span as the synthetic time series in the climate statistics space. The research challenge is to identify determination the 'perturbation' will be done while keeping the representativeness of the new scenario. In this paper, given an output PCP_t^{WGN} of the KNN weather generator and the perturbation is done by applying the following transformation to downscaled climate model outputs:

$$[2] \quad PCP_t^{RCM,TR} = PCP_t^{RCM} \frac{\underset{\substack{month=month(t)\\1986\leq year(t)\leq 2005}}{AVG}\left(PCP_t^{WGN}\right)}{\underset{\substack{month=month(t)\\1986\leq year(t)\leq 2005}}{AVG}\left(PCP_t^{RCM}\right)}$$

Where $PCP_t^{RCM,TR}$ is the perturbed RCM output, PCP_t^{RCM} is the original RCM output, and PCP_t^{WGN} one of the outputs of the KNN weather generator. The expectation is that after transformation, the mean of $PCP_t^{RCM,TR}$ would be closer to the mean and standard deviation of $PCP_t^{RCM,TR}$ on the historical period.

$$[3] \quad TMP_t^{RCM,TR} = \underset{\substack{month=month(t)\\1986\leq year(t)\leq 2005}}{AVG}\left(TMP_t^{WGN}\right) + \left(TMP_t^{RCM} - \underset{\substack{month=month(t)\\1986\leq year(t)\leq 2005}}{AVG}\left(TMP_t^{WGN}\right)\right) \frac{\underset{\substack{month=month(t)\\1986\leq year(t)\leq 2005}}{STD}\left(TMP_t^{WGN}\right)}{\underset{\substack{month=month(t)\\1986\leq year(t)\leq 2005}}{STD}\left(TMP_t^{WGN}\right)}$$

Where $TMP_t^{RCM,TR}$ is the perturbed RCM output, TMP_t^{RCM} is the original RCM output, and TMP_t^{WGN} one of the outputs of the KNN weather generator. The expectation is that after transformation, the mean of $TMP_t^{RCM,TR}$ would be closer to the mean and standard deviation of TMP_t^{WGN} on the historical period.

3 RESULTS AND DISCUSSION

3.1 Calibration and Validation of The Hydrological Model

The hydrological model was successfully calibrated and validated on the UNBB, with NS coefficient ranging from 0.6 to 0.93 in calibration, and from 0.45 to 0.9 in validation.

Table 2: Calibration and validation performance

Hydrometric	Latitude	Longitude	Nash–Sutcliffe efficiency coefficient (NS) (validation period)	(calibration period)
Kankan	10.38	-9.31	0.8	0.65
Baro	10.51	-9.72	0.73	0.68
Kouroussa	10.64	-9.88	0.78	0.68
Banankoro	11.68	-8.67	0.9	0.86
Sélingué aval barrage	11.64	-8.24	0.64	0.45
Koulikoro	12.85	-7.56	0.93	0.87
Kirango	13.69	-6.08	0.82	0.86
Ké-Macina	13.95	-5.36	0.91	0.9
Bougouni	11.39	-7.45	0.78	0.77
Pankourou	11.44	-6.58	0.6	0.74
Douna	13.21	-5.9	0.9	0.79
Mopti	14.49	-4.21	0.78	0.8
Akka	15.4	-4.24	0.84	0.87
Diré	16.27	-3.39	0.75	0.9

3.2 Climate Change Projections and Downscaling

The 40 downscaled climate time series were used to force the calibrated SWAT model. Results show an increase in temperature ranging from 1ºC to 5.9 ºC between 2041 and 2100 (Table 3). Projected changes in precipitation (resp. water volume in the IND) range from -29% to +27.5% (-45% to 104.2%).

Table 3: range of projected changes in temperature, precipitation and water volume in the IND

	Temperature (ºC)	Precipitation (mm)	Water volume in the IND (million m^3)
Average value in 1986-2005	28.6	1001	4671
Change in 2041-2070, RCP4.5	1.0 to 2.9	-116.0 (-11.6%) to 66.3 (6.6%)	-990.6 (-21.2%) to 4867.2 (104.2%)
Change in 2041-2070, RCP8.5	2.0 to 4.6	-146.6 (-14.6%) to 89.9 (9.0%)	-1204.9 (-25.8%) to 991.5 (21.2%)
Change in 2071-2100, RCP4.5	1.2 to 3.4	-115.8 (-11.6%) to 96.1 (9.6%)	-920.9 (-19.7%) to 4216.6 (90.3%)
Change in2071-2100, RCP8.5	2.6 to 5.8	-298.2 (-29.8%) to 275.1 (27.5%)	-2102.9 (-45.0%) to 2085.6 (44.6%)

The above results show that some rather drastic changes are projected by individual models, and raises the question of the credibility of these models. In the traditional 'top-down' approach, each model in the pool is given the same credibility.

3.3 Climate Statistics Space and Climate Response Function

Using stepwise regression (entrance *p*-value=0.01; exit *p*-value=0.05), the following variables were retained as potential predictors for the average volume of water in the IND: $AVG(CumPCP_t^{1,9,*})$, $AVG(CumPCP_t^{1,9,*})$, $AVG(CumPCP_t^{3,1,*})$, $AVG(CumPCP_t^{4,1,*})$, $AVG(CumPCP_t^{6,7,*})$, $AVG(CumPCP_t^{7,10,*})$, $AVG(AvTMP_t^{4,1,*})$, $STD(CumPCP_t^{1,9,*})$, $STD(CumPCP_t^{4,1,*})$, $STD(CumPCP_t^{3,1,*})$, $STD(CumPCP_t^{4,12,*})$, $AVG(CumPCP_t^{7,6,*})$ where $AVG(CumPCP_t^{R,M,*})$ is the average of cumulative precipitation between month 1 and month M on region R, $AVG(AvTMP_t^{R,M,*})$ is the average temperature between month 1 and month M over region R, and $AVG(CumPCP_t^{R,M,*})$ is the standard of cumulative precipitation between month 1 and month M on region R. The '*' as the end means that the average and standard deviation can be calculated on any of the three periods considered in the study (1986-2005, 2041-2070 or 2071-2100). Of these 11 potential predictors, the most correlated with the average volume of water in the IND were $AVG(CumPCP_t^{6,7,*})$ and $AVG(CumPCP_t^{7,10,*})$. They were therefore retained for inclusion in the CSS. The fit of the Climate response function (a linear regression equation linking the two selected predictors and the average volume of water in the IND is presented in Figure 5, panel a. The outputs of the RCMs and the KNN-WG in the CSS are presented on Figure 5, panel b. The empirical PDF derived from the weather generator outputs is presented on figure 6, panel a along with the undisturbed RCM output. It can be noticed that a significant number of RCM outputs are in the low probability area, suggesting that only five RCM outputs should actually be used.

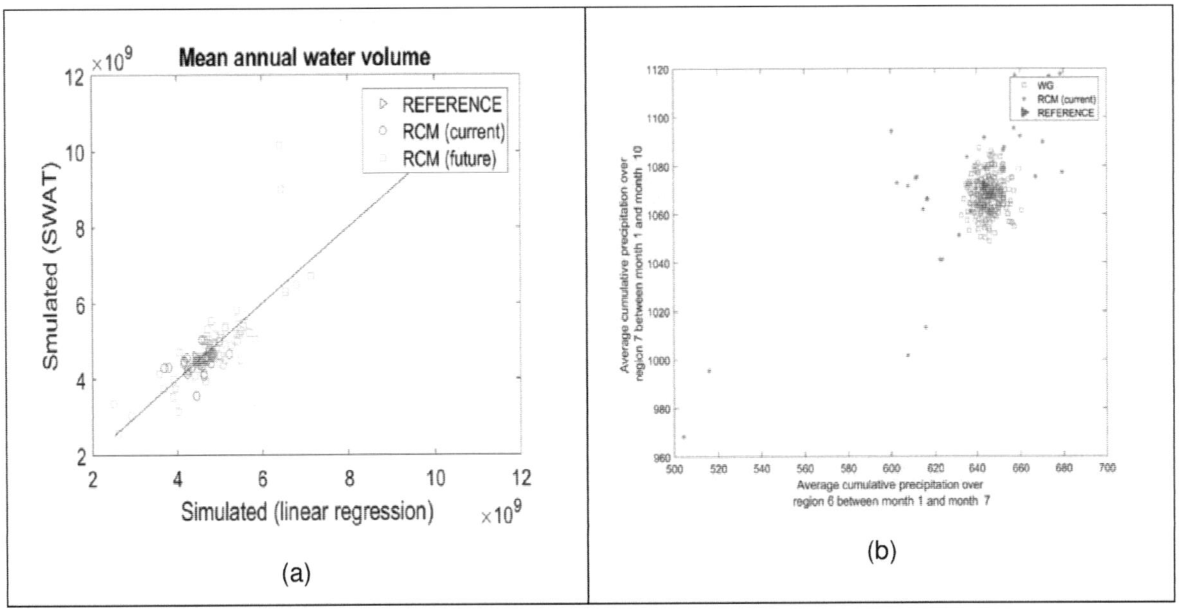

Figure 6: a) Selected Climate Response Function (Linear regression between the mean annual water volume in the IND and the two selected predictors) and b) representation of RCM outputs and synthetically generated time series in the DCC.

The magnitude of the empirical PDF can be used to set a credibility to RCM outputs both on the historical period. and future periods. This is already an improvement from the usual practice of assigning an equal weight to all RCM outputs. However, the low number of points in the high probability space may ne an issue. That's why the transformation described in section 2.7 was proposed to comb weather generator outputs and downscaled climate models outputs to generate high resolution projections. Figure 6.b shows the transformed RCM outputs after the transformation described in section 2.7. The number of projections went from 40 to 10,000 with a large number of them in the high probability area. This will allow a better coverage of potential future risks if the new time series are fed in the SWAT model of the UNBB or the Climate Response Function.

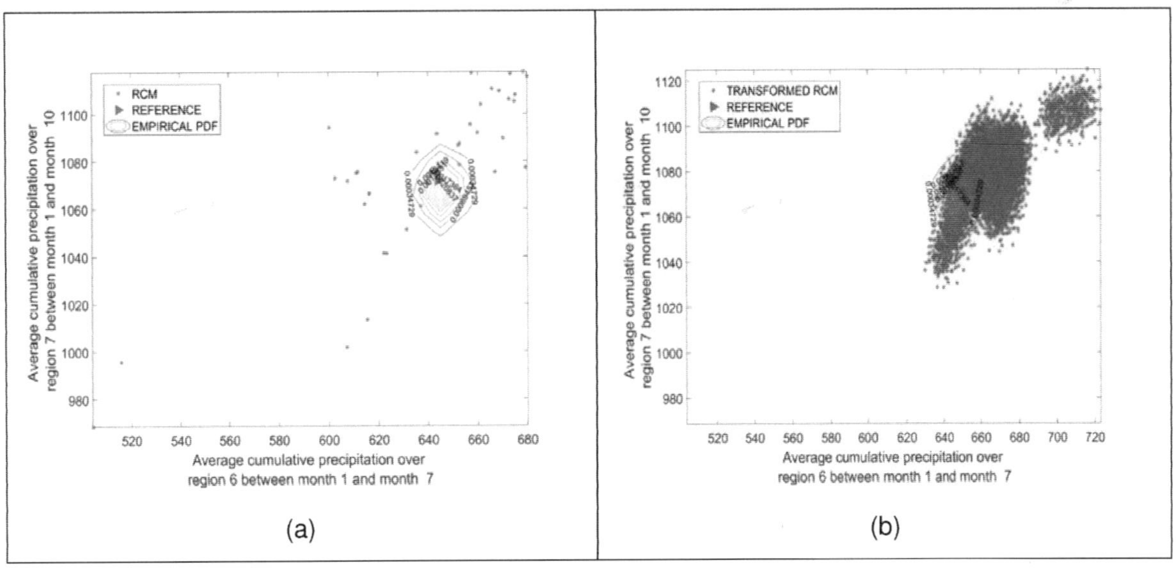

Figure 7: a) Empirical PDF and position of the RCM outputs in the CSS and b Empirical PDF and position of the transformed RCM outputs in the CSS.

Some aspects that need to be improved in the proposed methodology are a) the nature of the Climate Response Function and b) the method to combine weather generators outputs and RCM outputs so that the results cover the entire high probability region. Relationships in the climate system are usually highly nonlinear, and it is anticipated that the used of non-linear climate response functions (e.g. artificial neural networks) would lead to a better fit in the relationships between elements of the CSS and elements of the RPIS. In this paper, the cloud of transformed RCM outputs did not perfectly overlay with the cloud of the weather generator outputs in the CSS, leading to an imperfect coverage of the RPIS. Alternative transformation methods such as quantile mapping may lead to better results. Despite these limitations, the proposed methodology is considered promising and may lead to a better identification of risks associated with climate change, and better adaptation decision.

4 CONCLUSIONS

This paper highlighted the promises of the bottom-up approach to risk discovery compared to the traditional top-down approach. The limited number of RCM projections available to modellers and the lack of methodology to rank RCM outputs was identified as one of the major limitations to the application of the bottom-up approach. A new stochastic method is proposed to generate a large number of climate change projections by combining the outputs of weather generators to the outputs of regional climate models, and associate them with a likelihood value. The increased number of projections allows a better coverage of the risk space and while the likelihood value allows for a more realistic assessment of the plausibility of future risks.

References

Abbaspour, K.C. (2012). SWAT-CUP 2012: SWAT calibration and uncertainty programs. Eawag: Swiss Federal Institute of Aquatic Science and 8 Technology. Retrieved from http://www.neprashtechnology.ca.

Amadou, A., Gado Djibo, A., Seidou, O., Sittichok, K., Seidou Sanda, I. (2014). Changes to flow regime on the Niger River at Koulikoro under a changing climate. Journal of hydrological sciences. DOI: 10.1080/02626667.2014.916407

Arnold, J.G., Srinivasan, R., Muttiah, R. S. and Williams, J.R. (1998). Large area hydrologic modeling and assessment part i: model development. Journal of the American Water Resources Association, 34(1):73–89, doi: 10.1111/j.1752-1688.1998.tb05961.x.

Bartholom, E. and Belward, A.S. (2005). GLC2000: a new approach to global land cover mapping from Earth observation data. International Journal of Remote Sensing, 26(9):1959–1977, doi: 10.1080/01431160412331291297.

Brown, C, Y Ghile, M Laverty, and K Li. 2012. "Decision Scaling: Linking Bottom-up Vulnerability Analysis with Climate Projections in the Water Sector: decision scaling-linking vulnerability analysis." Water Resources Research 48 (9): doi:10.1029/2011WR011212.

FAO/UNESCO. Digital soil map of the world and derived soil properties [online]. rev. 1 cd rom. Available online at h ttp://www.fao.org/catalog/what new-e.htm, 2003.

Giorgi, F, C Jones, and G R Asrar. 2009. "Addressing Climate Information Needs at the Regional Level: The CORDEX Framework." Bulletin of the World Meteorologic Organization 58 (3): 175–83.

Jarvis, A., Reuter, H., Nelson, A., Guevara, E. (2008). Hole-flled srtm for the globe version 4, available from the cgiar-csi srtm 90m database (http://srtm.csi.cgiar.org).

Sharif M, Burn D H (2007) Improved K-nearest neighbor weather generating model. J Hydrol Eng 12(1):42–51

G. P. Weedon, G.P., Balsamo, G., Bellouin, N., Gomes, S., Best, M.J., and Viterbo, P. (2014). The WFDEI meteorological forcing data set: WATCH Forcing Data methodology applied to ERA-Interim reanalysis data. Water Resources Research, 50(9):7505–7514, doi: 10.1002/2014WR015638.

Stainforth, D A, T E Downing, R Washington, A Lopez, and M New. 2007. "Issues in the Interpretation of Climate Model Ensembles to Inform Decisions." Philosophical Transactions of the Royal Society A: Mathematical, Physical and Engineering Sciences 365 (1857): 2163–77. doi:10.1098/rsta.2007.2073.

Taylor, K E, R J Stouffer, and G A Meehl. 2012. "An Overview of CMIP5 and the Experiment Design." Bulletin of the American Meteorological Society 93(4): 485–98. doi:10.1175/BAMS-D-11-00094.1.

Building Tomorrow's Society
Bâtir la Société de Demain

Fredericton, Canada
June 13 to 16, 2018 / *13 au 16 juin, 2018*

FRP REINFORCED GLULAM BEAMS UNDER HIGH STRAIN-RATES

Lacroix, Daniel[1,3] and Doudak, Ghasan[2]
[1] Carleton University, Canada
[2] University of Ottawa, Canada
[3] Daniel.Lacroix2@Carleton.ca

Abstract: An experimental program investigating the potential for using fibre-reinforced polymers (FRP) as a strengthening option for glulam beams subjected to simulated blast loads was undertaken. A total of seven different retrofit configurations were investigated. Increases in resistance and maximum deflection at peak resistance in the range of 1.35-1.57 and 1.30-1.62, respectively, were obtained when FRP tension laminates with and without confinement were used. The addition of FRP confinement prevented premature de-bonding and significantly altered the failure mode from simple or splintering tension failure to a combination of brash tension and compression failure while limiting the damage to a small region. Whereas the use of unidirectional FRP as confinement did not contribute significantly to the post-peak resistance, the use of bidirectional FRP significantly improved the post-peak resistance above the 50% threshold for deflections corresponding to 1.57-3.40 that recorded at peak resistance.

1 INTRODUCTION

The increased presence of glulam in high-profile structures may put them at risk for deliberate attacks or accidental explosions. The available literature on the behaviour of glulam under such loading is scarce (e.g. Lacroix 2017). Also, while design guidelines in the Canadian blast standard (CSA 2012) exist on the strengthening and retrofitting of structural elements using fibre-reinforced composite materials for reinforced concrete and masonry, such provisions are lacking for wood elements. However, significant research has been conducted on the behaviour of retrofitted wood members under static loading.

Research conducted on the behaviour of FRP reinforced wood members under static loading have shown that increases in resistance and stiffness for wood reinforced with simple FRP tension reinforcement of up to 50% and 20%, respectively, can be attained (Buell and Saadatmanesh 2005, Hernandez et al. 1997, Johns and Lacroix 2000, Yang et al. 2016). It is important to note that these increases highly depend on the type and amount of reinforcement provided, reinforcement scheme (e.g. horizontal or vertical, sheets or bars), and type of wood tested (i.e. lumber, timber, EWP).

The main drawback in using simple tension reinforcement is the accompanied partial- or full-length de-bonding of the reinforcement when the outer wood tension layers fail (Dorey and Cheng 1996a, Dorey and Cheng 1996b, Hernandez et al. 1997). Buell and Saadatmanesh (2005) investigated the effect of wrapping timber beams with carbon FRP (CFRP), which was found to provide confinement to the wood, and thereby reducing the effects of defects. Johns and Lacroix (2000) investigated the effect of simple longitudinal tension reinforcement and half U-shaped longitudinal reinforcement on sawn lumber. It was observed that the presence of FRP material arrested the crack opening, confined local rupture, and bridged local defects naturally found in wood. Sonti et al. (1996) investigated the effect of partial-length reinforcement and found that providing such confinement enhanced the strength and stiffness in the order of 2.1 to 23.2% and 1.9 to 9.3%, respectively.

The current paper is part of a broader research program investigating the behaviour of glulam beams and columns subjected to simulated blast loadings (Lacroix 2017). Herein presented are some key experimental results of the test program involving the flexural retrofit of full-scale glulam beams using both unidirectional and bidirectional glass FRP (GFRP). Full-scale retrofitted beams were tested destructively under both static and dynamic loading.

2 EXPERIMENTAL PROGRAM

2.1 Description of Specimens

A total of sixteen 137 x 222 mm^2 24f-ES Spruce-Pine retrofitted glulam beams were tested to failure under simply supported static and dynamic four-point bending. The glulam layup consisted of six laminates in width and five laminates in depth and was manufactured according to ASTM D3737 (ASTM 2012) to meet the specifications of CSA O177 in Canada (2015). A total of seven different GFRP retrofit configurations, used to enhance the flexural behaviour of the beams were investigated. The experimental program involved investigating the effects of unidirectional and multi-directional fabrics on the post-peak behaviour of the beams. The test matrix for both the unretrofitted and retrofitted beams is presented in Table 1.

Table 1: Test matrix

Specimen type	Loading type	Specimens	FRP retrofit configuration
Unretrofitted	Static	B1-B4	-
	Dynamic	B5-B9	-
Retrofitted	Static	R1-1, R1-2	T[0]2
	Dynamic	R1-3, R1-4	T[0]2
	Static	R2-1, R2-2	T[0]2C[90]
	Dynamic	R2-3, R2-4	T[0]2C[90]
	Static	R3-1, R3-2	U[0]2C[90]
	Dynamic	R3-3, R3-4	U[0]2C[90]
	Dynamic	R4-A	U[0]2C[0/90]2
	Dynamic	R4-B	C[0/90]4
	Dynamic	R5-A	U[0]2C[±45]2
	Dynamic	R5-B	C[±45]4

Specimens are identified with letters B, or R, denoting whether the specimen is an unretrofitted beam, or retrofitted beam, respectively. For the retrofitted specimens, the stacking sequence is presented in terms of simple tension reinforcement (T) or U-shaped tension reinforcement (U) and confinement (C). Found in the square brackets is the fibre orientation of the fabric used while the subscript denotes the number of layers applied in cases where there were multiple layers applied.

Retrofit 1 consisted of two layers of GFRP tension reinforcement. Due to the known issue of de-bonding of the FRP reinforcement, a layer of unidirectional GFRP in the perpendicular to wood fibre direction was added in Retrofit 2. Investigated in Retrofit 3 is the effect of a tension U-shaped GFRP with partial-length confinement. Retrofit 4-A consisted of two unidirectional GFRP U-shaped layers along with two layers of 0/90 bidirectional GFRP whereas Retrofit 4-B consisted of four layers of 0/90 bidirectional GFRP. Similarly, Retrofit 5-A consisted of two unidirectional GFRP U-shaped layers along with two layers of ±45 bidirectional GFRP whereas Retrofit 5-B consisted of four layers of ±45 bidirectional GFRP. For both Retrofits 4 and 5, the reinforcement was applied in the direction parallel to the length of the beam such that the reinforcement on the tension side was continuous.

2.2 Static and Dynamic Test Setups

To establish the flexural behaviour of the specimens, static and dynamic four-point bending tests with simply supported boundary conditions and a clear span of 2,235 mm were conducted. Figure 1 shows the static and dynamic test setups where it can be seen that the same loading pattern, span, and boundary conditions were used to allow for a direct comparison between the two loading regimes.

(a) Static (b) Dynamic

Figure 1: Test setups

Data acquisition systems with sampling rates of 10 and 100,000 samples per second for the static and dynamic tests, respectively, were used to record the beams' response. Reactions, mid-span displacement, and tensile and compressive strain-time histories were recorded for the static tests. Similarly, the reflected pressure, mid-span displacement, reactions, and mid-span tensile and compressive strain-times histories were measured for the dynamic tests.

3 EXPERIMENTAL RESULTS AND ANALYSIS

3.1 Characterization of Failure Modes

Representative static and dynamic failure modes for Retrofits 1 through 3 are shown in Figure 2. Both the static (Fig. 2a) and dynamic (Fig. 2d) failure of Retrofit 1 was characterized by a failure in the outer wood tension laminates with extensive damage throughout the depth of the cross-section. Upon initial wood tensile failure, the outer tension laminates pushed outwards on the GFRP causing separation between the wood and reinforcement. The overall failure of the wood was dominated by simple or splintering tension failure that was influenced by the presence of knots or finger-joints.

(a) R1-1

(b) R2-1

(c) R3-1

(d) R1-4

(e) R2-4

(f) R3-3

Figure 2: Representative static and dynamic failure modes for Retrofits 1 through 3

The failure mode observed in Retrofit 2 significantly differed from that observed in Retrofit 1. A brash tension failure resulting in a straight cut through the beam width was observed under both static (Fig. 2b) and dynamic (Fig. 2e) loading. The addition of full-length confinement prevented the separation between the tension GFRP reinforcement and the wood tension face while also limiting the damage to the wood to a small section with significant compression failure. Furthermore, as anticipated in unidirectional FRP fabric, tearing in the FRP confinement perpendicular to the fibre direction was observed (Figs. 2b and 2e).

Representative static and dynamic failures for Retrofit 3 are shown in Figures 2c and 2f, respectively. While some separation of the FRP tension reinforcement was observed (Figs. 2c and 2f), it occurred at a deflection level that was beyond that obtained at peak resistance. The specimens investigated in Retrofit 3 were able to maintain some level of post-peak resistance. For example, specimen R3-3 deflected approximately 1.6 times more than the deflection corresponding to peak resistance without de-bonding of the FRP.

Figure 3a shows the final deflected shape of specimen R4-A, where the de-bonding occurred past the point of ultimate resistance. Specimen R4-B, which deflected to a maximum displacement corresponding to 4.15 times that recorded at peak resistance without any de-bonding, is shown in Figure 3b. While specimen R4-B did not reach complete failure (i.e. blowout) and recovered part of its displacement, local "buckling" of the FRP on the compression face near the mid-span region was observed. Specimen R5-A, shown in Figure 3c, was subjected to a pressure-impulse combination similar to that of specimen R4-A, and as it can be seen the permanent deflection as well as the associated damage in R5-A is significantly less. While Specimen R5-A did not reach ultimate failure, it recovered with a permanent deflection corresponding to sixty percent of the maximum displacement. Specimen R5-B (Figure 3d) was subjected to a pressure and impulse combination similar to that specimen R5-A was exposed to. The specimen deflected 3.6 times that recorded at peak resistance.

(a) R4-A (b) R4-B (c) R5-A (d) R5-B

Figure 3: Representative static and dynamic failure modes for Retrofits 4 and 5

3.2 Analysis of Experimental Test Results

3.2.1 Static

The key static test results for the retrofitted beams are presented in Table 2 along with the average of the unretrofitted beams obtained from previous work conducted by the authors (Lacroix 2017).

Figure 4 shows the static resistance curves for the retrofitted specimens as well as the average of the unretrofitted beams. Providing two layers of GFRP tension reinforcement (e.g. Retrofit 1) increased the ultimate resistance by a factor of 1.35, deflection at ultimate resistance by a factor of 1.30, and stiffness by a factor of 1.1 compared to the unretrofitted beams. Although the reinforcement enhanced the stiffness and strength of the beam on average, it did not provide significant post-peak resistance. The addition of GFRP confinement in addition to the tension reinforcement (e.g. Retrofit 2) increased the ultimate resistance, deflection at ultimate resistance, and stiffness by factors of 1.40, 1.39, and 1.13 relative to the unretrofitted beams, respectively.

Table 2: Static test results

Specimen	R_{max}[a] (kN)	Δ_{Rmax}[b] (mm)	K[c] (kN/mm)	$\varepsilon_{t,w-f}$[d] x 10^{-4} (mm/mm)	$\varepsilon_{t,w-sf}$[e] x 10^{-4} (mm/mm)	$\varepsilon_{t,FRP-f}$[f] x10^{-4} (mm/mm)	$\varepsilon_{c,w-f}$[g] x 10^{-4} (mm/mm)
B_{AVG}	*141.6*	*23.7*	*6.0*	*40.8*	-	-	*-52.4*
R1-1	196.7	32.9	6.4	39.2	37.5	40.2	-42.0
R1-2	186.2	28.9	6.7	40.8	28.0	41.8	-37.8
$R1_{AVG}$	*191.5*	*30.9*	*6.6*	*40.0*	*32.7*	*41.0*	*-39.9*
R2-1	206.3	32.6	6.7	44.1	37.6	45.1	-47.3
R2-2	189.8	33.2	6.8	49.0	35.0	51.4	-94.5
$R2_{AVG}$	*198.0*	*32.9*	*6.8*	*46.5*	*36.3*	*48.2*	*-70.9*
R3-1	214.3	38.5	6.6	43.6	-	94.2	-60.7
R3-2	230.3	38.6	6.8	48.3	33.8	50.2	-
$R3_{AVG}$	*222.3*	*38.5*	*6.7*	*46.0*	*33.8*	*72.2*	*-60.7*

[a] Maximum (peak) resistance

[b] Deflection at peak resistance

[c] Stiffness

[d] Strain at tensile rupture on tension face

[e] Strain at tensile rupture at 21 mm from tension face

[f] FRP maximum strain

[g] Maximum wood compressive strain

The confinement provided some post-peak resistance, albeit small in comparison to peak resistance. Retrofit 3 had a significant post-peak resistance (Fig. 4) of approximately 75% of the ultimate resistance. Retrofit 3 provided increases of 1.57, 1.62, and 1.12 in resistance, deflection at peak resistance, and stiffness, respectively, when compared to the unretrofitted beams.

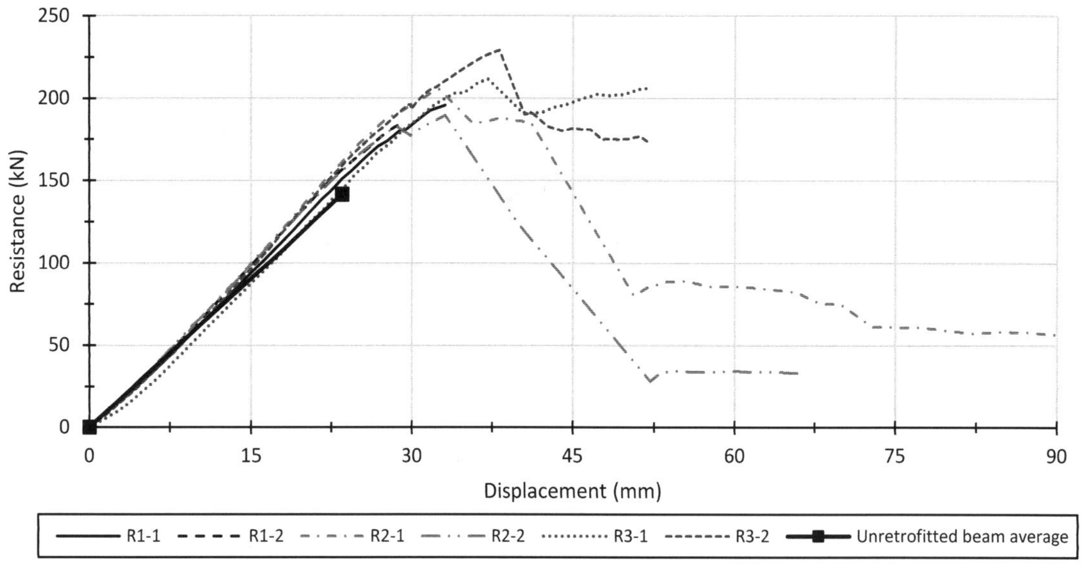

Figure 4: Static resistance curves of retrofitted beams compared to unretrofitted beam average

3.2.2 Dynamic

The key dynamic test results are summarized in Table 3 along with the results of the dynamically tested unretrofitted beams.

Table 3: Dynamic test results

Specimen	P_R[a] (kPa)	I_R[b] (kPa-ms)	R_{max-d}[c] (kN)	Δ_{Rmax-d} (mm)	K[e] (kN/mm)	ε_{f-T}[f] x 10^{-4} (mm/mm)	ε_{f-Ts}[g] x 10^{-4} (mm/mm)	ε_{f-FRP}[h] x 10^{-4} (mm/mm)	ε_{f-C}[i] x 10^{-4} (mm/mm)
B_{AVG}	53.0	561	153.6	28.4	5.9	37.6	-	-	-47.4
R1-3.1	52.2	577	156.1	26.7	6.9	39.5	32.5	88.9	-37.2
R1-4.1	68.7	727	216.1	36.9	6	49.1	40.5	51.9	-41.4
R2-3.1	62.3	718	198.0	32.7	6.0	42.8	34.7	61.1	-133.1
R2-4.1	68.8	757	208.0	34.6	6.0	N/A[j]	N/A[j]	N/A[j]	N/A[j]
R3-3.1	63.5	698	213.0	40.3	6.4	51.8	N/A	175.3	-113.9
R3-3.2	73.8	829	183.7	61.0	3.0	N/A[j]	N/A[j]	165.9	-73.6
R3-4.1	67.6	818	242.3	41.7	6.5	49.8	33.8	118.6	N/A[j]
R3-4.2	74.4	780	182.9	58.9	3.5	N/A[j]	N/A[j]	113.4	N/A[j]
R4-A	76.3	1,091	241.2	38.2	6.3	42.8	-	143.9	-164.3
R4-B	71.5	1,092	242.9	37.7	6.5	45.5	-	207.8	-69.9
R5-A	74.6	1,153	246.0	36.0	7.4	43.2	-	152.7	-54.7
R5-B	65.7	1,170	234.3	38.9	6.4	47.7	-	251.0	-59.7

[a] Reflected pressure
[b] Reflected impulse
[c] Beam dynamic resistance
[d] Deflection at peak resistance
[e] Stiffness
[f] Strain at tensile rupture on tension face
[g] *Strain at tensile rupture at 21 mm from tension face*
[h] FRP maximum strain
[i] Maximum wood compressive strain
[j] Strain gauge malfunction

A comparison of the dynamic resistance of Retrofits 1 to 3 (Table 3) to the static average of its respective retrofit (Table 2) shows an average increase of 1.15. The dynamic resistance was obtained by considering the dynamic equilibrium of the system using a single-degree-of-freedom approach (Lacroix 2017), The apparent increase in strength is associated with high strain-rate effects. The dynamic increase factor (DIF) obtained in this study compares well with the value of 1.14 found in the literature based on comparison between static and dynamic resistance of unretrofitted glulam beams (Lacroix 2017).

Figure 5 shows the dynamic resistance curves of the retrofitted specimens where it can be seen that in general significantly greater deflections were experienced by the specimens during the dynamic testing compared to the static testing.

From Table 3 and Figure 5, it can be seen that the various configurations of reinforcement provided similar peak resistance (R_{max}), but different post-peak behaviour. While the addition of confinement for Retrofits 2 and 3 improved the post-peak resistance in comparison to the unretrofitted beams and Retrofit 1 beams, the use of multi-directional fabric significantly improved the post-peak behaviour. Post-peak resistance, above the 50% threshold of peak resistance, was sustained for deflections ranging between 1.57 and 3.40 that recorded at peak resistance for Retrofits 4 and 5.

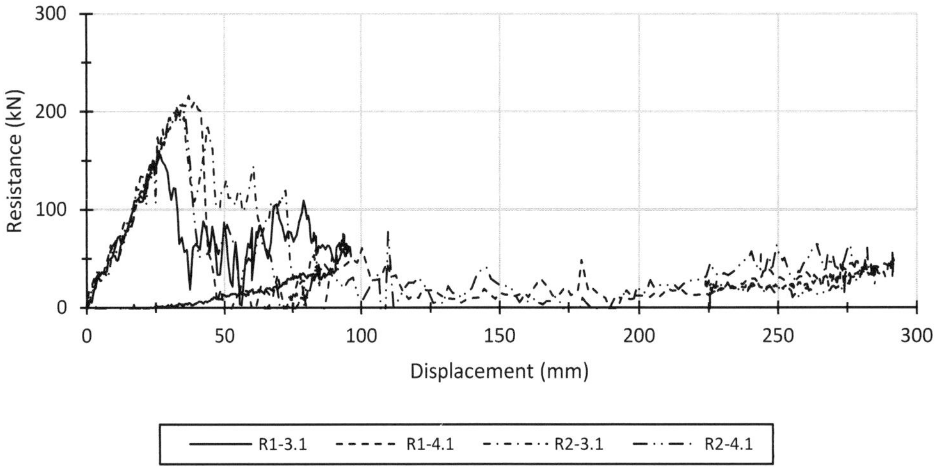

(a) Retrofits 1 and 2

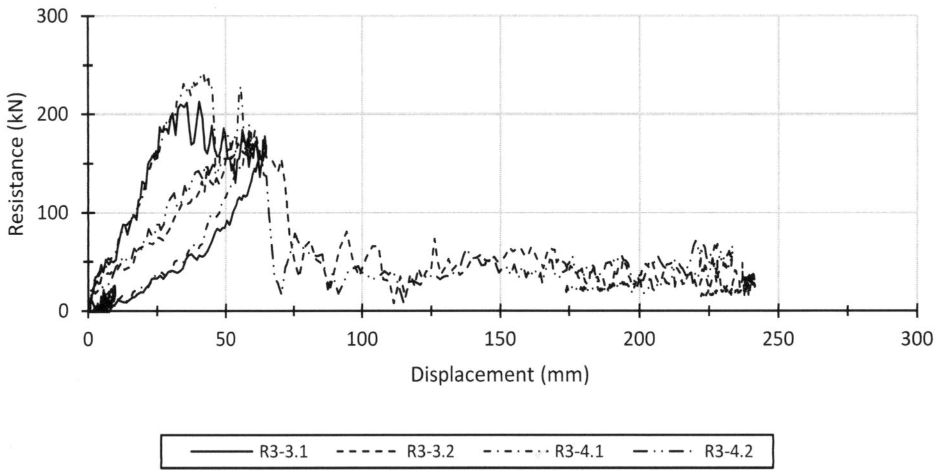

(b) Retrofit 3

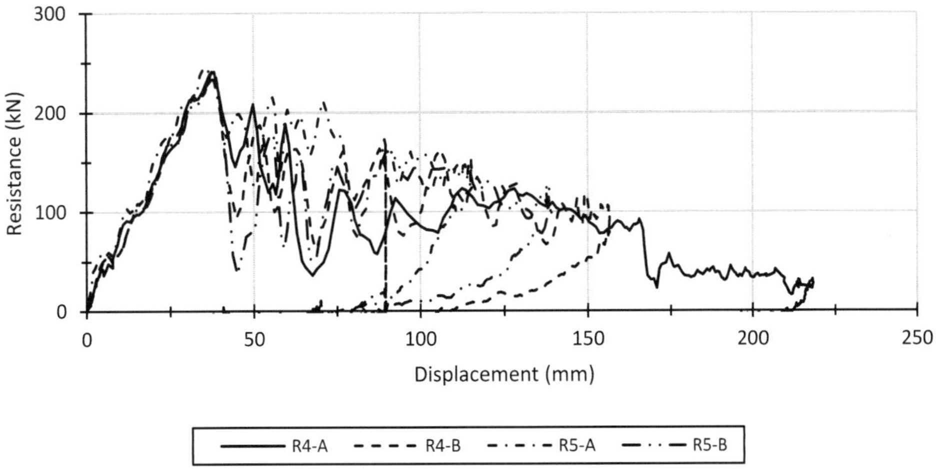

(c) Retrofits 4 and 5

Figure 5 Dynamic resistance curves of retrofitted beams

4 CONCLUSIONS

Results from an experimental program investigating the behaviour of retrofitted glulam beams using unidirectional and bidirectional FRP were presented. Increases in resistance and deflection at peak resistance in the order of 1.35-1.57 and 1.30-1.62, respectively, were observed for the reinforced beams relative to the unretrofitted beams. While adding simple tension reinforcement contributed to an increase in performance, this retrofit resulted in premature de-bonding between the FRP and the wood. Adding confinement prevented premature de-bonding and significantly altered the failure mode from simple or splintering tension failure to a combination of brash tension and compression failure while limiting the damage to a small region. Apart from Retrofit 3, none of the retrofit configurations investigated using unidirectional FRP significantly contributed to the post-peak behaviour. The use of bidirectional FRP in Retrofits 4 and 5 significantly increased the post-peak behaviour.

References

ASTM. 2012. Standard Practice for Establishing Allowable Properties for Structural Glued Laminated Timber. *ASTM D3737-12.* West Conshohocken, PA, ASTM International.

Buell, T.W., and Saadatmanesh, H. 2005. Strengthening timber bridge beams using carbon fiber. *Journal of Structural Engineering*, **131**(1):173-187.

CSA. 2012. Design and assessment of buildings subjected to blast loads. *CSA S850.* Mississauga, ON, Canadian Standards Association Group.

CSA. 2015. Qualification code for manufacturers of structural glued-laminated timber. *CSA O177.* Mississauga, ON, Canadian Standards Association Group.

Dorey, A.B., and Cheng, J.J.R. 1996a. The behavior of GFRP glued laminated timber beams. *Adv. Comp. Mat. in Bridges and Struct. II*, The Canadian Society for Civil Engineering, Montreal, Canada.

Dorey, A.B., and Cheng, J.J.R. 1996b. Glass Fiber Reinforced Glued Laminated Wood Beams. In *Canada-Alberta Agreement Documents.* Edmonton, Alberta: Natural Resources Canada, Canadian Forest Service, Northern Forestry Centre.

Hernandez, R., Davalos, J., F., Sonti, S.S., Kim, Y., and Moody, R., C. 1997. Strength and stiffness of reinforced yellow-poplar glued laminated beams. Madison, WI: U.S. Department of Agriculture, Forest Service, Forest Products Laboratory.
Johns, K.C., and Lacroix, S. 2000. Composite reinforcement of timber in bending. *Canadian Journal of Civil Engineering*, **27**(5):899-906.
Lacroix, D. 2017. Investigating the Behaviour of Glulam Beams and Columns Subjected to Simulated Blast Loading. Ph.D., Civil Engineering, University of Ottawa.
Sonti, S.S., GangaRao, H.V.S., and Superfesky, M.C. 1996. Rehabilitation and strengthening of glulam stringers for bridge superstructures. *First International Conference on Composites in Infrastructures*, University of Arizona, Tucson, Arizona.
Yang, H., Liu, W., Lu, W., Zhu, S., and Geng, Q. 2016. Flexural behavior of FRP and steel reinforced glulam beams: Experimental and theoretical evaluation. *Construction and Building Materials*, **106**:550-563.

Building Tomorrow's Society
Bâtir la Société de Demain

Fredericton, Canada
June 13 to 16, 2018 / *13 au 16 juin, 2018*

HOLISTIC APPROACH TO FLOOD RISK MITIGATION IN NOVA SCOTIA

Fernandez, Victoria[1,2], Wilson, Alexander[1]
[1] CBCL Limited, Canada
[2] vfernandez@cbcl.ca (corresponding author email)

Abstract: The third United Nation World Conference on Disaster Risk Reduction recognized the importance of understanding disaster risks as result of the combination of not only natural hazards and infrastructure capacity but also social vulnerability, exposure levels, and emergency preparedness. Understanding these factors and their combined effect is fundamental for assessing risks, preventing disasters, mitigating damages and preparing for emergencies. In the case of Canada, flooding has been identified as the leading cause of emergencies. In this context, risk management requires not only understanding the hydrologic and hydraulic processes that result in extreme events but also recognizing that protection of infrastructure does not necessarily translate into the protection of public safety. In Atlantic Canada, more than 40% of residents live in rural and low budget communities. Populations at risks include First Nation communities and coastal settlements located in the Bay of Fundy (home to the highest tides in the world). Holistic flood risk management approaches in this region have focused on determining the priorities and needs of stakeholders, identifying vulnerabilities, evaluating the suitability of approaches such as retreat, protection and adaptation; and enhancing capacity for resilience. The implementation of such approaches in Atlantic Canada needs to be placed in the context of limited budget availability, complex environmental settings, limited historical data and uncertain future climate. This paper discusses the implementation of holistic flood management approaches and softer flood mitigation techniques in the province of Nova Scotia.

1 INTRODUCTION

Rural and urban communities in the peninsula of Nova Scotia are located within 70 Km of the coast and exposed to complex and diverse environmental settings. In the South Shore, for example, a deep natural harbour allowed the establishment of Halifax, now a densely populated urban center. Coastal developments are exposed to waves and storm surges from the Atlantic Ocean and are at risk of permanent inundation, as a result of sea level rise. In the Bay of Fundy, the world's highest tides inundated the coastal areas, creating nutrient rich soils that attracted agricultural development. Communities along these shores reclaimed coastal flooding areas through the construction of dykes and tide gates (aboiteaux) for flood protection. In recent times, many of these communities have experienced major flooding events caused by the overtopping of the dykes or stormwater backup (Nova Scotia Environment 2015).Throughout the province, communities at risk of flooding are exposed not only to flooding hazards of different nature but also to diverse socio-economical settings. These settings are relevant in defining a community's vulnerability and exposure to flooding disasters (Thistlethwaite and Henstra 2017).

Communities in Nova Scotia have traditionally relied on structural defenses from hazards up to a specific magnitude. The standard design criterion is generally the 100-year rainfall event (an event with an annual exceedance probability of 1%). On October 10, 2016, Thanksgiving Day, Sydney NS experienced a major flooding after a precipitation event of 225 mm of rain in 24 hours, an amount close to the 2000-year rainfall calculated with historical data in the area (Figure 1). The occurrence of events of magnitude above the infrastructure design threshold shows that structural defenses alone may not be sufficient to achieve effective flood management and disaster mitigation. A broader or holistic flood management strategy treats flood risks as a combination of multiple factors including hazards, emergency preparedness, environmental settings and socio-economic conditions. This paper discusses elements to consider when implementing an integrated approach to flood studies in Nova Scotia including exposure (the proximity of the community to the hazard), vulnerability (the environmental, social, and physical conditions that make a community susceptible to a hazard), and the limitations of human intervention when providing full protection against a flooding hazard.

Figure 1 Flood in Sydney NS on October 12, 2016. Two days after the storm (CTV News 2016)

2 PARADIGM SHIFT

Communities living in floodplains and along coastal shorelines have traditionally relied on flow control and storage (dams and reservoirs), channel enhancement and diversion, and structural defenses (levees and dykes) for protection against floods (World Metereological Organization 2009). This approach provided populations with access to areas naturally prone to flooding, but at significant long-term cost and risk. These include:

- the continuous need for infrastructure maintenance funds (Instittue for Sustainable Development 2013);
- performance limited to designs based on probability of occurrence (FEMA 2007);
- limited adaptability in the face of climate change (Blair and Thistlethwaite 2012);
- environmental damages associated with disconnecting the rivers from the floodplain (Tockner and Stanford 2002); and
- hidden risks associated with failures and additional developments (Managers 2007).

Recent estimates of flood-related economic loss indicate that effective flood management requires considering the limitations of structural defenses and the consequences of flooding (Thistlethwaite and Henstra 2017)

In the year 2009, the World Meteorological Organization called for a "paradigm shift" from this traditional approach to an integrated recognition of the dynamic nature of river and coastal systems (World Metereological Organization 2009).

In 2015, the Sendai Framework of Disaster Risk reduction, the most recent holistic guidance for global disaster management, identified four fundamental elements in the implementation of an integrated approach to disaster mitigation (United Nations 2015):

1. Understanding the Risks;
2. Strengthening disaster risk governance to manage the risks;
3. Promote public and private involvement; and
4. Enhance emergency preparedness

In the context of flooding, understanding the risks requires not only knowledge of the natural processes that cause flooding (rainfall, tides, snow melt, ice jams, storm surges) but also knowledge of the degree of exposure to the hazard and the degree of vulnerability. In this context, flood studies and floodplain mapping can provide a comprehensive description of the natural hazard and the extent of flooding.

Strengthening disaster risk governance for flood management involves the implementation of programs, plans, guidance, and regulations directed at managing the exposure to risk of flood-prone areas, strengthening emergency preparedness and response, and fostering resilience and recovery (United Nations 2015).

Increasing public awareness of flooding risks and the potential effects of climatic change on those risks should be included in public and private involvement and investment, with the goal of preventing the loss of life, reducing losses, and facilitating recovery. This also includes providing a mechanism for the public to establish its priorities for protection and potential needs during recovery. Citizen data collection also provides a platform to collect historical flood event information and creates awareness about the magnitude of hazards.

Enhancing emergency preparedness aims at fostering effective emergency response and identifying opportunities for implementing improvements when restoring or replacing damaged infrastructure. Generally, municipal, provincial, and federal authorities rely on emergency management plans outlining response strategies. In the context of flooding, calibrated hydraulic and hydrologic models, using rainfall and temperature predictions can potentially identify areas at risk. This would allow authorities to notify the public of the risk and prepare for extreme events ahead of time.

The following sections evaluate the implementation of a holistic approach to flood studies and flood management in Nova Scotia using the four elements identified in the Sendai Framework of Disaster Risk Reduction

3 UNDERSTANDING FLOODING RISKS IN NOVA SCOTIA

Floodplain maps are an effective tool for anticipating, preparing, and mitigating flood damage when based on updated weather and land use information (Henstra&Thistlethwaite, 2017). Between 1975 and 1996, under the Flood Damage Reduction Program (FDRP) more than 900 communities throughout Canada developed 1 in 20 year and 1 in 100 year flood maps aiming to identify areas vulnerable to floods and regulate development in those areas (Environment and Climate Chance Canada 2013). A study for the National Floodplain Management Framework of Public Safety Canada identified that by the year 2014, approximately 50% of flood maps in Canada were produced before 1996 (MMM Group Limited 2014) .

Since the FDRP ended,20 years of historical weather and hydrometric observations have become available through Environment Canada. Global models have increased their resolution and capability to project future climates (Le Treut, et al. 2007); computer modelling has evolved to allow hydraulic modelling in two and three dimensions (2D and 3D); and accurate gauging instruments have become available. In addition to this, community engagement (Natural Resources Canada 2017)and emergency preparedness plans (Public Safety Canada 2018) have become relevant in the evaluation of flood risks and vulnerability.

Several municipalities within Nova Scotia have conducted recent flood studies recognizing the value of updating their flood map information. These flood studies were based on the computer modelling of complex hydraulic and sediment transport processes in the Bay of Fundy or on simulations of a complex system of lakes, wetlands, river and hydraulic structures in the Halifax Regional Municipality (CBCL Limited 2016). However, despite the potential of the technological platform, which allows the assessment of flooding hazards, flood mapping in Nova Scotia faces the following challenges:

3.1 Lack of Historical Data

Advances in technology have included more complex computer models, high-resolution aerial imagery, and Lidar information availability allowing the simulation of complex hydraulic processes with a much greater degree of accuracy. However, evaluating the frequency of flooding and the probability of occurrence of extreme events requires an assessment of the local historical weather with a typical minimum of 20 years of records. In Nova Scotia, and generally in Atlantic Canada, this is limited by the low availability of gauging stations with sufficient historical records of rainfall amounts, water levels, and flow rates to conduct statistical analysis. For example, Figure 2 shows that very few rainfall stations over the province feature more than 20 years of record data within the last 50 years.

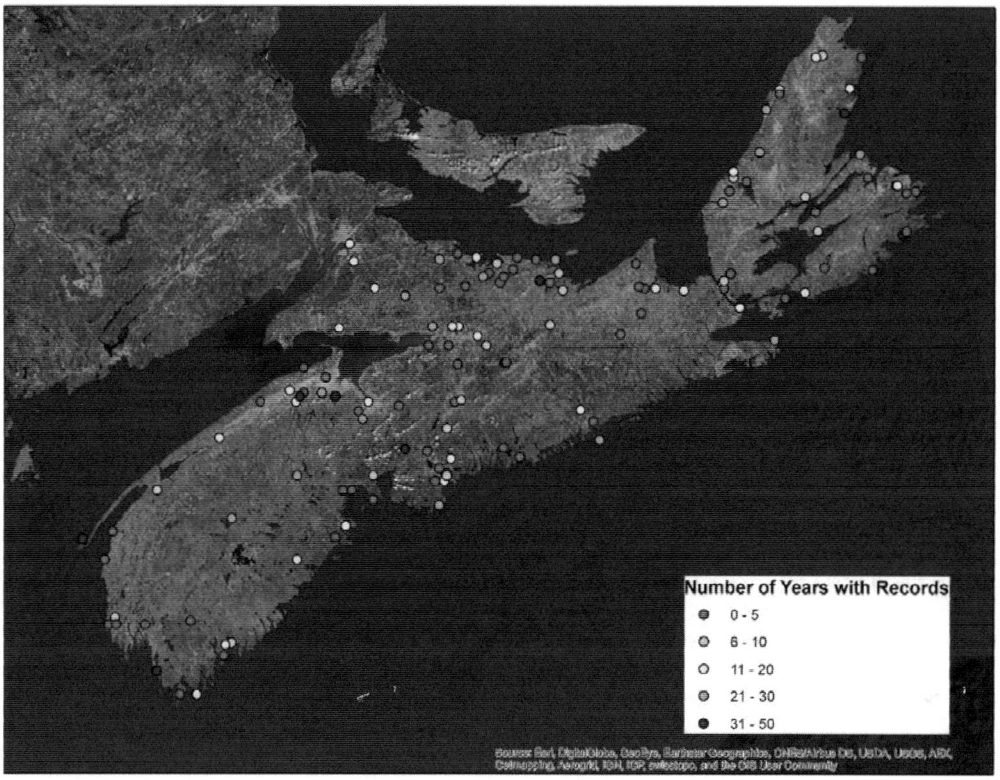

Figure 2 Environment Canada Rainfall Stations with Records since 1970

Overcoming the challenges associated with these information gaps requires a series of approximations and projections that may carry a certain degree of uncertainty and require additional validation efforts. For example, The Truro Flood Study required the development of a calibrated hydrodynamic model of the Bay of Fundy capable of projecting gauged water levels from Saint John Harbour to Truro (CBCL Limited 2014).

In the future, remote sensing records may offer an alternative to rainfall gauging stations for the calculation of rainfall intensity and duration curves (Marra, et al. 2017) which are typically used in the assessment of extreme stormwater flooding. In Nova Scotia, this methodology could be explored in the short term, as Environment Canada maintains rainfall radar data records, available from the year 2007 (Environment Canada 2018).

3.2 Complex Hydrodynamics

In the most recent flood studies in the region, advanced hydrodynamic modelling has provided insights into the complex combination of processes that result in flood risks (CBCL Limited 2016). For example, the interaction between the large tidal ranges in the Bay of Fundy with high rates of sediment transport and the network of dams and tidal gates (aboiteaux) result in temporal and spatial distribution of water levels, sediments, and flows that complicate field monitoring programs and model calibration. Figure 3 shows how more than one meter of sediment accumulated over a tide gauge during a field-monitoring program in the Bay of Fundy.

Figure 3 Sediment Accumulation Bay of Fundy at Tide Gauge Location (March 2013 Left, January 2014 Right). Credit: Alexander Wilson, CBCL Limited

3.3 Effect of Environmental Setting and Socio-Economic Conditions in Flood Mitigation Options

In the Bay of Fundy, sedimentation rates result in up to 2 meters of accumulation during the summer months and significant morphodynamical changes to river channels after the construction of hard infrastructure (Van Proosdij, et al. 2009). For example, the Truro Flood Risk study (CBCL Limited 2014) showed that under these conditions, the construction of protective infrastructure with limited life spans might require a significant capital investment to achieve minimal benefits, while increasing the risk of damage in case of failure. The study evaluated the implementation of more sustainable options including basin wide approaches that identify opportunities for avoiding exposure of key infrastructure to areas prone to flooding.

3.4 Limited implementation of nature-based approaches

Public awareness about the benefits of nature-based approaches for flood risk management is growing (World Bank 2017). The physical conditions of riverine and coastal systems are the result of the feedback between sediment transport, hydrodynamic processes and ecology. (Masselink and Hughes 2014). Nature-based solutions aim to reduce erosion issues, mitigate flood risks and improve water quality by mimicking processes of the natural environment. Benefits of this approach include adaptability, sustainability and low cost (NOAA 2015) .

Throughout the Province there are limited examples of flood mitigation approaches based on adaptation to natural processes. The capacity enhancement of Broad Brook in Yarmouth is one example of a flood mitigation approach that incorporates the functional features of adjacent wetlands and fish habitat (CBCL Limited 2016). The design is based on a system-scale perspective of the local socio-economic and environmental conditions after evaluating more than 10 combinations of infrastructure upgrade and stream enhancement. The project also includes considerations for adaptive management based on long-term monitoring.

4 FLOOD MANAGEMENT GOVERNANCE

In Nova Scotia, provincial and municipal governance includes the implementation of by-laws that regulate development within the 100-yearfloodplain (Nova Scotia Municipal Affairs 2002). Some municipalities such as the Halifax Regional Municipality require new developments to maintain peak runoff rates equal to those discharging from the site before development. Nova Scotia Environment regulates the environmental impacts that flood protection infrastructure may exert over adjacent wetlands and water courses (Nova Scotia Environment 2015). For approaches involving the construction of dams and water storage, Nova Scotia Environment requires a safety assessment based on the Safety Guidelines of the Canadian Dam Association. At the Federal level, the Department of Fisheries and Ocean requires the inclusion of fish passage considerations in the design and construction of any hard infrastructure within a watercourse. Therefore, an integrated approach to flood studies in Nova Scotia requires an understanding of the regulatory framework that governs the implementation of flood mitigation approaches. These regulations limit the implementation of traditional approaches that relied on extensive channel dredging, wetland reclamation, and river diversion.

5 PUBLIC AND PRIVATE INVOLVEMENT

A recent survey conducted by the University of Waterloo showed that about 94 percent of Canadians living in designated flood risk areas do not know they are at risk (Thistlethwaite et al. 2017). Effectively involving community residents in the floodplain mapping process is a critical component in establishing community-level awareness of flood risk (White, Kingston, & Barber, 2010).

Flooding studies for the Town of Mahone Bay in Lunenburg and for the Town of Truro, included public consultation at several stages of the project. The meetings unfolded through open house sessions that included "stations" to capture public feedback. Community members were directed to large maps where they could discuss their flooding experiences and concerns with the consultation team, illustrating their issues with markers and pens on the maps provided. The final station provided an opportunity for community members to rate priorities for flood protection. In both cases, this approach provided the community with updated information about flooding risks and allowed the development of mitigation recommendations taking into consideration the vision, values, needs, and priorities of each community.

6 FLOOD RISK MITIGATION AND EMERGENCY PREPAREDNESS

Flood risk mitigation in Nova Scotia requires recognizing the limitations of infrastructure-based protection approaches and recognizing the uncertainties associated with estimating the time and magnitude of the impacts of climate change. Reducing the risk of flooding under the complex environmental conditions of the Bay of Fundy, the socio-economic conditions of urban centers such as Halifax and Sydney and the physical characteristics of rural areas such as Yarmouth, may require combining strategies such as the following (Will and Leeds 2012, IPCC 2014):

- **Accommodate:** This approach involves changing land uses in exposed areas using infrastructure that can tolerate flooding. Examples of this approach include raising or flood proofing vulnerable infrastructure and using floating structures.

- **Retreat:** This long-term approach consists of relocating people and infrastructure away from hazardous coastal areas. Managed retreat involves selecting what to relocate and mitigating the environmental impacts of leaving infrastructure exposed to natural processes. Abandonment is another type of retreat that does not involve planning for relocation or for the impacts that abandonment may cause on the environment. Abandonment is not a beneficial adaptation strategy but may be necessary in cases of emergency.

- **Avoid:** This approach prevents development in hazardous coastal and riparian areas and locates critical infrastructure such as hospitals and emergency services in areas where risks of flooding are negligible.
- **Holding the line:** This approach maintains land uses and development in the vulnerable areas and relies in the design and maintenance of hard infrastructure to resist extreme events to a set level.

In Nova Scotia, selecting a suitable flood management approach requires an understanding of the regional setting that influences the scale of flooding risks, the characteristics of the population and the value of the infrastructure at risks. Flood mitigation options in areas of negative growth may require balancing the community outlook with the needs of their current residents. In addition, the occurrence of major flooding events such as the Thanksgiving storm in Sydney, shows the fundamental role of emergency preparedness as part of a integrated flood risk management strategy.

6.1 Expanding Emergency Preparedness in Nova Scotia

A wide range of factors, including availability of financial resources, physical constraints, and long-term unpredictability of the magnitude of the threat limits the effectiveness of human intervention in flooding risk protection (World Metereological Organization 2011). Therefore, emergency preparedness is a fundamental element in reducing the risk of life and infrastructure losses.

Flood mapping creates awareness about areas at risks and allows emergency management officers to implement general response plans. However, flooding maps are insufficient to account for the variability of timing, location, and magnitude of flood hazards. Therefore, the public and emergency response teams can benefit from the implementation of a flood forecasting and warning system.

Implementation of an effective flood forecasting and warning system requires consideration of multiple elements (World Metereological Organization 2011) including the type of expected flood (urban, seasonal, flash, coastal, estuarine, etc.); the size of the area to monitor, type of model (lumped or distributed), hydrological and coastal considerations. In Nova Scotia a fundamental consideration includes the simultaneous effect of multiple processes (rainfall, snowmelt, high tides, waves, and in some cases sediment transport) which can be better understood through computer modelling.

The availability of remotely sensed weather prediction, calibrated hydrodynamic models, and platforms for citizen data collection can potentially allow the anticipation of where and when a forecasted weather event is likely to cause more damage. Across Canada flood forecasting systems range from water level and flow monitoring to complex systems involving calibrated computer models.

In Nova Scotia, recent flood studies have developed calibrated computer models for a wide range of communities (CBCL Limited 2016). Some of these models can support real-time updates using weather and storm surge forecast. Environment Canada provides radar rainfall forecasting data from the High Resolution Deterministic Prediction System (HRDPS), a 2.5 Km resolution climatic model with up to 48-hour forecast period. NOAA and Deltares provide global and regional storm surge forecast. Webtide provides tidal predictions for the Atlantic. However, emergency preparedness programs based on flood forecasting using complex hydrodynamic models are not yet available in the Province (Natural Resources Canada 2014)

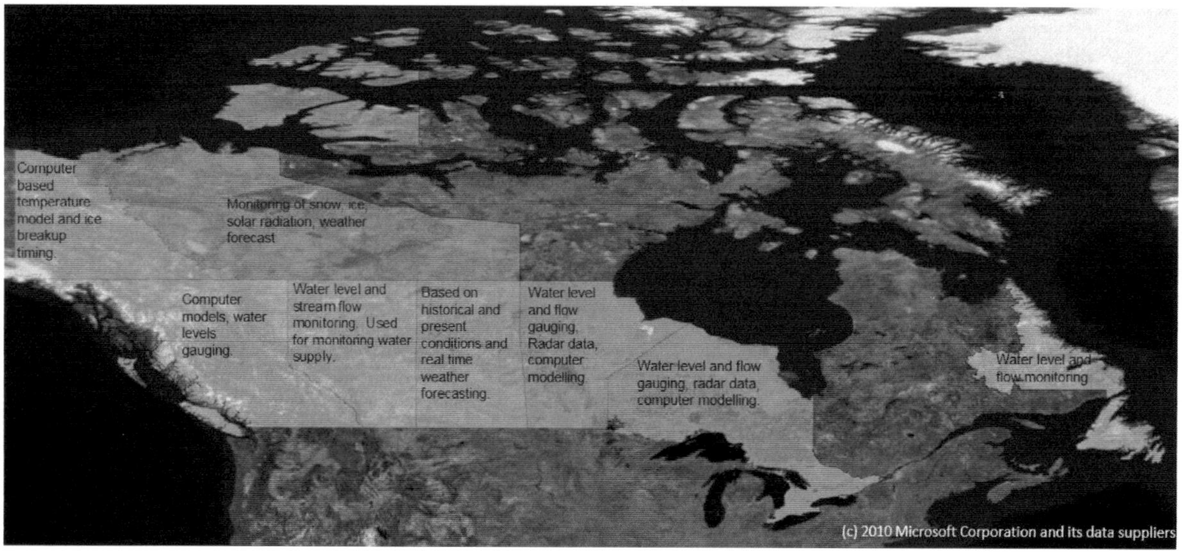

Figure 4 Flood Forecasting in Canada (Natural Resources Canada 2014)

7 CONCLUSIONS

Nova Scotia is no exception to the complexities of flood risk management that have been recognized and acknowledged worldwide in the face of climate change. The United Nations have developed a framework for disaster risk mitigation that aims to mitigate life and infrastructure loss and foster resilience. In the context of flooding, this framework and a holistic approach to flood risk studies can be implemented in Nova Scotia to define effective risk management strategies in vulnerable communities. An integrated approach to flood mitigation acknowledges the limitation of an infrastructure-based approach, takes into consideration the degree of uncertainty associated with climate change projections, recognizes that flooding events are the result of natural processes at a large scale, and identifies opportunities for improving the performance of the system as a whole.

Recent flooding studies in the province have used this approach in the evaluation and mapping of flooding risks under current climatic conditions and under the projected effects of climate change. These studies have provided recommendations that recognize the limitations of using hard infrastructure alone and have recognized the importance of community engagement as a source of information about historical events and the priorities and needs of vulnerable communities. The most recent studies have relied on complex computer modelling for producing flood maps identifying vulnerable areas. The availability of these models along with remote sensing information and weather forecast data provide an emergency preparedness strategies in the Province.

References

Blair, Feltmate, and Jason Thistlethwaite. 2012. Climate Change Adaptation: A Priorities Plan for Canada. Waterloo: University of Waterloo.

CBCL Limited. 2016. Assessment of Broad Brook Improvements. Yarmouth, NS: CBCL Limited.

CBCL Limited. 2016. Status of Flood Maps in the Atlantic Canadian Provinces. Halifax: CBCL.

CBCL Limited. 2014. "Truro Flood Risk Study for the Joint Flood Advisory Committe (County of Colchester, Town of Truro and Milbrook First Nation)." Halifax.

CTV News. 2016. Sydney, N.S., after Hurricane Matthew flooded the area, on Oct. 12, 2016. 10 13. Accessed 04 01, 2018. https://www.ctvnews.ca/canada/i-ve-lost-everything-insurance-coverage-in-question-after-atlantic-storm-1.3113324.

Environment and Climate Chance Canada. 2013. Flood Damage Reduction Program. 07 17. (https://ec.gc.ca/eau-water/default.asp?lang=En&n=0365F5C2-1).
Environment Canada. 2018. 1 11. http://climate.weather.gc.ca/radar/index_e.html.
FEMA. 2007. Design Guideline for Improving Critical Facility Safety from Flooding and High Winds. Washington DC: FEMA.
Instittue for Sustainable Development. 2013. Climate Change Adaptation and Canadian Infrastructure. Review of Literature. Winnipeg: The International Institute of Sustainable Development.
IPCC. 2014. "Climate Change 2014: Impacts, Adaptation, and Vulnerability. Part A: Global and Sectoral Aspects. Contribution of Working Group II to the Fifth Assessment Report of the Intergovernmental Panel on Climate Change." By V.R. Barros, D.J. Dokken, K.J. Mach, M.D. Mastrandrea, T.E. Bilir, M. Chatterjee, K.L. Ebi, Y.O. Estrada, R.C. Genova, B. Girma, E.S. Kissel, A.N. Levy, S. MacCracken, P.R. Mastrandrea, and L.L. White (eds.) C.B., 1132. Cambridge, United Kingdom and New York: Cambridge University Press.
Le Treut, H., R. Somerville, R. Cubasch, Y. Ding, and C. Mauritzen. 2007. Historical Overview of Climate Change. In: Climate Change 2007: The Physical Science Basis. Contribution of Working Group I to the Fourth Assessment Report of the Intergovernmental Panel on Climate Change. Cambridge, United Kingdom and New York, NY, USA: Cambridge University Press.
Managers, Association of State Floodplain. 2007. "Levees: the double-edged sword. ." Madison, WI.
Marra, Fracesco, Efrat Morin, Nadav Peleg, Yiwen Mei, and Emmanouil Anagnostou. 2017. "Intensity–duration–frequency curves from remote sensing rainfall estimates: comparing satellite and weather radar over the eastern Mediterranean." Hydrology and Earth System Sciences 2389-2404.
Masselink, Gerhard, and Michael Hughes. 2014. Introduction to Coastal Processes and Geomorphology. New York: Routledge.
MMM Group Limited. 2014. National Floodplain Mapping Assessment. Ottawa, ON: Public Safety Canada.
Natural Resources Canada. 2017. Federal Floodplain Mapping Framework. Ottawa: Natural Resources Canada.
—. 2014. Flood forecasting centres across Canada. 07 03. https://www.canada.ca/en/environment-climate-change/services/water-overview/quantity/floods/forecasting-centres.html.
NOAA. 2015. A Guide to Assessing Green Infrastructure Costs and Benefits for Flood Reduction. Charleston: NOAA Office for Coastal Management.
Nova Scotia Environment. 2015. "Guide to Altering Water Courses." Nova Scotia: Nova Scotia Environment.
—. 2015. Nova Scotia Flood Events from 1992 to 2015. December 15. https://fusiontables.google.com/DataSource?docid=1UwMP_E8K47cl2gcEpDfz660xa7Q3kit2czO6j7Lo#rows:id=1.
Nova Scotia Municipal Affairs. 2002. "Model Land Use By-Law." Halifax, July.
Public Safety Canada. 2018. "Canada's National Disaster Mitigation Strategy." Department of Public Safety and Emergency Preparedness . 01 31. https://www.publicsafety.gc.ca/cnt/rsrcs/pblctns/mtgtn-strtgy/mtgtn-strtgy-eng.pdf.
Thistlethwaite, Jason, and Daniel Henstra. 2017. "Municipal Flood Risk Sharinf in Canada: A policy instrument analysis." Canadian Water Resources Journal 349-363.
Tockner, Klement, and Jack Stanford. 2002. "Riverine Flood Plains: Present State and Future Trends." Environmental Conservation 308-330.
United Nations. 2015. Sendai Framework for Disaster Risk Reductoin . New York: United Nations.
Van Proosdij, Danika, Timothy Milligan , Gary Bugden, and Karl Butler. 2009. "A Tale of Two Macro Tidal Estuaries: Differential Morphodynamic Response of the Intertidal Zone to Causeway Construction." Journal of Coastal Research 772-776.
Will, McBain, and Arup Leeds. 2012. "Twenty-frist Centruy Flood risk Management." In Flood Risk: Planning, design and management of flood defence infrastructure, by ed. P.B. Sayers, 7-21. Oxfordshire: ICE.
World Bank. 2017. Implementing Nature-Based Flood Protection: Principles and Implementation Guid. Washington: The World Bank Group.
World Metereological Organization. 2009. Integrated Flood Management. 2009: World Meterological Organization.

World Metereological Organization. 2011. Manual on Flood Forecasting and Warning. Geneva: World Metereological Organization.

Building Tomorrow's Society
Bâtir la Société de Demain

Fredericton, Canada
June 13 to 16, 2018 / *13 au 16 juin, 2018*

IMPACT OF CLIMATE VARIABILITY ON FLOW AND SEDIMENT TRANSPORT OF THE MAGDALENA RIVER THROUGH THE ANALYSIS OF HISTORICAL SERIES

Y. Amaya, F.[1,2], O. Castro, M.[1], Chang, P.[1], V. Upegui, J.[1]
[1] *Universidad Nacional de Colombia Sede Manizales*, Colombia
[2] fayaraa@unal.edu.co

Abstract: The Colombian Hydrological, Meteorological and Environmental Science Institute (IDEAM) highlights that Colombia is susceptible to be impacted by the effects of climate change. Current trends show that the hydrological conditions of various catchment basins are being affected. The Magdalena river basin has experienced an upsurge in erosion processes. This is the largest river artery and the most important in Colombia. The river runs approximately 15 km along the urban perimeter of La Dorada (Caldas) township where dynamic erosion and sediment transport processes can readily be observed. This study focused on the analysis of historical hydrological series in the region, seeking to correlate flow data, sediment bed transport and suspended concentration with the NOAA indexes allowing one to quantify the sediment transport processes that may occur over periods of extreme climatic events (i.e. El Niño and La Niña). Through a field campaign, bed and suspended sediment transport was quantified and considered in relation to river discharge, such data was then analyzed against available historical series over the last 40 years. The results allow one to infer the validity of various NOAA indexes as long term indicators considering the most favorable and unfavorable climate change scenarios and its consequence in terms of predicted sediment transport for the river. Series analysis showed that the BEST and MEI indexes have a higher incidence on the variables analyzed, with a significant correlation from 20% up to 53%.

1 INTRODUCTION

Over the last decades, Colombia has observed an increase in the magnitude and frequency of extreme climatic events and the Magdalena River catchment basin has seen a steady increase in precipitation over the years. Recently a number of anomalies in rainfall events have occurred and are being more recurrent, impacting populations and productive sectors of Colombia (Hurtado, 2010). Figure 1a and figure 1b show the disastrous effect of such extreme events as observed in La Dorada, Caldas.

Figure 1a. Extreme weather events during El Niño, (La Patria, Jan. 2016).

Figure 1b. Extreme weather event during La Niña, Yara. F., Apr. 2017.

La Dorada is located midway upriver along a sharp curve of the Magdalena. The city is situated in a wide valley amongst the central mountain range of the Andes at 178 meters above mean sea level with a population of 76 574 inhabitants as of 2005 (DANE, 2015). The meander located in this area is limited in its width by a geological formation which restricts its evolution. Secondary flow processes are actively deteriorating the outer curve of the river bank producing significant erosion and deposition.

Over the years the city limit has expanded significantly with limited urban planning and zoning restrictions being imposed. As a consequence impoverished neighborhoods have reached the river embankments and with 15 km of the urban perimeter exposed to the channel, the situation has reached a breaking point, exposing housing, public and private property to imminent tumbling or collapse and threatening the local population. Also, intense precipitation has caused flooding and has affected local neighbourhoods aggravating the situation. Figure 2 illustrates the general urban area and city limits.

Figure 2. Bing Maps 2018. Meander of the Magdalena River in La Dorada – Caldas.

The catastrophic events of November and April 2011 are exemplary of how the situation has deteriorated over time. On such occasion, precipitation intensity exceeded any historical previously recorded rainfall in the country and the limited mitigation measures that had been implemented collapsed under pressure.

The level of the Magdalena River exceeded historic flow on record with 6509 m^3/s registered. The event affected 10 neighborhoods and nearly 4000 families. Overall 750 houses were damaged, of which 150 were totally destroyed. In addition, 918 people were evacuated to temporary shelters. (El Tiempo.com newspaper, 2011). The municipality did not have flood drainage canals or levees to evacuate the waters to lower areas. Furthermore, as 5 kilometers of the river embankment was exposed to erosion and scour, the flooding was further aggravated.

Internationally different statistical techniques have been used to establish multivariable correlations to determine the relationship between the ENSO phenomenon and the discharge of a catchment basin (Martelo, 2003).

The effect of the extreme phases of the ENSO (El Niño and La Niña) in a region depend on the time of year, affecting the climate over a whole region and impacting in particular, the magnitude and components of the hydrological cycle (Bedoya et al, 2010). The consequences of such events is a function of the conditions of vulnerability of the considered area. In the case of Colombia, its geographic position in the Andean zone results in a distinct bimodal behavior, with two rain and dry periods during the year (Inzunza, 2001).

In order to better understand the characteristics of flood events in relation to historical data and current climate change scenarios, a formal hydrological event series correlation analysis was undertaken.

2 METHODOLOGY

2.1 Data Series

Historical data series for the region were analyzed using the IDEAM (*Instituto de Hidrología, Meteorología y Estudios Ambientales*) gauge station #23037010 Puerto Salgar - La Dorada, that monitors channel discharge, water depth as well as suspended sediment transport. The following annual data series have been analyzed:

- Daily average discharge 1946 – 2015 excluding 1958 and 1980.
- Monthly average discharge 1946 – 2014, excluding 1958.
- Maximum monthly discharge 1946 – 2014, excluding 1958 and 1980.
- Minimum Monthly discharge 1946 – 2014, excluding 1958 and 1980.
- Daily average waterline 1960 – 2014, excluding 1980.
- Monthly average waterline 1960 – 2014.
- Maximum monthly waterline 1960 – 2014.
- Minimum monthly waterline 1960 – 2014.
- Total transport 1946 – 2015.
- Suspended solid discharge 1946 – 2015.

2.2 Treatment of the data series

A qualitative and quantitative analysis using different estimation time windows was completed on the historical series with a view to classify and organized the recorded data.

The series were examined in relation to the climate series of the National Oceanic and Atmospheric Administration (NOAA) which include: the ONI index, the BEST index, the MEI index and finally the SOI index, establishing the various correlation with the related variables of the Magdalena River in La Dorada – Caldas. The analysis was performed using the Southern Oscillation index for El Niño (ENSO) and the R^2 coefficient.

The analysis strived to determine the best correlation index that can be related to historical discharge data. This variable generally allows for higher correlation values being a persistent variable in hydrology. The data series were smoothed using a mobile average over 3 months to filter higher frequencies, before establishing correlation values.

3 RESULTS

The correlation analysis was completed based on the following consideration (Trenberth, 2001):

El Niño: characterised by an ONI positive value, greater or equal to +0.5ºC

La Niña: characterised by an ONI negative value, smaller or equal to -0.5ºC

3.1 ONI Index

Figure 3a presents the ONI index results as they relate to channel flow over time. One can observe that positive sharp rise of the index match reported El Niño events, in the same way as negative spikes or sharp decline of the index are associated with reported La Niña events in the area with intense rainfall.

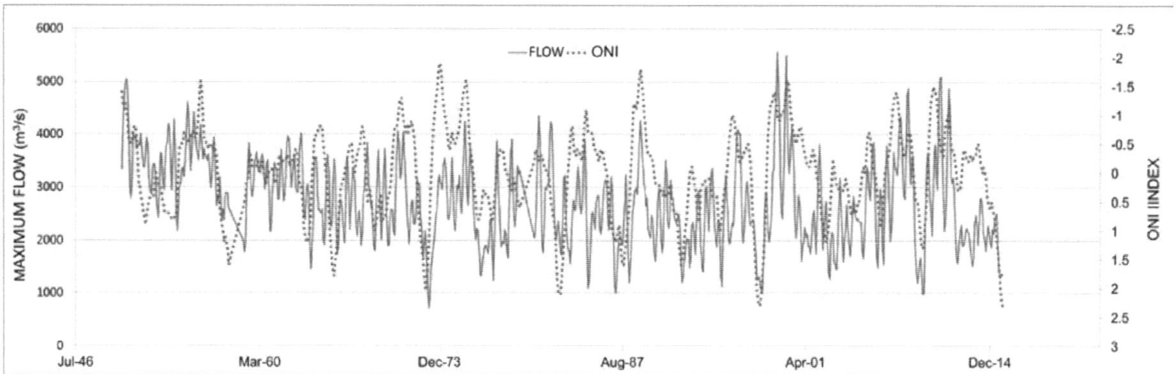

Figure 3a. Maximum discharge correlation to ONI index

The best correlation was found to be 48% for a time window estimate of 3 months, which tends to indicate that maximum flow discharge is related to the information as provided by the ONI index, i.e. the surface temperature of the ocean recorded in the equatorial Pacific Ocean 5 ° N-5 ° S, 170 ° W - 120 ° W, (known as El Niño 3.4) (Tootle, 2008).

The higher correlation between the ONI index and flow indicates that the behavior of the latter in the area, is mostly influenced by the extreme events of El Nino and La Nina. This is partly due to the fact that the registered flows depend on the contributions of the tributary sub-basins to the Magdalena River. As evidenced in figure 3a, La Nina events tend to develop in the months of March and April, presenting the highest intensities in October and November. El Niño events tend to develop in the period of May and June, reaching its maximum level during the period of December and February. (Poveda et al, 2001).

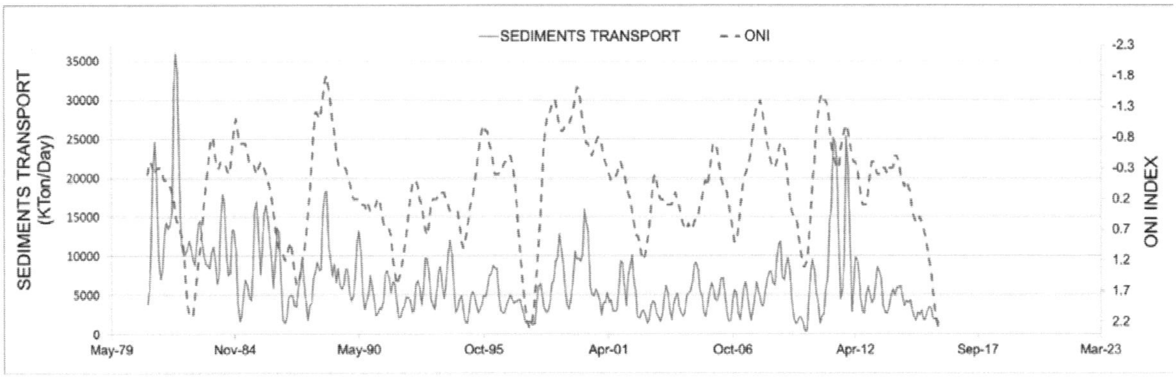

Figure 3b. Sediments transport correlation to ONI index.

In the case of sediment transport (Figure 3b), the correlation with the ONI index is low, averaging approximately 27%, this can be attributed to the number of variables that are involved and affect the transport process. In addition, it was observed that sediment transport presents the least number of months that are significantly related to the index.

3.2 BEST Index

The BEST index combines the SOI index with given temperature at sea level for the El Niño 3.4 (SST) region (Yan, 2011). Figure 4a presents the BEST index results as they relate to channel flow over time. Here positive values can be associated with El Niño events, while negative values can be related to La Niña. The BEST index with 51% over a 2 month estimation time window presents the highest correlation that this index was able to determine. One has to recall that the available data series was provided monthly were the BEST index is determined as a monthly and trimestral average which tend to display a better representation of the phenomena.

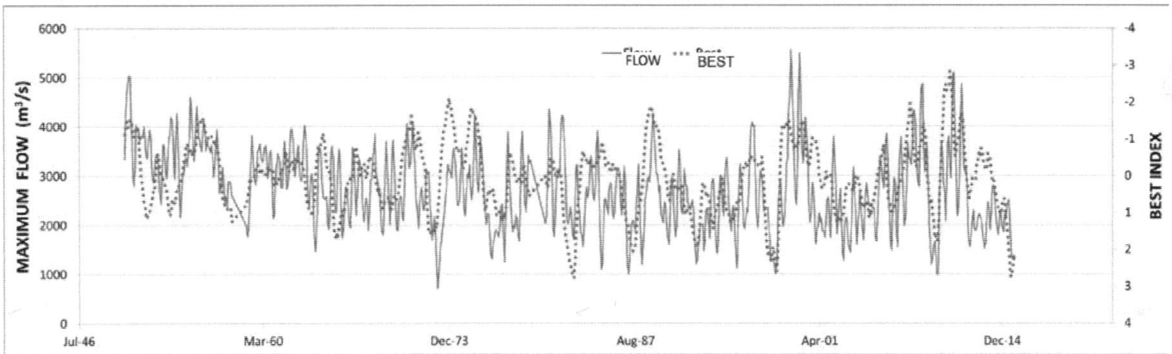

Figure 4a. Maximum discharge correlation to BEST index

For this particular case, there are substantially more months (analysis over a two-month estimate) where the historical series for flow correlates with the BEST index contrasting with the ONI index mentioned previously. Again, La Nina events tend to develop in March and April, and present the highest intensities in October and November. El Niño events tend to develop over the period May and June and reaching its maximum level during the months of December and February.

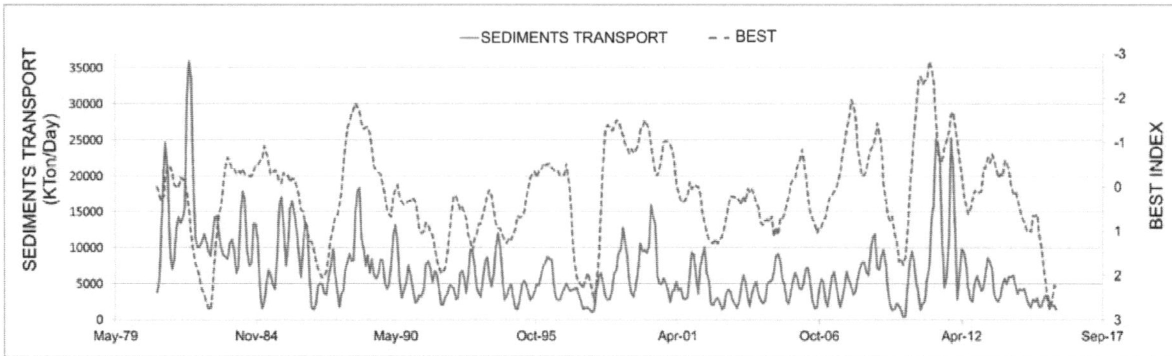

Figure 3b. Sediments transport correlation to BEST index.

Considering Figure 4b, sediment transport is related to the index with calculated correlation of approximately 30%. It may be inferred that sediment transport is influenced by an increase in river stage and increased discharge. In the case of the Magdalena River, the series indicate that the months of March-April and October-November are historically rainy periods, which may explain why the correlation of this variable appears greater during these time periods. It was evidenced that a time window estimate over 3 months of the BEST index shows a greater correlation with extreme El Niño and La Niña events.

3.3 MEI Index

Figure 5a presents the MEI index results. The best correlation encountered between discharge and the MEI index was found to be 49 % considering a 3 month time window. The MEI index is determined considering six monitored main variables of the Pacific tropics including: atmospheric pressure at sea level, zone and meridian components of the surface wind, sea surface temperature, surface air temperature, and total cloudiness fraction of the sky (Wolter, 1987).

In this case, the greatest correlations were obtained relating the MEI index with the minimum and average flow, with 52% and 53% respectively. Furthermore it was observed that peak correlation were in better agreement with the La Niña events. In general, the best correlation values were obtained using estimation over 1 and 3 month time period.

Data analysis indicate that in the case of El Niño, the MEI index was not significantly correlated to the analyzed variables contrary to what was observed with La Niña events.

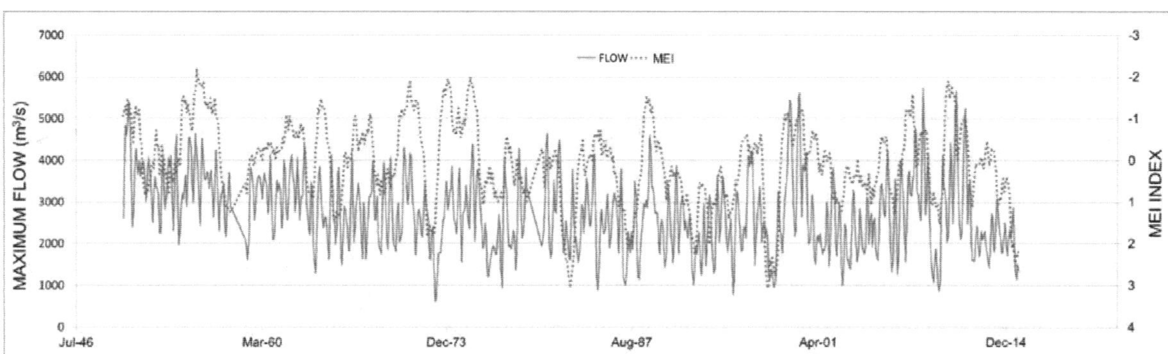

Figure 4a. Maximum discharge correlation to MEI index

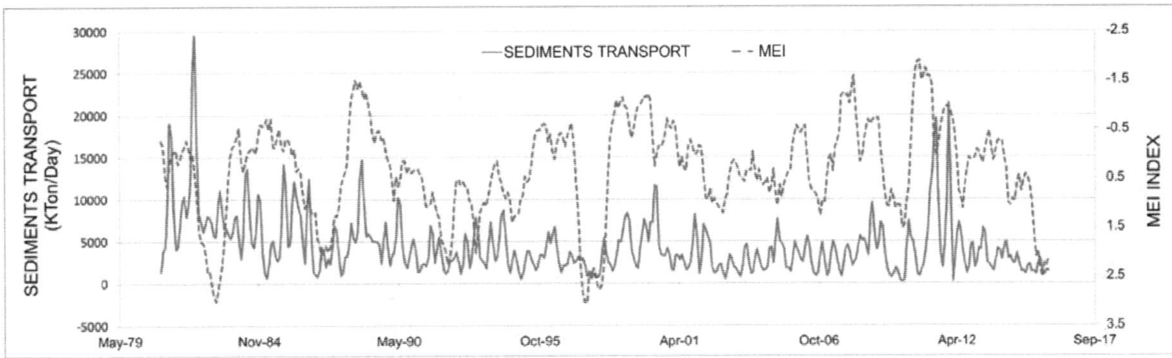

Figure 5b. Sediments transport correlation to MEI index.

Sediments transport analysis for the MEI index (Figure 5b), determined a 27% correlation. As sediment transport is influenced by an increase in discharge, the MEI index displayed specific peaks during the rainy period of March-April and October-November, as was also observed with the BEST index previously.

3.4 SOI Index

In the case of the SOI index, the best correlation was found to be 33% for a 2 month time window estimate as it relates to maximum discharge. The SOI index relates observed abnormalities between mean monthly pressure found in Tahiti (French Polynesia) and Darwin (Northern Australia), (INOCAR, 2001). Figure 6a illustrates the maximum discharge to SOI index results.

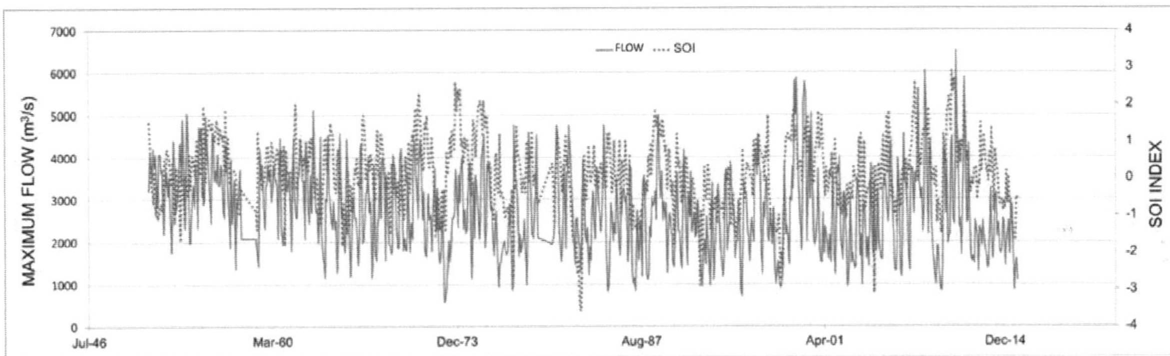

Figure 6a. Maximum discharge correlation to SOI index

It may be suggested that the SOI index displays poor correlation values as it is determined monthly whereas other indexes are generally evaluated over a two or three-month period.

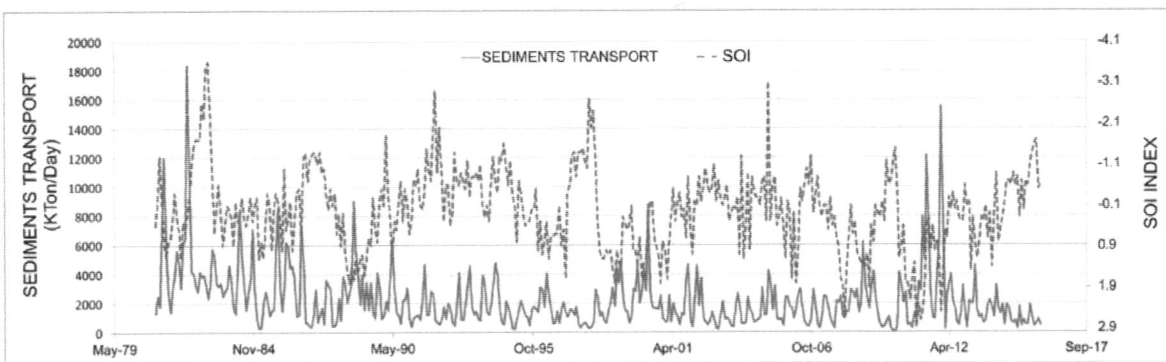

Figure 6b. Sediments transport correlation to SOI index.

In the case of sediment transport (Figure 6b), the correlation is low, averaging approximately 19%. Again, as the SOI index is determined through a monthly average this may explain the lower correlation values in this case. As the data series are considered for a similar time window estimate a greater correlation variability is to be expected.

4 ANALYSIS

The overall results of the various indexes are presented and compared in table 1, illustrating the maximum correlation obtained using the Pearson product moment correlation coefficient of the data series. The latter being interpreted as the proportion of the variance of one variable attributable to the variance of the other.

Table 1. Maximum correlation for each studied variable

MAXIMUM CORRELATION (%)					Time Window for Max. Correlation (Month)
PARAMETER	ONI	BEST	MEI	SOI	
Suspended Solids Tpt. (Kg/m^3)	-23.15	-20.25	-20.65	12.46	3
Bed Tpt. (Kton/Day)	-27.41	-30.13	-27.01	19.02	3
Level min (m)	-43.52	-42.27	-39.31	30.74	2
Level mean (m)	-47.27	-46.56	-42.49	33.4	2
Level max (m)	-47.74	-47.33	-42.81	32.37	2
Flow min (m^3/s)	-43.24	-48.83	-52.31	34.45	3
Flow mean (m^3/s)	-47.05	-52.03	-52.78	35.32	3
Flow max (m^3/s)	-48.34	-51.28	-49.07	33.37	2

The MEI index provided the best results of the analysis with a 52% and 53% correlation to minimum and mean discharge respectively. Nonetheless the BEST index was found to be a better indicator of sediment transport with a 30% correlation factor.

5 CONCLUSIONS

This study has evaluated the historical series available from the IDEAM #23037010 Puerto Salgar gauge station with the NOAA indexes. Generally, the data series related to transport and discharge in relation to evaluated indexes has been found to present a satisfactory fit. Significant correlations have been determined and the persistence of the data has been found to be adequate.

The BEST and MEI indexes showed consistently a higher incidence on the variables analyzed, with a significant correlation coefficient from 20 up to 53%. This would indicate that such indexes are better indicators of hydrometeorological events occurring in the zone. In general, the best correlation results were determined using a 3 month time window estimate. It was also observed that when analyzing the historical series with the NOAA indexes, greater anomalies were observed and lesser predictivity was determined when relating sediment transport to La Niña events.

In all cases, accuracy of the correlation is related to the homogeneity of the available data set. Most historical events related to discharge and sediment transport have shown an acceptable intersect; although little is known of the accuracy of the data provided by the monitoring gauge station, further probabilistic evaluation of the data is limited.

Nonetheless, one can infer that the monitoring station data series would provide for a qualitative and long term planning and management tool relating liquid and solid discharge to reported El Niño and La Niña events in the La Dorada, Caldas region and providing for a better understanding of the hydrological alteration presently occurring in Colombia.

Finally considering the various NOAA indexes (NOAA, 2018) studied, extreme hydrological events resulting from climate change do impact suspended solid and bed transport rates over the months of May and November in particular displaying greater correlation values for the region of La Dorada Caldas.

References

Bedoya, M., Contreras, C. y Ruiz, F. (2010). Capítulo 7. Alteraciones del régimen hidrológico y de la oferta hídrica por variabilidad y cambio climático. Estudio Nacional del agua 2010. Instituto de Hidrología Meteorología y Estudios Ambientales (IDEAM), Colombia, pp. 282-320.

DANE. (2015). Censo General 2005. Bogotá: Imprenta Nacional de Colombia. Recovered from: https://www.dane.gov.co/files/censo2005/PERFIL_PDF_CG2005/17380T7T000.PDF.

El Tiempo. (2011). Inundaciones dejan sin clases a 18 mil estudiantes en La Dorada. El tiempo.com. Recovered from: http://m.eltiempo.com/colombia/eje-cafetero/invierno-deja-estudiantes-sinclases-en-la-dorada/9223160

Martelo, M. "Influencia de las Variables Macroclimáticas en el Clima de Venezuela". Dirección de hidrología, meteorología y oceanología – dirección general de cuencas hidrográficas – Ministerio del Ambiente y los Recursos Naturales. Caracas, Venezuela, 2003

NOAA. (2018). PSD Climate and Weather Data. Physical Sciences Division. Available: https://www.esrl.noaa.gov/psd/data/.

Hurtado, G. "El Clima: Origen y Aplicaciones". Memorias: I congreso nacional del clima 2010 "El desarrollo económico de Colombia bajo un nuevo escenario climático". Instituto de Hidrología, meteorología y Estudios Ambientales (IDEAM). Bogotá, Julio 13 a 15 de2010.

INOCAR. (2001). Atlas oceanográfico 2011. CAPÍTULO 7: Océano y Clima. Documentación IDEAM Colombia (ideam.gov.co). Recovered from: http://documentacion.ideam.gov.co/openbiblio/bvirtual/002342/inocar/pages/oceanografia/capitulo7/int_cap7_imp.html.

Inzunza, J. "Tema 7. Cambio Climático Global". Pedagogía en Ciencias Naturales y Carreras de la infancia. (2001). Universidad de Concepción. Chile. Recovered from: http://www2.udec.cl/~jinzunza/infancia/tema7.pdf

La Patria Newspaper. (2017). Con el Magdalena se secan las ilusiones. lapatria.com. Recovered from: http://www.lapatria.com/node/249441

Poveda, J. G.; Jaramillo, R., A.; Gil, M., M.; Quiceno, N.; Mantilla R., I. Seasonality in ENSO related precipitation, river discharges, soil moisture and vegetation index (NDVI) in Colombia. Water Resources Research 37:2169-2178. 2001.

Support.office.com. (2018), United States. RSQ function, applies to: Excel 2016, Excel 2013, Excel 2010, Excel 2007. Recovered from: https://support.office.com/en-us/article/rsq-function-d7161715-250d-4a01-b80d-a8364f2be08f

Tootle, A.; Piechota, T.C.; Gutiérrez, F. The relationships between Pacific and Atlantic ocean sea Surface temperature and Colombian streamflow variability. Journal Hydrology 349:268-276. 2008.

Trenberth, K., E. The definition of El Niño. Bulletin of the American Meteorological Society. 78:2771-2777. 1997.

Wolter, K. (1987). The Southern Oscillation in surface circulation and climate over the tropical Atlantic, Eastern Pacific, and Indian Oceans as captured by cluster analysis. J. Climate Appl. Meteor., 26, 540-558. Available from the AMS.

Yan, H.; Sun, L.; Wang, Y. (2011). A record of the Southern Oscillation Index for the past 2,000 years from precipitation proxies. Nature Geoscience 4: 611–614.

Building Tomorrow's Society
Bâtir la Société de Demain

Fredericton, Canada
June 13 to 16, 2018 / *13 au 16 juin, 2018*

IMPROVED EARTHQUAKE RESILIENCE WITH PERFORMANCE-BASED SEISMIC DESIGN IN CANADA

Dutrisac, Hélène[1,2] and Saatcioglu, Murat[1,3]
[1] University of Ottawa, Canada
[2] helene.dutrisac@uottawa.ca
[3] murat.saatcioglu@uottawa.ca

Abstract: The focus of current Canadian traditional code-based seismic design is to primarily ensure life safety of building occupants. This objective is currently achieved by satisfying prescriptive criteria expressed in terms of drift and strength requirements as prescribed by the National Building Code of Canada or the Building Code in effect under the applicable jurisdiction. However, earthquakes in the last decade such as Chile (2010) and New Zealand (2011) have clearly demonstrated that these natural disasters can have devastating social and economic impacts. Many of the large urban centres in Canada are located within zones of moderate to high seismicity and are susceptible to the effects of major seismic events. Whereas conventional code-based design focuses on meeting prescriptive requirements, Performance-Based Seismic Design (PBSD), a relatively new and emerging seismic design methodology, provides various stakeholders with a better appreciation of the expected performance of a building during a seismic event, often expressed for multiple levels of ground shaking. This paper provides an overview of the state of practice in performance-based seismic design of buildings in North America, with an emphasis on reinforced concrete shear wall buildings. Topics explored include current Canadian code-based seismic design procedures, existing North American performance-based seismic design guidelines, targeted building performance levels, hazard levels, performance objectives and acceptance criteria. This methodology is then briefly examined for adaptation to a Canadian design context. Needs for future research are also identified.

1 INTRODUCTION

Whereas conventional code-based design relies on prescriptive requirements which focus primarily on life safety of the building occupants, Performance-Based Seismic Design (PBSD), a relatively new and emerging seismic design methodology, considers seismic design at different intensities of ground shaking with various performance criteria. This approach provides stakeholders with a better appreciation of the expected performance of a building during a seismic event, often expressed for multiple levels of ground shaking. This paper outlines the state-of-practice in performance-based seismic design of buildings, with an emphasis on reinforced concrete shear wall buildings. Topics explored include a brief review of current Canadian code-based seismic design procedures, an overview of performance-based seismic design methodology and targeted building performance level as well as a short review of the latest available North American guidelines on PBSD. Needs for future research are also identified.

2 OVERVIEW OF PERFORMANCE-BASED SEISMIC DESIGN

Over the last two decades, several guidelines have emerged in the United States with a PBSD framework. Although PBSD was originally developed for the assessment and evaluation of existing buildings (ATC 1996, FEMA 1997, FEMA 2000 and ASCE 2017b), there has now been a shift in the last decade towards incorporating this framework for the next generation analysis and earthquake design of new buildings (PEER 2017 and LATBSDC 2017).

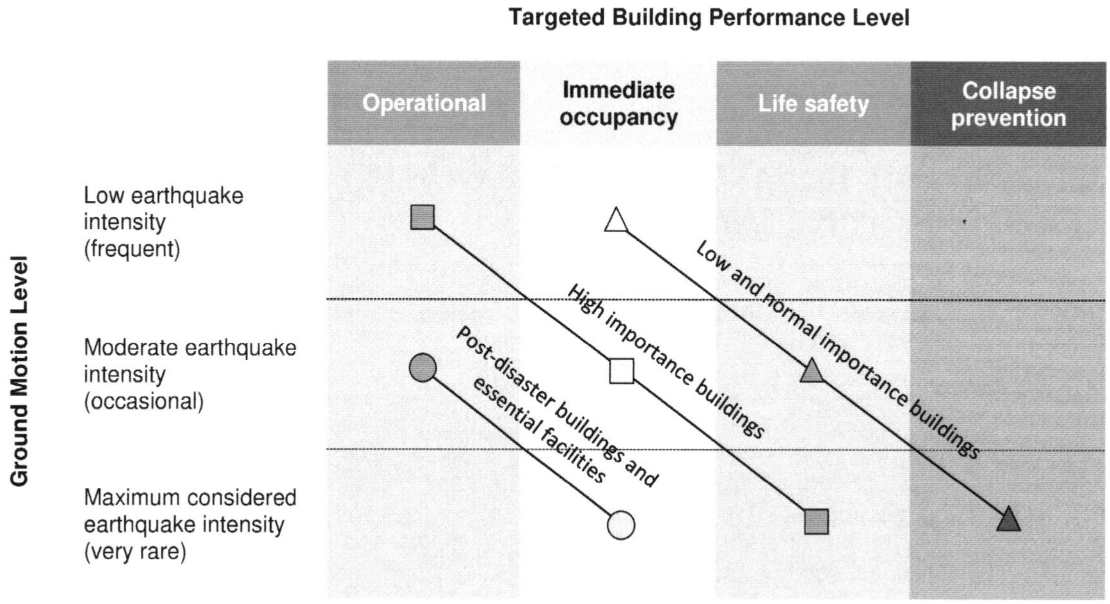

Figure 1: Performance-Based Seismic Design Concept
Adapted from FEMA 450 (FEMA 2004)

As illustrated in Figure 1, a performance-based seismic design framework consists of evaluating the seismic performance of a building for multiple levels of ground shaking. Acceptance criteria are set for each targeted building performance for each ground motion level. Also, different targeted building performance levels apply at a given intensity of shaking as a function of a design category (shown here as a function of the building importance category). This emerging methodology aims to provide various stakeholders with a better appreciation of the expected performance of a building during a seismic event for a given design category. In addition, this framework also enables building owners to choose a design approach which aims at attaining a better performance than the minimum prescribed by Code. Given the complexity of this procedure over traditional code-based methods, it requires extensive knowledge in the characterization of ground motions (tectonic environment, earthquake magnitude, type of faulting/sources, ground attenuation relationships and local site conditions), the nonlinear behaviour of structural materials under reverse cyclic loading (post-yield and post-capping), dynamic modeling and the response evaluation of the structural components.

2.1 Targeted Performance Levels

In the PBSD approach developed for the seismic assessment and rehabilitation of existing buildings (ATC 40, FEMA 273, FEMA 356 and ASCE 41), targeted building performance levels and corresponding objectives were established with the pairing of a structural performance level with a non-structural performance level. Commonly recognized structural performance levels in the above noted publications include immediate occupancy, damage control, life safety, limited safety, structural stability and not considered. The objectives covered under these levels relate to the performance of the structural components and are expressed qualitatively. For concrete shear wall SFRSs, these objectives are mostly described in terms of expected concrete cracking, spalling and/or crushing as well as damage to boundary elements. On the other hand, non-structural performance levels generally consist of operational, position retention, life safety and not considered. These levels cover qualitative objectives related to non-structural components such as partitions, ceilings, mechanical and electrical equipment and other architectural ornamentations.

While several combinations of structural and non-structural levels are identified, it is inferred (ASCE 2017b) that only a select few pairings are considered to provide a balanced design approach. The main targeted building performance levels are Operational, Immediate Occupancy, Life Safety and Collapse Prevention with objectives expressed qualitatively in terms of drifts, strength / stiffness retention and function of the gravity load-carrying elements post-earthquake. Buildings designed to the highest building performance level (Operational Level) are expected to retain most of their pre-earthquake strength and stiffness and have no permanent drifts. Although very light damage may occur, it is noted (ASCE 2017b) that this should not impede on the operation and functionality of the building. At the other end of the spectrum, buildings designed to the lowest performance level (Collapse Prevention) are expected to sustain significant and extensive damage during high intensity ground shaking. Despite significant loss in strength / stiffness of the SFRS elements and expected large permanent drifts, the load-bearing structural elements should still be able to support the gravity loads. It is noted that buildings designed under this performance level are essentially expected to be near collapse and it is recommended to prohibit occupancy in these buildings post-earthquake (ASCE 2017b). As illustrated in Figure 2, the higher the selected performance, the lower the anticipated loss is.

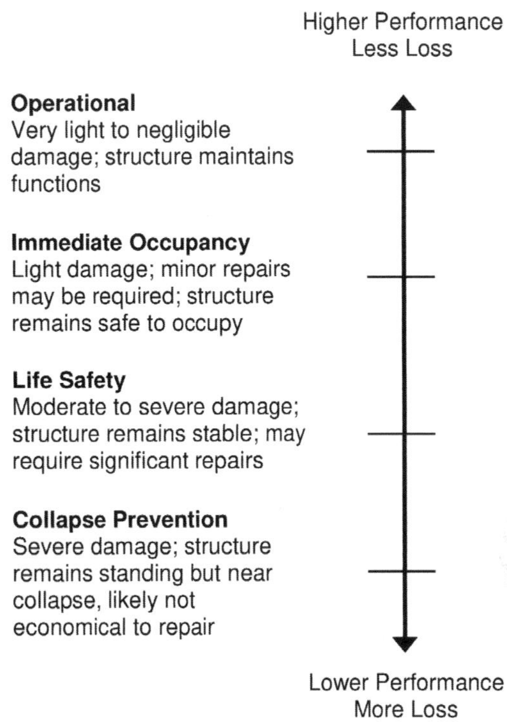

Figure 2: Expected Post-Earthquake Damage State for Various Building Performance Levels (Adapted from Table C2-1 of ASCE 2017b)

In essence, performance levels identify what to expect from a structure as it responds to various intensities of ground shaking. Although performance objectives have remained relatively unchanged over the years, it is worth noting that the initial edition ASCE 41-06 and its predecessor FEMA 273 provided insights in terms of expected drifts, which are summarized in Table 1 for concrete frames and shear wall SFRSs. These target indicators were removed in subsequent editions. A generic force-deformation response is also illustrated schematically in Figure 3.

Table 1: Structural performance level drifts

Seismic Force Resisting System (SFRS)	**Type of drift**	**Expected drifts**		
		Immediate Occupancy	**Life Safety**	**Collapse Prevention**
Concrete frames	Transient Permanent	1% negligible	2% 1%	4% 4%
Shear walls	Transient Permanent	0.5% negligible	1.5% 0.5%	2% 2%

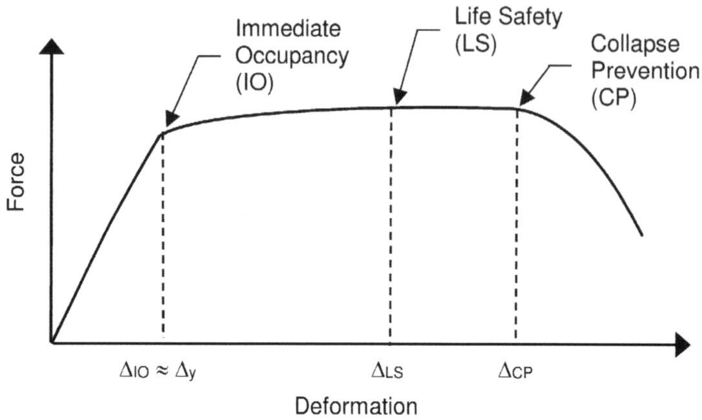

Figure 3: Schematic of structural performance levels and associated deformations

It is generally considered the norm currently in Canada to design buildings for a targeted building performance level of "Life Safety" at the maximum considered level of ground shaking; however, buildings of greater importance such as hospitals are required to remain functional post-earthquake. As such, these buildings are typically designed for a higher performance level to remain functional at the same level of ground shaking.

2.2 Hazard Levels

Current traditional code-based seismic provisions only focus on one hazard level. In Canada, the NBCC requires consideration to a hazard level corresponding to a probability of exceedance of 2% in 50 years. In the US, the requirements of Chapter 16 of ASCE 7-16 on nonlinear response history analysis (NLRHA) specify a design earthquake level which corresponds to the Maximum Considered Earthquake (MCE_R) where MCE_R is based on a risk adjusted uniform hazard with probability of exceedance of 2% in 50 years and deterministic ground motions. Other types of seismic analyses in ASCE 7 consider a design level earthquake which is taken as 2/3 MCE_R. For the analysis and design of new buildings, recent PBSD guidelines for tall buildings (PEER 2017 and LATBSDC 2017) recommend the use of two hazard levels corresponding to probabilities of exceedance of 50% in 30 years and an ASCE 7-16 MCE_R level, for service level and MCE_R evaluations, respectively. On the other hand, various hazard levels are considered in ASCE 41-17: ASCE 7-16's MCE_R, 2/3rds MCE_R, 5% in 50 years, and 20% in 50 years.

3 CURRENT CODE DESIGN PROCEDURE

Building regulations provide minimum requirements that need to be met for the construction of new buildings. In Canada, building codes mandated by municipalities and provinces are typically based on the adoption of Canada's National Model Construction Code, the National Building Code of Canada (NBCC), with some modifications in certain jurisdictions. Canada adopted its first building code in 1941 with an empirical formula for seismic design covered under the Code's appendix. The very simplistic static force formulation consisted of computing the product of the building weight and a constant expressed as a function of the soil bearing capacity, with the resulting horizontal force applied at the center of gravity of the structure.

Although different, the current 2015 NBCC (NRCC 2015) continues to provide code users with a traditional code-based static approach for earthquake design. This conventional approach, known as the Equivalent Static Force Procedure (ESFP), consists of computing the elastic earthquake demand based on uniform hazard values having a probability of exceedance of 2% in 50 years and adjusting it to account for the anticipated ductility (R_d) and overstrength (R_o) of the chosen seismic force resisting system (SFRS). Where permitted, buildings designed under the ESFP are required to resist a lateral earthquake force, V, distributed over the height of the building as a function of the ratio of the product of storey weight and height relative to the summation of this product over the entire height of the building. The lateral force is also adjusted for soil type, higher mode effects and to account for the type of building using an importance factor.

Despite that the ESFP is permitted in certain cases, the default method of analysis in the 2015 NBCC is the Dynamic Analysis Procedure. Under this procedure, buildings may be designed using linear dynamic analysis (either the Modal Response Spectrum Method or the Numerical Integration Linear Time History Method) or Nonlinear Dynamic Analysis. Where Linear Dynamic Analysis is used, the analysis elastic base shear, V_e, is to be modified to account for the importance category of the building and the ductility of the SFRS by multiplying the resulting analysis shear by the importance factor, I_E, and dividing it by R_dR_o to obtain the design base shear, V_d. For regular and irregular structures, the minimum base shear determined from the Linear Dynamic Analysis, may not be taken less than 80% and 100%, respectively, of the base shear, V, determined from the ESFP described above. The elastic storey shears, storey forces, member forces and deflections obtained from a Linear Dynamic Analysis are also to be multiplied by the ratio V_d/V_e to obtained the design values.

Very little information is provided in the current NBCC with respect to linear and nonlinear time history analyses. For linear analysis, the code provisions specify that ground motion histories used must be compatible with the code-specified response spectrum; whereas, for nonlinear analysis, the code simply indicates that a special study is required. Despite the limited information in the actual body of the NBCC, a new Annex on the "Selection and Scaling of Ground Motions" has been introduced in Commentary J of the User's Guide – 2015 NBC Structural Commentaries (NRCC 2017). This Annex provides guidance on how to develop the target response spectrum, determine the applicable period range and the number of ground motions required as well as methods for scaling the ground motions to be used for time history analyses to ensure adequate and target response spectrum compatible motions are used. Guidance is also provided in the Commentary on determining force and deformation demands. A summary of selected requirements of the Annex is shown in Table 2. These are also discussed in greater detail in the Section 4.

Except for restrictions, design limitations and other design requirements, two acceptance criteria are noted in the NBCC: a global drift criterion and a general strength criterion. The global acceptance criterion specifies an interstorey drift limit of 1% for post-disaster buildings, 2% for high importance buildings and 2.5% for other types of buildings. The strength criterion simply specifies that the factored resistance of structural components must be greater than the factored demand. Whereas the NBCC specifies requirement for design loads, the structural resistance and detailing requirements for buildings typically fall under the material CSA design standards. The latest seismic design and detailing requirements for concrete buildings are specified under Clause 21 of the CSA A23.3-14 design standard, Design of Concrete Structures (CSA 2014). These requirements are consistent with a capacity design approach where pre-determined elements of the SFRS are designed and detailed to allow energy dissipation to occur with the remainder of the structural elements designed for amplified forces to ensure that the targeted energy dissipating mechanism is obtained. For example, flexural shear walls are designed with an intended flexural hinge mechanism typically located at the base. These walls are then required to be capacity protected in shear by amplifying the shear design force in consideration to the actual flexural capacity of the shear wall.

4 REVIEW OF NORTH AMERICAN NONLINEAR RESPONSE HISTORY ANALYSIS PROVISIONS AND PERFORMANCE-BASED SEISMIC DESIGN GUIDELINES

Several documents exist on the topic of performance-based seismic design and nonlinear response history analysis (NLRHA). This section of the paper specifically focuses and provides a brief overview of the following documents:

- "Chapter 16 – Nonlinear Response History Analysis" – ASCE 7-16,
- "Seismic Evaluation and Retrofit of Existing Buildings" – ASCE 41-17,
- "TBI Guidelines on the Performance-Based Seismic Design of Tall Buildings" – PEER (2017), and
- "National Building Code of Canada 2015 Nonlinear Time History Analysis" – NRCC (2015) and "User's Guide – NBC 2015 Structural Commentaries, Commentary J" – NRCC (2017).

The overarching goals, ground motion characterization procedures, modeling parameters and acceptance criteria are discussed below, summary of which is included in Table 2. It is important to note that the table does not cover all requirements and/or exceptions. For a comprehensive list of requirements, the reader is referred to the publications.

4.1 Performance Objectives

Performance objectives vary amongst the above noted publications. The intent of the PEER (2017) guidelines, as noted in this document, is to provide designers with a performance-based seismic design approach for the design of tall buildings as an alternative to the prescriptive code-based procedures of ASCE 7-16. It is noted that buildings designed in accordance with these guidelines are expected to remain essentially elastic under the service level earthquake (SLE) and to respond to the MCE_R intensity shaking without the loss of the gravity load system (i.e. collapse prevention). The scope of these guidelines is however limited to tall buildings, which is defined as those having a lateral fundamental period greater than 1.0s, a high mass participation and lateral response in higher modes of vibration and a SFRS with a slender aspect ratio. Although not explicitly expressed in the PEER (2017) guidelines; it is implied that the goals are aligned with those of ASCE 7-16 by invoking the “alternative design” method permitted by both ASCE 7 and the IBC. The overarching goals identified in ASCE 7-16, expressed as targeted maximum probabilities of structural collapse, P[C], given an MCE_R intensity level of ground shaking, are 10% for Risk Categories (RC) I and II, 5% for Risk Category III and 2.5% for Risk Category IV. It is noted in PEER (2017) that these acceptable collapse probabilities stem from the work reported in FEMA P695. Chapter 16 of ASCE 7-16 on NLRHA is specified as a function of the risk-targeted Maximum Considered Earthquake (MCE_R), whereas the traditional code design process specifies a design load intensity corresponding to 2/3 MCE_R, which is referred to as the Design Earthquake (DE). Other implicit objectives of seismic design under ASCE 7-16 are expressed in qualitative terms such as the ability to withstand the DE load without significant hazards to lives (life safety objective) and to withstand more frequent lower intensity earthquakes with limited damage. It is further noted in the PEER (2017) guidelines that the expected level of damage under service level intensity hazard may include minor cracking of concrete and yielding of steel in a limited number of structural elements but that the damage should be easily repairable and not compromise the ability of the building structure to withstand an earthquake at the MCE_R intensity.

ASCE 41-17 contains multiple performance objectives as a function of the targeted building performance and the building category for various levels of shaking.

No targeted maximum probabilities of structural collapse are identified in the National Building Code of Canada. However, NRCC (2017) notes in qualitative terms that the primary objective of seismic design addressed by the NBCC is to provide an acceptable level of safety. Under strong ground shaking (which is defined as a ground motion having a probability of exceedance of 2% in 50 years), it is expected that buildings in general may sustain substantial structural and non-structural damage but they should have sufficient strength to support gravity loads. Buildings of greater importance such as hospitals are required to be designed to higher loads and are expected to be functional post-earthquake. This is indirectly achieved by amplifying the design forces by an importance factor.

The PEER (2017) guidelines require evaluations at two levels of ground shaking: one under an intensity of shaking corresponding to a probability of exceedance of 50% in 30 years termed Service Level Earthquake (SLE) evaluation and a second evaluation at MCE_R. Chapter 16 of ASCE 7 requires evaluation only at MCE_R and the NBCC prescribes one evaluation with an intensity of shaking corresponding to a probability of exceedance of 2% in 50 years.

4.2 Ground Motion Characterization

An important aspect of NLRHA yet complex process, as noted in all the above noted publications, is the proper characterization of time history ground motions, which involves proper selection and scaling of ground motion histories to a compatible target response spectrum. It is suggested to select motions that are compatible with the tectonic regime with similar magnitude, fault distance, source mechanisms and site soil conditions.

Table 2: Selected summary of nonlinear response history analysis provisions

Response History Analysis Components	Design / Assessment Method for NLRHA			
	ASCE 7–16 Chapter 16 ASCE (2017a)	**ASCE 41-17 ASCE (2017b)**	**PEER TBI PBSD Guidelines PEER (2017)**	**NBCC 2015 NRCC (2015) & (2017)**
General				
Explicit goals	P[C] < 10%, 5% and 2.5% for RC I or II, RC III and IV, respectively	Varies based on targeted building performance	Intended to be at least equal to ASCE 7-16 by invoking alternative solution	Not specified
Ground motion hazard analysis	Probabilistic and deterministic	Probabilistic and deterministic	Probabilistic and deterministic	Probabilistic
Ground motion level	MCE_R	Multiple	SLE, MCE_R	2% in 50 yrs
Target spectrum	UHS or multiple spectra with risk adjustment	UHS with or without risk adjustment	UHS or multiple spectra with risk adjustment	UHS or multiple spectra
Ground motion selection (at maximum considered earthquake intensity)				
Number of motions	≥ 11 pairs per target spectrum	≥ 11 pairs per target spectrum	≥ 11 pairs total ≥ 5 records per source contributing more than 20% to hazard but not less than 11 total ≥ 5 records pulse or no pulse subsets for sites where pulse-type motions are considered	<u>Method A</u> ≥ 5 records per suite ≥ 11 records per source ≥ 11 records (total for all suites) <u>Method B1 or B2</u> ≥ 11 records for each scenario target spectrum
Scaling / modifications of motions to match target spectrum				
General approach	Amplitude scaling or spectral matching	Amplitude scaling or spectral matching	Amplitude scaling or spectral matching	Amplitude scaling or spectral matching
Amplitude scaling	Match records on average to target, enforce 90% suite floor	Match records on average to target, enforce 90% suite floor	Match records on average to target, enforce 90% suite floor	Match records on average to target, enforce 90% suite floor
Spectral matching	Match records to target considering a 110% suite floor	Match records to target considering a 110% suite floor Not to be used with Method 2,	Match records to target considering a 110% suite floor	Match each record on average to target, enforce 110% suite floor
Period range	min[0.2T, $T_{90\%}$] – 2.0T	$0.2T_{min}$ – max[1,1.5T_{max}] where T_{min} and T_{max} are the smallest and largest 1st mode periods for the two principal directions	min[0.2T, $T_{90\%}$] – 2.0T	min[0.15T,$T_{90\%}$] – max[2.0T, 1.5s]

Modeling and analysis				
Modeling parameters of nonlinear elements	Consistent with ASCE 41 or applicable laboratory test data	Based on experimental evidence	Use hysteretic models that adequately account for strength and stiffness deterioration under imposed deformations and cyclic loading; physical test data	Representative of actual nonlinear cyclic load test data
Design seismic demand	Use mean values; where unacceptable response is permitted, use 120% median value but not less than mean value of acceptable responses	Where component response is independent of direction of action, use mean of maximum absolute responses, where unacceptable response is permitted, use 120% median value but not less than mean value of acceptable responses	Use mean values; where unacceptable response is permitted, use 120% median value but not less than mean value of acceptable responses	Use mean value of all ground motions when only 1 suite is used When 2 or more suites are used – use largest of the mean of each suite where each suite has at least 11 ground motions; otherwise, use mean of the n highest values of all suites where n is the average number of records in all suites
Response metrics and acceptance criteria (at maximum considered earthquake intensity)				
Peak storey drifts	μ < twice code typical limit;	No limit specified	Suite mean of absolute peak transient story drifts ≤ 0.03	μ of analysis using lower-bound strength properties < typical code limit
Residual storey drifts	No limit specified	No limit specified	Suite mean of absolute residual drifts ≤ 0.01	No limit specified
Deformation-controlled actions	μ < ASCE 41 CP limit x $1.0/I_e$, or $Q_u < 0.3Q_{ne}/I_e$ for critical and $0.5Q_{ne}/I_e$ for non-critical	μ < limit specified for the targeted performance level as per Chapter 10 for concrete SFRSs	δ of any response history analysis to be less than the ultimate deformation capacity, δ_u, of the component; with some exceptions	μ of analysis using lower-bound strength properties < CSA limit x $1.0/I_E$ or ASCE 41 LS limit x $0.7/I_E$ in absence of CSA limit or test data x $0.5/I_E$
Force-controlled actions	$\gamma I_e(Q_u - Q_{ns}) + Q_{ns} \leq Q_e$ where γ is taken as 2.0, 1.5 and 1.0 for critical, ordinary and non-critical actions, respectively	$\gamma\chi(Q_{uf} - Q_g) + Q_g \leq Q_{CL}$ where γ is taken as 1.3, 1.0 and 1.0 for critical, ordinary and non-critical actions, respectively	For well-defined yield mechanisms: $(1.2 + 0.2S_{MS})D + 1.0L + E_M \leq \phi_s R_n$ $(0.9 - 0.2S_{MS})D + E_M \leq \phi_s R_n$	μ of analysis using upper-bound strength properties < resistance determined from CSA standard

		and χ is taken as 1.0 for CP and 1.3 for LS and IO and $\gamma\chi \leq 1.5$	For other mechanisms: $(1.2 + 0.2S_{MS})D + 1.0L + 1.3I_e(Q_T\text{-}Q_{ns}) \leq \phi_s B R_n$ $(0.9 - 0.2S_{MS})D + 1.3I_e(Q_T\text{-}Q_{ns}) \leq \phi_s B R_n$	
Treatment of collapse or unacceptable response cases	1 unacceptable response permitted for RC I & II where spectral matching is not used	1 unacceptable response per 11 analyses permitted for LS or lower performance levels	Where ϕ_s is as per ACI 318 for critical elements, and 0.9 and 1.0 for ordinary and non-critical elements, respectively	Where spectral matching is not used, 1 unacceptable response permitted if suite has at least 11 ground motions and additional evaluations are carried out

The estimation of the ground shaking is typically done either deterministically or probabilistically. A probabilistic approach accounts for uncertainties by considering weighted alternatives of various earthquake sources, possible magnitudes, multiple viable ground attenuation relationships (ground motion prediction equations) and earthquake recurrence. The cumulative probabilities of all considered scenarios is then used to quantify the seismic hazard. In a deterministic approach, the seismic hazard is simply defined as the maximum response resulting from the considered scenarios.

Whereas ASCE 7-16 Chapter 16, PEER (2017) and ASCE 41-17 typically define the seismic hazard at the MCE_R intensity using a probabilistic approach with a deterministic cap, the NBCC hazard stems from a purely probabilistic approach. Unlike the US, there is no deterministic cap applied to the hazard assessment in Canada. PEER (2017) specifies the use of a probabilistic approach for the characterization of the Service Level Earthquake (SLE) Evaluation ground shaking hazard.

Selected ground motion histories are then scaled to match a target response spectrum to ensure adequacy of the adjusted histories for use in analysis. ASCE 7-16, PEER (2017) and NRCC (2017) indicate that the target spectrum can be based on a uniform hazard spectrum or a spectrum constructed from multiple scenarios.

Ground motion records can be either amplitude scaled or matched spectrally, over a defined period range, to a minimum specified level. The period range is typically specified to span from a minimum period $T_{90\%}$, which corresponds to the lowest period required to capture at least 90% of the mass participation to a maximum taken as 2 times the lateral fundamental period of the structure, T. An absolute minimum of 0.2T is specified for the range in PEER (2017) and ASCE 7-16. On the other hand, Commentary J of the NBC User's Guide (NRCC 2017) specifies an absolute minimum of 0.15s regardless of the $T_{90\%}$ value in determining the period range and a minimum upper range of 1.5s. In general terms, records are scaled over the period range such that the suite meets 90% and 110% of the targeted spectrum for amplitude scaling and spectral matching, respectively.

In addition to properly scaled ground motions, an adequate number of records is needed to properly predict and quantify the response of a structure subjected to ground shaking. PEER (2017) specifies the minimum number of ground motions to be used as a function of the type of assessment to be undertaken. For SLE evaluation, a minimum of 3 and 7 pairs of ground motion histories are specified for linear and nonlinear analyses, respectively. For MCE_R evaluation, a minimum of 5 records are to be used for each source contributing more than 20% to the hazard but not less than 11 records in total. A minimum of 5 pulse-type records is also specified for sites where pulse-type motions are to be considered. The minimum of 11 pairs at MCE_R evaluation aligns with ASCE 7 Chapter 16 requirements.

In Canada, the minimum number of records is expressed as a function of the method used to define the target response spectrum. Where Method A is used, minimum requirements specify 5 records per suite, 11 records per source and an absolute minimum of 11 records. Where Method B1 or B2 is used, a minimum of 11 records for each scenario target spectrum is specified.

Contrary to code-based approaches which only consider one level of shaking, ground motions used for PBSD are to be characterized for each level of shaking considered.

4.3 Modeling and Analysis

Proper modeling and analysis of building structures is another important aspect of NLRHA and performance-based seismic design. PEER (2017) recommends modeling of structural components based on expected properties to provide an unbiased estimate of the expected response of the building. It specifies the expected concrete strength as 1.3 times the specified concrete strength. It is also noted that the 1.3 multiplier may be different depending on the strength of the concrete, the use of fly ash and other additives as well as the type of aggregates used. NRCC (2017) specifies the use of lower-bound strength properties and upper-bound strength properties for modelling of non-linear SFRS elements to determine demands for deformation-controlled and force-controlled actions, respectively.

For modeling of nonlinear components, all publications indicate that proper material and hysteretic models that adequately account for strength and stiffness deterioration under imposed deformation and cyclic loading should be used. It is also inferred that model parameters should be validated by experimental data.

The PEER (2017) guidelines require the use of well-defined three-dimensional mathematical models which appropriately reflect the spatial distribution of the mass and the stiffness of the building structure. The models are also to be representative of the anticipated effective stiffness and damping under the level of shaking under consideration. The NBCC does not have any explicit requirements with respect to three-dimensional modeling for use with nonlinear time history analysis.

4.4 Acceptance Criteria

As a final step in response history analysis, the seismic demand is compared and checked against acceptance criteria. Requirements regarding the determination of the seismic demand vary depending on the level of evaluation under consideration and the type of analysis used. Under SLE, PEER (2017) indicates that typical traditional code-based response modification factors are not to be applied to response spectrum or linear history analysis results as the intent is to have the buildings remain essentially elastic under this level of shaking. Where nonlinear analysis is used, PEER (2017) generally specifies the use of average values to determine the seismic demand for both SLE and MCE_R evaluations. In Canada, Commentary J of the NBC User's Guide specifies that mean values of all ground motions may be used where only 1 suite is used. When 2 or more suites are used, the largest of the mean of each suite, where each suite has at least 11 ground motion may be used; otherwise, the mean of the n highest values of the response parameters (where n is the average number of ground motions in all suites) is to be used.

All publications identify unacceptable responses. In general, these include an analysis which fails to converge, cases where the predicted demand exceeds the range of modelling or capacity of the component and demands which exceed deformation limits at which members can no longer support gravity loads. One unacceptable response is permitted in some cases as highlighted in Table 2.

Response metrics and acceptance criteria are specified in terms of global and component acceptance criteria. Global acceptance criteria include drift limits which are further subcategorized into peak storey drifts and residual storey drifts. PEER (2017) specifies an overall peak storey drift limit of 0.005 for SLE evaluation and a mean absolute peak transient storey drift limit of 0.03 for each suite for the MCE_R evaluation. A mean absolute residual story drift limit of 0.01 is also specified for MCE_R evaluation. Peak storey drift limits for response time history analysis under ASCE 7-16 and NBCC 2015 are noted as twice the typical code drift limits and the typical code drift limits, respectively. These documents do not list any requirements/limits for residual storey drifts.

For component checks, all publications require the classification of actions into either deformation-controlled actions or force-controlled actions. Deformation-controlled actions are noted as those actions that are expected to undergo nonlinear behavior in response to earthquake shaking whereas as force-controlled actions are not expected to undergo nonlinear behavior. A second tier of classification (critical, ordinary and non-critical) further distinguishes the consequence of failure of the force-controlled actions in PEER (2017), ASCE (2017a) and ASCE (2017b). The acceptance criteria for deformation- and force-controlled actions are listed in Table 2. As noted, demands or acceptance criteria for these components are generally adjusted to reflect the importance category of the building under evaluation. For force-controlled actions, the US publications in general require that the portion of the seismic only demand be amplified to obtain the design demand for these components which is then compared to the resistance of the component using either expected or lower-bound material properties. In Canada, the demand for force-controlled components are to be computed based on the probable resistance of the SFRS or using an analysis which considers upper-bound strength properties for the nonlinear elements.

Given the complexities involved with nonlinear time response history analysis, all four documents require review by an independent third-party engineer with experience in nonlinear analysis and seismic design.

5 ADAPTATION TO CANADIAN CONTEXT AND NEED FOR FUTURE RESEARCH

As described in this paper, most of the documentation to date on performance-based seismic design has been developed for the assessment and retrofit of existing buildings. To the authors' knowledge, few guidelines exist (other than PEER 2017 and LATBSDC 2017) on PBSD of new buildings, with these being focused on tall buildings. Other than the limited information introduced in Commentary 'L' of the NBC User's Guide (NRCC 2017), the authors are unaware of any PBSD information that is specific to Canada. It is also noted that the NBCC does not quantify the target performance intended by the Code in terms of probability of collapse as does ASCE 7.

On that basis, the following is identified as needs for future research in Canada:

- elaboration / development of performance-based seismic design framework for assessment of existing buildings and design of new buildings (levels of ground shaking, performance levels, performance categories), and
- elaboration / development of nonlinear modelling parameters and acceptance criteria for use within a PBSD framework for various levels of shaking consistent with the minimum detailing requirements outlined in the respective CSA material design standard.

It is important to note that in Canada, the National Building Code of Canada prescribes design forces whereas the referenced CSA material design standard specifies the material resistance. In general, there is limited guidance provided with respect to the mathematical and computer modeling of building structures. This is a gap that also needs to be addressed, especially considering the availability of modeling software tools and the shift to dynamic analysis as the default method. Although the information stems from a US methodology perspective, the PEER (2017) guidelines and ASCE 41-17 provide valuable insights on this.

6 CONCLUSION

This paper explored the North American state-of-the practice in NLRHA and performance-based seismic design of buildings with an emphasis on reinforced concrete seismic force resisting systems. As noted, PBSD provides various stakeholder with a better representation of the expected performance of a building under earthquake shaking. Recent guidelines developed by PEER have also been explored. However, these guidelines strictly focus on the PBSD of tall buildings and stem from a US methodology perspective. As it can be deduced from the above noted literature review, Canada is significantly lagging the US in the development of Performance-Based Seismic Design methodology. Just as linear dynamic analysis is now the default method in current Canadian code-based seismic design, it is anticipated that performance-based seismic design will be the next generation seismic design methodology in the future.

References

ASCE. 2005. Seismic Evaluation and Retrofit of Existing Buildings. ASCE/SEI 41-06. American Society of Civil Engineers, Reston, Virginia.

ASCE. 2017a. Minimum Design Loads for Buildings and Other Structures. ASCE/SEI 7-16. American Society of Civil Engineers, Reston, Virginia.

ASCE. 2017b. Seismic Evaluation and Retrofit of Existing Buildings. ASCE/SEI 41-17. American Society of Civil Engineers, Reston, Virginia.

ATC. 1996. "Seismic evaluation and retrofit of concrete buildings". ATC-40 Report, Applied Technology Council, Redwood City, California.

CSA. 2014. "Design of Concrete Structures", CSA A23.3-14, Canadian Standards Association, Toronto, Ontario.

FEMA. 1997. NEHRP Guidelines for the Seismic Rehabilitation of Buildings, FEMA 273, Applied Technology Council, Redwood, California.

FEMA. 2000. Prestandard and Commentary for the Seismic Rehabilitation of Buildings, FEMA 356, American Society of Civil Engineers, Reston, Virginia

FEMA. 2004. NEHRP Recommended Provisions for Seismic Regulations for New Buildings and Other Structures. FEMA 450. Building Seismic Safety Council & National Institute of Building Sciences, Washington, DC.

LATBSDC. 2017. An Alternative Procedure for Seismic Analysis and Design of Tall Buildings Located in the Los Angeles Region. Los Angeles Tall Buildings Structural Design Council, Los Angeles, California.

NRCC. 2015. National Building Code of Canada 2015. Canadian Commission on Building and Fire Codes, National Research Council of Canada, Ottawa, Ontario

NRCC. 2017. User's Guide – NBC 2015 Structural Commentaries (Part 4 of Division B). Canadian Commission on Building and Fire Codes, National Research Council of Canada, Ottawa, Ontario

PEER. 2017. Guidelines for Performance-Based Seismic Design of Tall Buildings. PEER Report 2017/06, Pacific Earthquake Engineering Research Center, University of California, Berkeley, California

Building Tomorrow's Society
Bâtir la Société de Demain

Fredericton, Canada
June 13 to 16, 2018 / *13 au 16 juin, 2018*

INTEGRATION OF SEISMICALLY INDUCED SITE EFFECTS TO THE EVALUATION OF THE SEISMIC VULNERABILITY OF BRIDGES THROUGH A GIS PLATFORM

Farzam, Azarm[1,2], Nollet, Marie-José[1] and Khaled, Amar[1]
[1] Department of Construction Engineering, École de technologie supérieure, Montréal, Canada
[2] azarm.farzam.1@ens.etsmtl.ca

Abstract: This paper proposes an approach to integrate seismically induced site effects to the evaluation of the seismic vulnerability of bridges through a GIS platform. The evaluation of the seismic vulnerability of bridges aims to assess the potential damage to structures in order to prioritize the interventions following an earthquake or to plan mitigation strategies. Evaluation procedures are generally index-based, and consider structural vulnerability and seismic hazard, which is usually defined in terms of the probable acceleration that could integrate the amplification effect of seismic waves by the site. A few index-based procedures take into account induced site effects, such as landslides, rock falls or soil liquefaction. Quantification of amplification and induced site effects requires site-specific geotechnical information. However, seismic risk assessment studies are usually performed on a regional scale for a large number of infrastructures scattered on a large territory. The lack of site-specific geotechnical information makes the evaluation of the seismic vulnerability of these infrastructures less representative. This study is based on the development of susceptibility scales to amplification effect, landslides, rock falls and soil liquefaction using geographical information systems (GIS). These susceptibility scales are defined documenting and characterizing the general context of geomorphology, geology and hydrography. The resulting susceptibility maps are then used to integrate susceptibility levels to seismically induced site effects into an index-based procedure for a better evaluation of the seismic risk of bridge networks.

Keywords: amplification, landslide, rock falls, seismic vulnerability, risk assessment, scoring procedure, bridge

1 INTRODUCTION

Southeastern Canada experiences about 600 earthquakes every year (Government of Canada, 2011). While seismic hazard is generally moderate in eastern Canada, the density of the population in urban areas makes it the second-largest seismic hazard area in Canada. In the southeastern Canada, two seismic events with magnitude 5 (on a Richter scale) or more were felt in the last ten years (Government of Canada, 2018b). These events induced landslides such as in Val-des-Bois (2010), or dike damages in south of Bowman and damages to chimneys and houses due to local soil amplification (Gouvernement du Canada, 2010). During the 1988 Saguenay earthquake (Mw=5.9), rock falls, landslides and liquefaction, were observed along the Lowlands as far as 200 km from the epicentre (Lamontagne, 2002). More than 20 landslides occurred due to the presence of marine clay (Lefebvre et al., 1992; Locat, 2008). Moreover, damages, due to local amplification, were triggered up to a distance of 340 km on Montreal Island (Paultre et al., 1993).

Seismic risk assessment is an essential step towards effective mitigation measures and emergency planning. As the number of infrastructures and installations exposed to seismic hazard increase, interventions must prioritize the most vulnerable structures. In the province of Quebec, the total number of bridges and overpasses under the jurisdiction of the province is approximately 9600 (MTMDET, 2017); out of this total number, 70% were built between 1960 and 1980, when seismic design provisions were not as stringent as today (MTMDET, 2017). Bridges and overpasses of municipal networks experience similar situation. The province of Quebec counts approximately 10600 bridges and overpasses among which 48% are located in the region of study (Government of Canada, 2018a).

The first multi-criterion procedure for prioritization of bridges was developed by the California Department of Transportation following the San Fernando earthquake in 1971 (Basöz et Kiremidjian, 1995; Small, 1999). This method was the inspiration for several other procedures in the US (Kim, 1993; NYSDOT, 2002), Canada (Liu, 2001) and in the province of Quebec (Tinawi et al., 1993). In Quebec, the most recent procedure applied by the *Ministère des Transports, de la Mobilité durable et de l'Électrification des transports* (MTMDET) is based on the scoring procedure developed for the Quebec City bridges (Hida, 2009). It has been since modified and revised for a better identification of the most vulnerable bridges by considering the socioeconomic index to quantify value at risk (Lemaire, 2013). As in similar procedures, it considers the seismic hazard, the structural vulnerabilities of the bridges and the amplification effect. The potential of liquefaction of soil is sometimes taken into consideration when information is available. However, the contribution of those site effects to the seismic risk is frequently approximated due to the lack of site-specific information. Ground mass movements such as landslides or rock falls are rarely taken into consideration (Davi et al., 2011).

Seismic risk is composed of three main parameters: hazard, vulnerability and value at risk (UNDRO, 1991). This study focuses on the hazard aspects of the seismic risk in rapid screening procedures for bridges considering induced site effects. The proposed methodology is based on the development of maps characterizing susceptibility scales to amplification effect, landslides, rock falls and soil liquefaction using geographical information systems (GIS) based on hydrological, topographic and geological cartographic data (Farzam et al., 2016; Farzam et al., 2018). To validate the susceptibility scale to amplification effect, ambient noise measurements have been performed at the site of bridges and on the structures. The landslide susceptibility scale was validated using a landslide inventory in the St. Lawrence Valley (Quinn et al., 2008). The objective of this paper is to demonstrate how induced site effects could be integrated into seismic assessment scoring procedures for bridges using a GIS platform. An application to the seismic procedure developed by Lemaire (2013) for the bridges in the province of Quebec illustrates the impact of considering seismically induced site effects on the prioritization of bridges.

2 METHODOLOGY FOR SUSCEPTIBILITY MAPPING OF SITE EFFECTS

Seismically site effects are governed by geotechnical factors taking place in a geological context. There are many ways to characterize the susceptibility to seismic site effects including prospecting site by site with in situ geotechnical measurements and laboratory testing, and documenting and characterizing the general context of geomorphology, geology, hydrography and even climate factors (Theilen-Willige, 2010). Although a best estimate of seismically site effects hazards is obtained from site-specific prospecting, the information from the general context allow mapping large scale phenomena influencing site effect at smaller scale.

2.1 Susceptibility Mapping to Amplification

Seismic amplification is usually characterized by the attribution of a seismic site class as defined by codes and regulations. This process also called microzonation is mainly based on the calculation of the time-averaged shear-wave velocity V_{s30} over 30 metres (NRCC, 2015). However at a local scale, in the absence of specific shear wave velocities data, information about thickness of quaternary deposits and surficial geology can be correlated to seismic site classes using geostatistical relations obtained from existing microzonation maps for other regions with similar geological characteristics. This methodology is based on the hypothesis that soil column is similar for a same region depending on the surficial geology and the thickness of quaternary deposits (Braganza et al., 2016). Consequently, all sites with same surficial geology and same thickness of quaternary deposits should have equal V_{s30} and thereby the same seismic site class. For the region of study and its vicinity, microzonation have been completed for three cities: Quebec city (Leboeuf et al., 2013), Montreal (Rosset et al., 2015) and Ottawa-Gatineau (Motazedian et al., 2011). The geostatistics conducted for these three cities provide the probabilities to belong to a seismic site class for a specific surficial geology and thickness of quaternary deposits. However, in some cases, probabilities are not strictly defining one seismic class and intermediate levels are introduced. Seven levels of susceptibility to amplification ranging from very very low (VVL) to very high (VH) are introduced in this study as presented in Table 1.

Table 1: Susceptibility scale for amplification depending on attributed seismic site class

Attributed seismic site class from surficial geology and thickness deposits	Susceptibility scale
A and B	Very very low (VVL)
B - C	Very low (VL)
C	Low (L)
C - D	Low to moderate (LM)
D	Moderate (M)
D - E	High (H)
E	Very high (VH)

To validate the approach by geostatistical relations ambient vibration measurements were conducted on 50 different sites. Knowing the thickness of quaternary deposits and the average shear wave velocity of bedrock at the sites, resonance frequencies of soil were extracted and V_{s30} were computed. The results were then compared with the seismic site class attributed to the sites from the information on the surficial geology and thickness deposits. The final susceptibility scale is presented in Table 2 for the surficial geology and thickness of deposits in the region.

Table 2: Susceptibility scale for amplification depending on surficial geology and thickness of deposits

Surficial geology\Thickness of deposits	0 to 5 [m]	5 to 10 [m]	10 to 20 [m]	20 to 30 [m]	>30 [m]
Alluvial deposits	LM	LM	LM	H	H
Colluvial and mass-wasting deposits	H	H	H	VH	VH
Eolian deposits	LM	L	L	LM	LM
Glaciofluvial and glaciolacustrine deposits	VVL	L	L	L	L
Anthropogenic deposits	L	L	L	L	N/A
Lacustrine deposits	VVL	LM	LM	M	M
Marine deposits	VL	L	LM	H	VH
Organic deposits	VVL	VL	VL	H	VH
Till-Glacial deposits	VVL	VL	VL	LM	LM
Undifferentiated deposits	VVL	L	L	LM	N/A

The map presented in Figure 1 shows the distribution of the susceptibility to amplification for the Lowlands of the Saint-Lawrence Valley. Very high levels of susceptibility of amplification are mostly found in the centre of the region of study where thicknesses of quaternary deposits are particularly high. The susceptibility map for amplification has a resolution of about 405 m, calculated from the root mean square of surficial geology layer with a resolution of 500 m (Parent et al., 2018) and thickness of quaternary deposits with a resolution of 275 m (Gouvernement du Canada, 2004).

To validate this susceptibility map for amplification, a comparison was done with the regional microzonation maps of Montreal and Quebec City: seismic site classes assigned from geostatistical relations were similar to the regional microzonation for 75% of the area covered by the two cities.

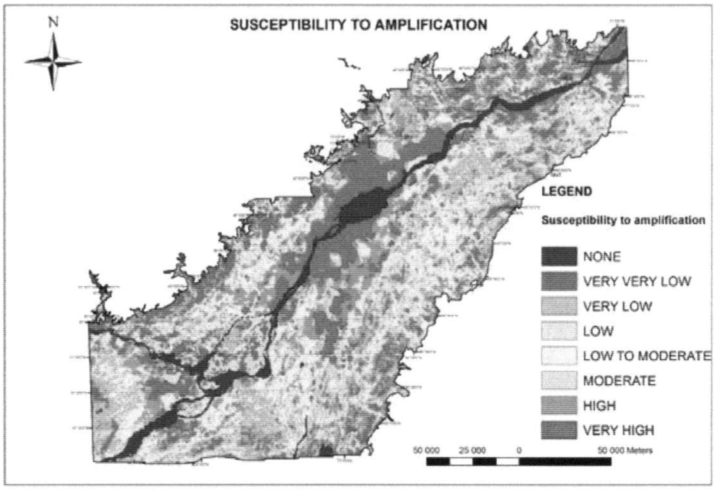

Figure 1: Susceptibility to amplification map (red = high, blue = low)

2.2 Susceptibility Mapping to Ground Mass Movements

The methodology used in this study to characterize ground mass movements (landslides and rock falls) hazard is highly inspired by susceptibility scale developed from experts judgments in the context of seismic hazard loss estimation (HAZUS) of the Federal Emergency Management Agency of US (FEMA, 2012). However, the particular sensitivity to fracture and erosion of Leda clay in the region of study is taken into account by introducing a new parameter that characterizes the proximity to watercourses. Six susceptibility levels to ground mass movement, ranging from none to very high, are developed based on: (a) the 12 types of surficial geology present in the Lowlands of the Saint-Lawrence Valley, in terms of their composition and state of consolidation divided in three groups (Group A: strongly cemented rocks, Group B: weakly cemented rocks and soils and Group C: argillaceous rocks) (b) the slope of the ground divided in 6 ranges (0 - 10°, 10° - 15°, 15° - 20°, 20° - 30°, 30° - 40°, > 40°), (c) the groundwater table (below or above 10 m), and (d) proximity to watercourses (< 100 m). For example, a site with till outcrop (Group B), with ground slope of 5°, on wet soil (groundwater table close to the surface (<10 m) and close to watercourse (<100 m) will have a high susceptibility to landslide. Rock falls hazards can be differentiated from landslides hazard by the presence of strongly cemented rock such as bedrock. Since bedrock outcrops is present in only 5% of the region of study, therefore rock falls hazard is only present in this 5%. Landslides and rock falls susceptibilities are mapped together in the Figure 2. The susceptibility map for ground mass movements has a resolution of 355 m, computed from the root mean square of surficial geology layer with a resolution of 500 m (Parent et al., 2018) and the slope derived by the digital elevation model with a resolution of approximately 30m (Farr et al., 2007). Landslide susceptibility map has been validated using an inventory for the region of study provided by Quinn (2009): 68% of landslides from the inventory are located on areas identified as highly or very highly susceptible. The method of attribution of susceptibility level is well correlated to the inventory of landslides.

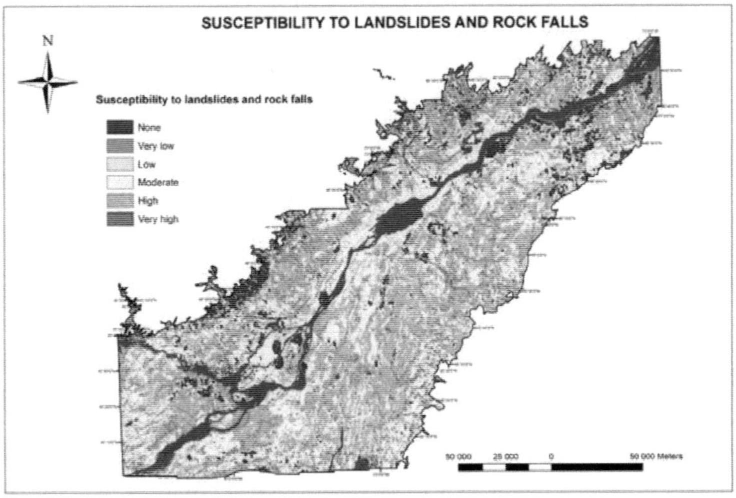

Figure 2: Susceptibility to landslides and rock falls (red = high, blue = low)

2.3 Susceptibility Mapping to Liquefaction

Liquefaction of soil may happen during an earthquake when saturated soils under multiple cycles of loadings lose their shear strength. It is known as a very complex and destructive phenomenon (Priestley, 1996). However, experts from FEMA (2012) classified lithologies in six levels of susceptibility to liquefaction (from none to very high) with respect to their type of deposit and their state of consolidation through their age. The same classification is adopted in this project. If groundwater table is close to the surface (< 10 m), final liquefaction levels of susceptibility are corresponding to those from lithologies classification, i.e. if groundwater table is lower than 10 m from the surface susceptibility level of liquefaction are classified from low to very low. The map in Figure 3 shows the distribution of susceptibility level to liquefaction in the region of study.

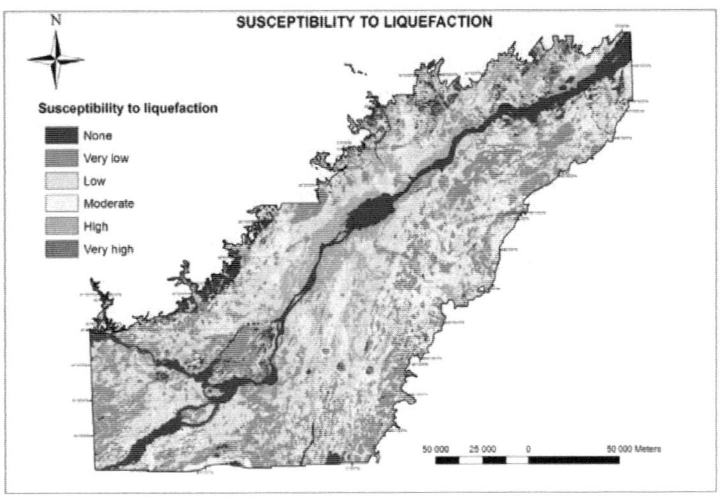

Figure 1: Susceptibility to liquefaction map (red = high, blue = low)

3 INTEGRATION OF SEISMICALLY INDUCED SITE EFFECTS TO SEISMIC RISK ASSESSMENT PROCEDURES FOR BRIDGES

To demonstrate how induced site effects could be integrated into seismic assessment scoring procedures for bridges using a GIS platform, the proposed susceptibility maps are used with the seismic scoring procedure developed by Lemaire (2013). This procedure allows prioritizing bridges in the province of Quebec according to their relative seismic risk. The bridges seismic vulnerability index ($SVI_{MTQ2013}$) proposed by (Lemaire, 2013) evaluates the seismic risk considering the structure vulnerability, the seismic hazard and the amplification effect at the site independently. As a result, the probability of each event intersected with the probability of the others and the resulting probability is given by the product of the respective probability of each independent event. In a rapid scoring procedure, this could be expressed by the product of three independent indices combined to obtain the final seismic risk or vulnerability index (SVI), as in the following equation:

$$[3]\ SVI_{MTQ2013} = V_{structural\ vulnerability} \times F_{hazard} \times F_{site}$$

The $SVI_{MTQ2013}$ varies from 0 to 100, with a maximum score for the most vulnerable bridges. The structural vulnerability index, $V_{structural\ vulnerability}$, varying from 1 to 32, is defined from structural characteristics and identification of seismic deficiencies of the bridges. The weight of this index has been carefully established from simulation scenarios of damage to the bridges of the inventory. The seismic hazard index, F_{hazard}, is computed from seismic hazard maps in terms of peak ground acceleration (PGA) (Halchuk et al. (2015)). PGA is one of the most widely used seismic hazard parameter for the evaluation of the seismic risk of bridges in regional studies (Basöz et Kiremidjian, 1995; Davi et Schmitt, 2003; Kim, 1993; Kiremidjian et al., 2007). F_{hazard} varies from 1 to 2. Local amplification effect is taken into consideration through the index F_{site}. It is defined from seismic site categorizations (NBC 2010) and varies from 1 to 1.25. Moreover, F_{site} is also taking into account liquefaction susceptibility. A maximum value of 10 is attributed to F_{site} in the case of a very high potential of liquefaction. As a consequence, the final SVI value can increase significantly which reflects the destructive capacity of this particular phenomenon and its impact on structures.

The present study introduces into this procedure susceptibility to ground mass movements including landslides and rock falls. These latter are inconsistent events; meaning that landslides cannot happen at the same time as rock falls. On the contrary, ground mass movements can be triggered during a seismic event at the same time as a local amplification of seismic waves. Therefore, the probability of landslides and rock falls are considered as the union of sets and their indices ($F_{landslides}$ and $F_{rock\ falls}$) are arithmetically added. It is also considered that the probability of ground mass movements intersects with the probability of local amplification (F_{site}) leading to a product of their respective indices, as shown in the revised seismic risk index, SRI (Eq. 4):

$$[4]\ SRI = V_{structural\ vulnerability} \times (F_{hazard} \times F_{site} \times (F_{landslides} + F_{rock\ falls}))$$

In order to preserve a maximum score of 100 and the same weight for hazards and vulnerability, F_{hazard} is set to vary between 1 and 2 instead of 2.5 while the site index F_{site} is kept between 1 and 1.25, depending on local amplification. It is defined from the susceptibility map to amplification (Figure 1) converting the susceptibility scale to scores from 1 to 1.25 (Very Very Low = 1, Very Low = 1.025, Low = 1.05, Low to Moderate = 1.1, Moderate = 1.15, High = 1.2 and Very High = 1.25).

The resulting index, combining landslides ($F_{landslides}$) and rock falls ($F_{rock\ falls}$) hazards, is defined from the susceptibility map to ground mass movements (Figure 3) converting the susceptibility scale to scores from 1 to 1.25 (None = 1, Very Low = 1.05, Low = 1.1, Moderate = 1.15, High = 1.2 and Very High = 1.25). For example a bridge presenting several structural vulnerabilities ($V_{struc} = 32$), $F_{hazard} = 2$ and $F_{site} = 1.25$ located on a site with low level of susceptibility to landslide ($F_{landslides} = 1$) will have its SRI increased from 80 to 100 if located on a site with very high level of susceptibility to landslide ($F_{landslides} = 1.25$).

Therefore bridge with no or few structural deficiencies ($V_{structural} < 16$) but located on a site of maximum seismic hazard should be maintained with a final score under 50, considered to be very low priority.

It should be noted that the modified seismic risk index SRI also takes into account the susceptibility to liquefaction by assigning a score of 2.0 in the case of high susceptibility to liquefaction and a value of 10.0 in the case of very high susceptibility. These relatively high score values are attributed to F_{site} to reflect the destructive capacity of this particular phenomenon and its impact on structures (Priestley, 1996).

4 APPLICATION TO A BRIDGE NETWORK AND DISCUSSION

To illustrate the impact of considering seismically induced site effects on the prioritization of bridges, the modified SRI described in the previous section is computed for 450 bridges in the region of study. Using a GIS platform, susceptibility levels to amplification, ground mass movements (landslides and rock falls) and liquefaction extracted at bridges sites are integrated in the computation of the respective SRI. It should be emphasized that to preserve confidentiality of the network analyzed, bridges from an existing network were relocated. The new geographical locations of the bridges are as shown on Figure 4, in an area where the seismic hazard index, F_{hazard}, is established to vary between 1.52 and 1.73.

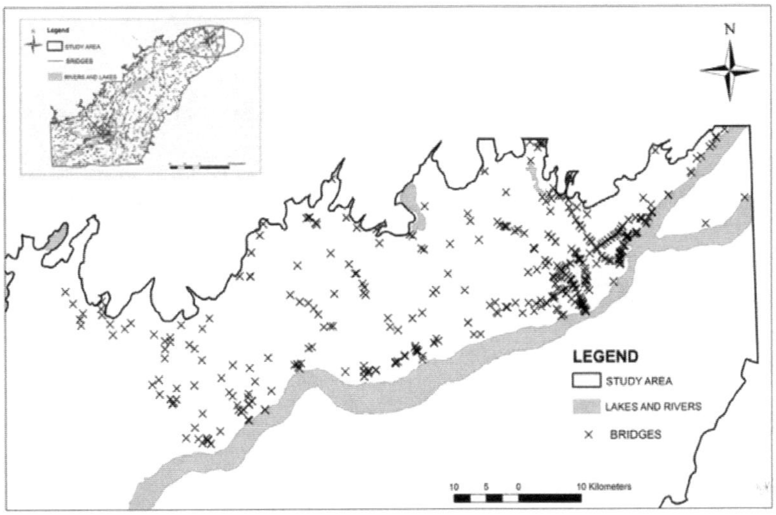

Figure 2: Geographical distribution of database bridges

The seismic risk indices were computed without and with consideration of site effects using susceptibility maps to amplification, ground mass movement and liquefaction (Figures 1 to 3). The SRI varies from 6 to 44 when taking none of the site effects. When considering all site effects, 186 bridges (41.3%) have their SRI increasing above 50, which is considered as higher priority. Furthermore, 11 (2.5%) bridges are located on sites with high susceptibility to liquefaction (F_{site} = 2) and their SRI reach the maximum score of 100.

In order to better represent the three aspects of the seismic risk (hazard, vulnerability and socioeconomic value of bridges), a socioeconomic index (SEI) is also assigned to bridges according to their importance on the network. This index varies between 0 and 100 and a value higher than 50 indicates an important bridge for safety and evacuation purposes. Both seismic risk and socioeconomic indices are combined in 4 categories to prioritize bridges for planning emergency measures or to define mitigation interventions (see Table 3).

Table 3: Category of prioritization according to values of seismic risk index and socioeconomic index

Category of prioritization	Seismic Risk Index	Socioeconomic Index (SEI)
I	SRI ≥ 50	SEI ≥ 50
II	SRI ≥ 50	SEI < 50
III	SRI < 50	SEI ≥ 50
IV	SRI < 50	SEI < 50

Figure 5 shows the distribution of bridges within the 4 categories of prioritization when SRI are computed without and with consideration of site effects using susceptibility maps. The results show that 19.3% of the bridges move to category I and 22.0% to category II, when taking into account the susceptibility to the 4 site effects (amplification, landslides, rock falls and liquefaction).

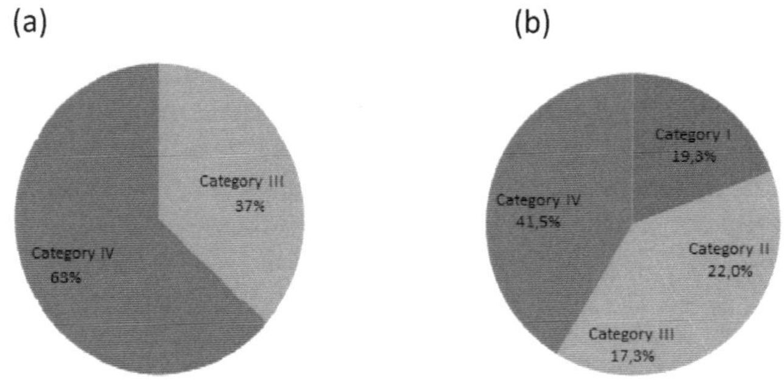

Figure 3: Distribution of the bridges into categories of prioritization: (a) without site effects, (b) with site effects

New indices are well distributed in the 4 categories of prioritization and allow a better classification of bridges taking into account all the components of the seismic hazards for the region of study. The weight of the structural vulnerability in the seismic risk index is preserved, which insure that a bridge with no structural deficiencies will be assigned a maximum SRI of 50 if exposed to a maximum seismic hazard.

5 CONCLUSION

Seismic site effects present in the region of study are amplification, liquefaction, landslides and rock falls. Their susceptibility levels were previously mapped by documenting and characterizing the general context of geomorphology, geology and hydrography. This approach to mapping site effects is conducted on a GIS platform and susceptibility levels can be easily extracted at bridge sites. These maps were then used to illustrate how seismically induced site effects could be integrated into seismic assessment scoring procedures for bridges using a GIS platform. The modified seismic risk index was then applied to a network of 450 bridges in the province of Quebec. Ground mass movements' susceptibilities are included in the new seismic risk index according to set theory. Liquefaction is also taken into consideration by increasing the site index significantly to reflect the destructive capacity of this particular phenomenon and its impact on structures. Results show that integrating susceptibility to seismically induced site effects into a scoring procedure allows a better identification of the most vulnerable structures which lead to improvement of the prioritization process. A GIS approach to susceptibility mapping increases the efficiency of scoring methods in the absence of site-specific information.

Acknowledgements

The authors wish to acknowledge the financial support provided by École de Technologie Supérieure and Natural Resources Canada for the scientific contribution to the project. Thanks also to Peter Quinn for giving us the access to his landslides inventory data base and to the Ministère des Transports, de la Mobilité durable et de l'Électrification des transports for information on its bridges.

References

Basöz, NI, et AS Kiremidjian. 1995. « Prioritization of bridges for seismic retrofitting ». *Technical Report The National Center for Earthquake Engineering Research*, nº 95-0007.

Braganza, S, GM Atkinson, H Ghofrani, B Hassani, L Chouinard, P Rosset, D Motazedian et J Hunter. 2016. « Modeling site amplification in eastern Canada on a regional scale ». *Seismological Research Letters.*

Davi, D, M Kahan, F Légeron, P Marchand, B Portier, J Resplendino, P Schmitt, C Thibault et A Vivier. 2011. « SISMOA: Evaluation préliminaire du risque sismique sur les ouvrages d'art existants ». *Service d'étude sur les transports, les routes et leurs aménagements.*

Davi, D, et P Schmitt. 2003. « SISMOA-Une methode d'estimation rapide de la vulnerabilite des ponts aux seismes ». *OUVRAGES D'ART*, n° 43.

Farr, TG, PA Rosen, E Caro, R Crippen, R Duren, S Hensley, M Kobrick, M Paller, E Rodriguez et L Roth. 2007. « The shuttle radar topography mission ». *Reviews of geophysics,* vol. 45, n° 2.

Farzam, A, M-J Nollet et A Khaled. 2016. « Integration of site conditions information using geographic information system for the seismic evaluation of bridges ». In *Resilient Infrastructure.* (London Ont). Canadian Society for Civil Engineering: CSCE / SCGC.

Farzam, A, M Nollet et A Khaled. 2018. « Susceptibility Modelling of Seismic Induced Effects (landslides and rock falls) integrated to rapid scoring procedures for bridges using GIS tools for the Lowlands of the Saint-Lawrence Valley ». *Geomatics, Natural Hazards and Risk (In Review)*

FEMA. 2012. *Hazus–MH: Multi-hazard Loss Estimation Methodology Earthquake Model.* Technical manual, Federal Emergency Management Agency,National Institute of Building Science Washington,D.C,USA.,2012.

Gouvernement du Canada, RNC. 2004. « Géologie urbaine et environnementale de la vallée du Saint-Laurent - Épaisseur des dépôts meubles».< http://ftp.geogratis.gc.ca/pub/nrcan_rncan/vector/geology/urb-geo-St.Lawrence/thickness.shp.zip >.

Gouvernement du Canada, RNC. 2010. « Compte rendu de séisme (2010-06-23) ». < http://www.seismescanada.rncan.gc.ca/pprs-pprp/pubs/GeoFacts/GEOINFO_ValdesBois2010.pdf >.

Government of Canada. 2018a. « Transport networks in Canada - CanVec – Transport Features ». < http://ftp.maps.canada.ca/pub/nrcan_rncan/vector/canvec/shp/Transport/ >.

Government of Canada, NRC. 2018b. « Earthquakes Canada, GSC, Earthquake Search (On-line Bulletin), ». < http://earthquakescanada.nrcan.gc.ca/stndon/NEDB-BNDS/bull-eng.php >.

Halchuk, SC, JE Adams et TI Allen. 2015. *Fifth generation seismic hazard model for Canada: grid values of mean hazard to be used with the 2015 National Building Code of Canada.* Commission géologique du Canada, Dossier public 7893, 26 p. < https://doi.org/10.4095/297378 >.

Hida, H. 2009. « Évaluation de la vulnérabilité sismique des ponts et viaducs ». Montréal, École de technologie supérieur, 175 p. Primo.

Kim, SH. 1993. *A GIS-based regional risk analysis approach for bridges against natural hazards.* State university of New York at Buffalo, 396 p.

Kiremidjian, A, J Moore, YY Fan, O Yazlali, N Basoz et M Williams. 2007. « Seismic risk assessment of transportation network systems ». *Journal of Earthquake Engineering,* vol. 11, n° 3, p. 371-382.

Lamontagne, M. 2002. « An overview of some significant eastern Canadian earthquakes and their impacts on the geological environment, buildings and the public ». *Natural Hazards,* vol. 26, n° 1, p. 55-68.

Leboeuf, D, D Perret, M-J Nollet, L Lamarche, M Nastev et M Parent. 2013. *Microzonage sismique des Villes de Québec-Ancienne-Lorette et Réserve indienne de Wendake (Catégories d'emplacement).* Commission géologique du Canada, Dossier Public 6875, Échelle 1:50,000, 1 feuille, doi : 10.4095/292640.

Lefebvre, G, D Leboeuf, P Hornych et L Tanguay. 1992. « Slope failures associated with the 1988 Saguenay earthquake, Quebec, Canada ». *Canadian Geotechnical Journal,* vol. 29, n° 1, p. 117-130.

Lemaire, A. 2013. « Étude de sensibilité des paramètres pour l'évaluation de la vulnérabilité sismique des ponts en vue de la calibration de la méthode d'évaluation du Québec ». Montréal, École de technologie supérieure, 194 p. Consulté le 2016.

Liu, Z. 2001. « Seismic prioritization of highway bridges in Canada ». Montréal, McGill University, 195 p.

Locat, J. 2008. « Localisation et magnitude du séisme du 5 février 1663 (Québec) revues à l'aide des mouvements de terrain ». In *the 4th Canadian Conference on Geohazards: From Causes to Management.* . (Québec). Presse de l'Université Laval.

Motazedian, D, JA Hunter, A Pugin et H Crow. 2011. « Development of a Vs30 (NEHRP) map for the city of Ottawa, Ontario, Canada ». *Canadian Geotechnical Journal,* vol. 48, n° 3, p. 458-472.

MTMDET. 2017. *Rapport annuel de gestion 2016-2017*. Bibliothèque et Archives nationales du Québec: Gouvernement du Québec: MINISTÈRE DES TRANSPORTS, DE LA MOBILITÉ DURABLE ET DE L'ÉLECTRIFICATION DES TRANSPORTS 154 p. < https://www.transports.gouv.qc.ca/fr/ministere/organisation/rapport-annuel/Documents/rag-2016-2017.pdf >.

NRCC. 2015. *National Building Code of Canada (NBCC)*. Ottawa, NRCC.

NYSDOT. 2002. *Seismic Vulnerability Manual. Structure Design and Construction Division - Bridge Safety Assurance Unit.* New York State Departement: New York State Departement, 44 p.

Parent, M, O Caron, H Dubé-Loubert et J Rivard. 2018. « Géologie des formations superficielles des Basses-terres centrales du Saint-Laurent et des régions adjacentes (Ottawa à Québec)/Surficial Geology of the Central St ». *Lawrence Lowlands and adjacent regions (Ottawa-Quebec City)]. Geological Survey of Canada, Open File (In Review).*

Paultre, P, G Lefebvre, JP Devic et G Côté. 1993. « Statistical analyses of damages to buildings in the 1988 Saguenay earthquake ». *Canadian Journal of Civil Engineering,* vol. 20, n° 6, p. 988-998.

Priestley, MJN. 1996. *Seismic design and retrofit of bridges* (1996). Wiley-Interscience.

Quinn, P. 2009. « Large landslides in sensitive clay in eastern Canada and the associated hazard and risk to linear infrastructure ». Kingston, Ontario, Canada, Queen's University, 437 p.

Quinn, P, DJ Hutchinson, MS Diederichs, RK Rowe et J Alvarez. 2008. « Susceptibility mapping of landslides in Champlain clay from a digital landslide inventory ». In *Proceedings of the 4th Canadian Conference on Geohazards, Québec City, Que.* p. 20-24.

Rosset, P, M Bour-Belvaux et L Chouinard. 2015. « Microzonation models for Montreal with respect to VS30 ». *Bulletin of Earthquake Engineering,* vol. 13, n° 8, p. 2225-2239.

Small, EP. 1999. « Development of a methodology for the integration of seismic risk mitigation activities in project and network level bridge management systems ». New York, State University of New York at Buffalo.

Theilen-Willige, B. 2010. « Detection of local site conditions influencing earthquake shaking and secondary effects in Southwest-Haiti using remote sensing and GIS-methods ». *Natural Hazards and Earth System Sciences,* vol. 10, p. 1183-1196.

Tinawi, R, S Tremblay et A Filiatrault. 1993. *Méthode rapide de classement séismique des ponts et viaducs du Québec* (1993). Coll.«Rapport». École Polytechnique de Montréal. Canada.

UNDRO. 1991. *Mitigating Natural Disasters Phenomena, Effects and Options* (1991). New York, 164 p.

Building Tomorrow's Society
Bâtir la Société de Demain

Fredericton, Canada
June 13 to 16, 2018 / *13 au 16 juin, 2018*

LANDSLIDE GENERATED TSUNAMIS: NEW LABORATORY EXPERIMENTS AND THEORETICAL DEVELOPMENTS

Ryan P. Mulligan[1,2], W. Andy Take[1], and Gemma K. Bullard[1]
[1] Department of Civil Engineering, Queen's University, Canada
[2] ryan.mulligan@queensu.ca

Abstract: Landslide generated tsunamis are major hazards on mountainous coasts that can cause natural disasters due to vast inundation of coastal communities. To investigate the complex process of tsunami generation by landslides, experimental observations are obtained by releasing material and capturing the wave behavior with high speed digital cameras and wave probes in a large laboratory landslide flume. These detailed observations are used to validate a new expression that describes the wave amplitude in the near-field zone. The theoretical relationship that governs the maximum wave amplitude is derived by considering the transfer of momentum from the sliding mass to the water. This work provides a new method for predicting the landslide wave amplitude that is easily applied in practice. The results can be used in the near-field evaluation of tsunamis or to define boundary conditions for numerical models to determine the extent of wave propagation and run-up from these major coastal geo-hazards.

1 INTRODUCTION

Tsunamis have caused some of the worst natural disasters in the world, with energetic waves that can inundate vast areas of land with hazardous debris-laden flows. Recent events have resulted in high loss of human life and high costs to rebuild infrastructure such as the earthquake-generated tsunami in Japan in 2011 that resulted in over 15,000 casualties and cost over $200B (Mimura et al. 2011). Tsunamis triggered by landslides in mountainous regions have immense local impacts with run-up over 100 m above shorelines. Recent examples include the massive slide into Icy Bay Alaska that resulted in run-up over 190 m in 2015 (Lynett et al. 2016), and the rapid rock avalanche into Karrat Fjord in Greenland with 45 Mm^3 of material entered the fjord along less than 1 km of shoreline over a period of only 2–3 min (Gauthier et al. 2018). In the latter case, the huge wave wiped out the small coastal community of Nuugaatsiaq.

To predict the waves generated by landslides, laboratory experiments can be carefully performed to make detailed measurements (e.g., Heller and Hager 2010; Miller et al. 2017; Bullard 2018), and the data used to develop empirical formulae that describe the wave characteristics from fitting parametric expressions to the observed data (e.g., Heller and Hager 2010). Alternatively, the data can be used to validate analytical solutions. For example, Mulligan and Take (2017) considered the momentum flux from a landslide to a water body, and derived idealized theoretical relationships for the maximum wave amplitude in the near-field zone. The problem is depicted in Figure 1, where the tsunami generation process necessitates understanding of the landslide, impact with water, momentum transfer, and near-field wave transformation due to wave breaking.

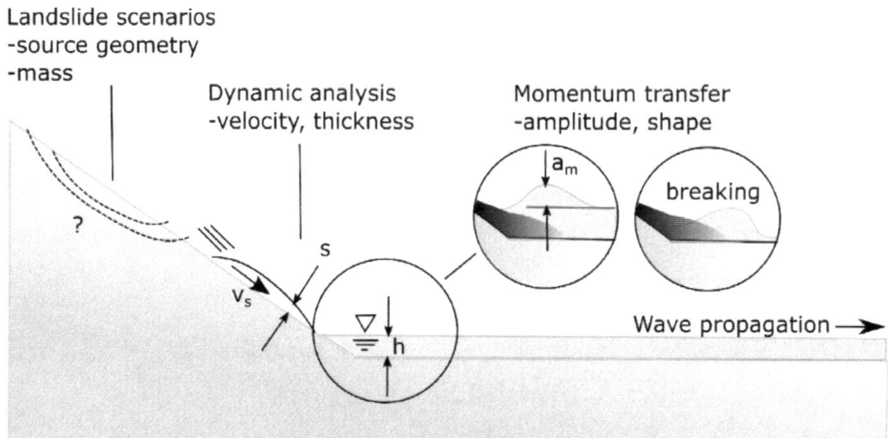

Figure 1: Conceptual diagram of the landside tsunami generation process: landslide, impact with water, momentum transfer, wave transformation and propagation.

In the present paper we describe the results of new laboratory experiments performed by Bullard (2018) in Section 2, discuss results from a predictive equation developed by Mulligan and Take (2017) in Section 3, and provide conclusions in Section 4 on the generation of tsunamis from landslides.

2 EXPERIMENTAL OBSERVATIONS

Laboratory experiments are performed in a landslide flume consisting of an 8.23 m long slope inclined at α = 30° to the horizontal to gravitationally accelerate landslides into a 33.0 m long and 2.1 m wide horizontal wave flume described in detail by Miller et al. (2017). Material is released from a source volume box at the top of the slope, accelerates down the landslide slope and impacts the water where it generates waves. The waves transform and propagate along the flume, run-up the slope at the end of the flume, and reflect back.

In the present study we examine a sub-set of the data collected by Bullard (2018). In these experiments, a triangular source volume V of water (ρ_s = 1000 kg/m^3) is used to generate a highly mobile slide with high impact velocity. The landslide and near field wave properties are observed using a system of high-speed cameras. The water surface is also measured using nine capacitive probes (P1-P9) along the flume at x = [2, 3, 6, 9, 11, 17, 21, 25, 33 m] that sample at 100 Hz. The water depth h is varied from 0.15-0.65 m and in total 41 tests are conducted, with a sub-set of preliminary data from 11 tests presented here.

The range of source volumes tested include V = [0.1, 0.2, 0.3, 0.4 m^3], which results in a range of landslide velocities v_s = 5.08-6.51 m/s and slide thicknesses s = 0.02-0.06 m at impact with the reservoir. The submarine part of the slide has effective length L_e= 0.8-2.4 m and effective time Δt_e= 0.44-0.83 s from the point of impact until the wave speed exceeds the decelerating slide speed and the wave detaches. An example of wave generation from the point of impact to wave detachment from the submarine part of the slide is shown in Figure 2 for the case with V = 0.3 m^3 and h = 0.30 m. The corresponding time series of water level elevations measurements at the wave probes are shown in Figure 3.

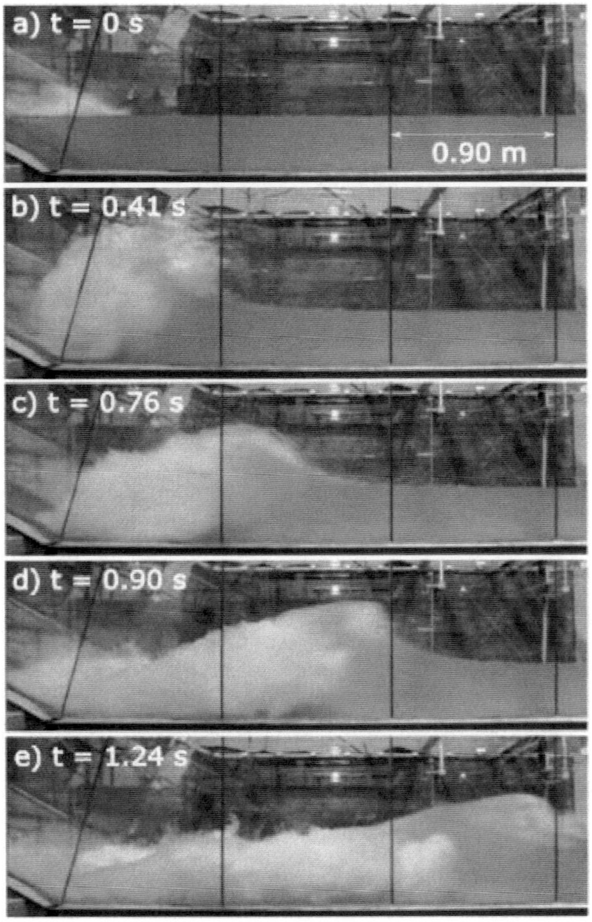

Figure 2: Near-field image time series of landslide wave generation in the laboratory.

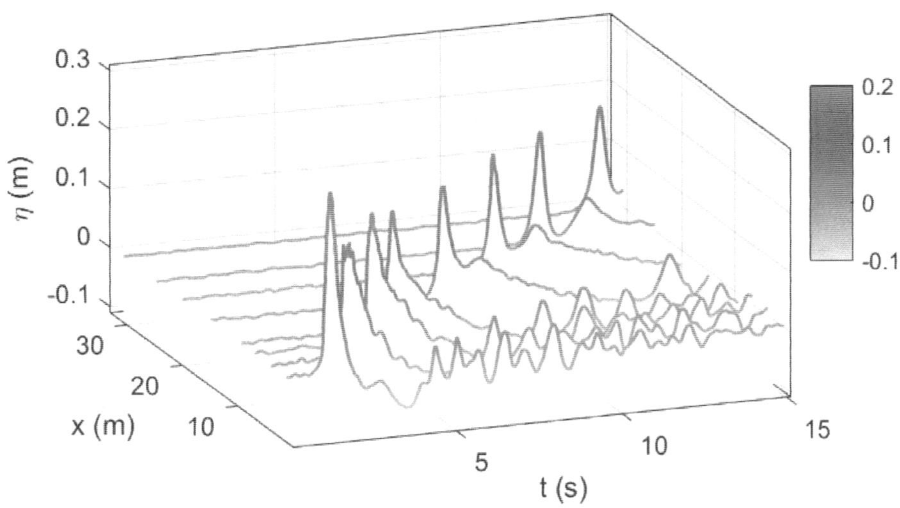

Figure 3: Time series of water level (η) observations at wave probes P1-P8 during landslide wave transformation and propagation.

3 THEORETICAL CONSIDERATIONS

Mulligan and Take (2017) consider a 1-dimensional approach to analyze the state of the bulk fluid at landslide impact. In this scenario, momentum that is transferred to the water is expressed by the hydrostatic pressure gradient generated by a near-instantaneous vertical water level displacement over a short time scale Δt_e and length scale L_e. The initial pressure gradient is evaluated from the change in water surface elevation $\eta(x,t)$, that has a maximum positive wave amplitude a_m above the still water depth h between two points in the landslide impact zone and the zone unaffected by the slide. The analysis yields the equation:

[1] $$a_q = \sqrt{h^2 + \frac{2\rho_s s v_s cos\alpha L_e}{\rho g \Delta t_e}} - h$$

where a_q denotes the maximum near-field wave amplitude derived from a quadratic equation, ρ is the fluid density and g is gravitational acceleration.

The observations and predictions of relative wave amplitude (a_m/h) are shown at several locations along the flume in Figure 4, where perfect agreement with Eq. [1] results in points along the 1:1 line. At the near-field wave probe (P1 at x =2 m) the agreement is good but not perfect, however at the far field sensors (e.g., P7 at x =22 m) the maximum wave amplitude is very well predicted for non-breaking waves with low values of a_m/h. This is due to depth-limited wave breaking, an important consideration for these highly turbulent flows, which causes an immediate reduction in amplitude as a function of h. The effect of breaking is indicated by the change in relative amplitude from near-field to far-field sensors. The overall agreement between data and model results suggests that the range of validity for the theoretical equations developed for granular landslides can be extended to slides with higher mobility. The predictive framework can therefore be a useful method of defining the boundary conditions for numerical models used to simulate the water surface (e.g., Mulligan et al. 2016) and flow velocity (e.g., Mulligan et al. 2018) evolution of landslide tsunamis.

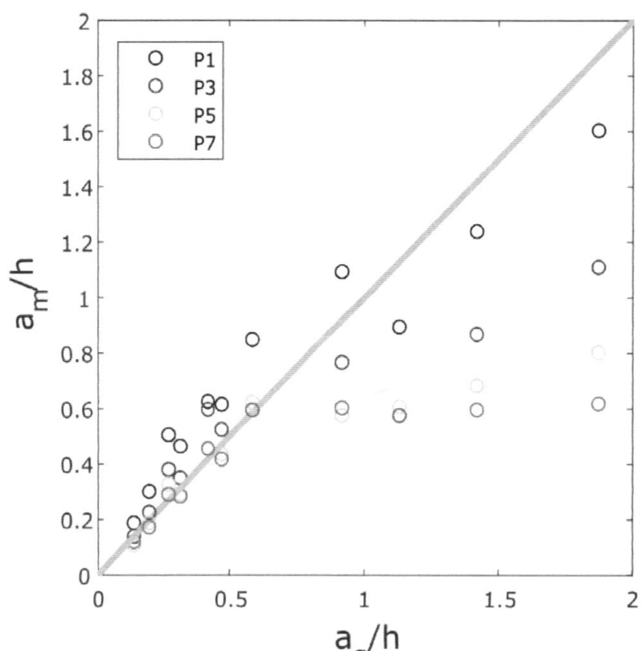

Figure 4: Comparison of observations (circles) for sub-set of tests from Bullard (2018) with the theoretical relationship for momentum balance from Eq. [1] at selected wave probes.

4 SUMMARY AND CONCLUSIONS

New insight into the generation of waves by landslides is acquired by conducting experiments in a large scale flume and developing a momentum-based predictive framework. In this paper, we extend the range of validity for a theoretical equation developed for granular landslides (Mulligan and Take 2017) to slides with higher mobility based on preliminary data from new experiments (Bullard 2018). The predictions of maximum wave amplitude are validated by tests for the highly mobile slides, and the results are in good agreement with observations except where the amplitude is reduced due to depth-limited wave breaking. The overall agreement suggests that the range of validity for the momentum-based equation developed for granular landslides can be extended to slides with higher mobility.

Acknowledgements

Funding for the research has been provided by the Natural Science and Engineering Research Council of Canada (NSERC) Discovery Grant programs of the first and second author. Funding for the design and construction of the large-scale landslide flume from the Canada Foundation for Innovation (CFI) is gratefully acknowledged.

References

Bullard, G.K. 2018. Wave characteristics of tsunamis generated by landslides of varying size and mobility, *Ph.D. Thesis*, Queen's University, Kingston, Canada.

Gauthier, D., Anderson, S.A., Fritz, H.M. and Giachetti, T. 2018. Karrat Fjord (Greenland) tsunamigenic landslide of 17 June 2017: initial 3D observations, *Landslides*, **15**:327–332.

Heller, V. and Hager, W.H. 2010. Impulse product parameter in landslide generated impulse waves, *Journal of Waterway, Port, Coastal and Ocean Engineering*, **136**(3): 145–155.

Lynett, P.J., Weiss, R., Mattox, A., Skanavis, V., Ayca, A., Tang, H., Higman, B.M., Stark, C.P. and Keen, A. 2016. The tsunami generated by the October 17th, 2015 Taan Fjord landslide, *American Geophysical Union Fall Meeting 2016*, San Francisco, USA, NH51D-05.

Miller, G.S., Take, W.A. and Mulligan, R.P. 2017. Tsunamis generated by long and thin granular landslides in a large flume, *Journal of Geophysical Research Oceans*, **122**: 1-16.

Mimura, N., Yasuhara, K., Kawagoe, S., Yokoki, H. and Kazama, S. 2011. Damage from the great east Japan earthquake and tsunami-a quick report. *Mitigation and Adaptation Strategies for Global Change*, **16**: 803–818.

Mulligan, R.P., Take, W.A. and Miller, G.S. 2016. Propagation and run-up of tsunamis generated by gravitationally accelerated granular landslides, *Proc. Coastlab16*, Ottawa, Canada, May 10-13.

Mulligan, R.P. and Take, W.A. 2017. On the transfer of momentum from a granular landslide to a water wave. *Coastal Engineering,* **125**: 16-22.

Mulligan, R.P., Take, W.A. and Bullard, G.K. 2018. Modelling tsunami from landslides with different mobility, *Proc. Coastlab18*, Santander, Spain, May 22-26.

Building Tomorrow's Society
Bâtir la Société de Demain

Fredericton, Canada
June 13 to 16, 2018 / *13 au 16 juin, 2018*

MODELLING OF MEANDERING RIVER DYNAMICS IN COLOMBIA: A CASE STUDY OF THE MAGDALENA RIVER

Guarín, L. I[1], Vargas, J[1,2], Barreneche, S[1], and Chang, P.[1]
[1] Universidad Nacional de Colombia Sede Manizales, Colombia
[2] liguarinm@unal.edu.co

Abstract: The Magdalena River located in Colombia is a large 1550 km long river basin draining the country from South to North with an average width of 275m and an annual mean discharge of 7200 m^3/s. The river displays high water level fluctuation and is characterized by a series of meanders midway upriver. The township of La Dorada has been affected over the years by serious flooding as the meander in this region is evolving at a steady pace endangering a great number of neighborhoods. This study has strived to correctly model the flow characteristics of the river in this region in order to evaluate various scenarios and mitigation measures related to erosion control and provide decision makers with a forecasting tool. Two field campaigns were completed over the dry and rainy seasons including extensive topographical and channel survey using Topcon GR5 DGPS and Sontek River Surveyor ADCP. Also, in order to characterize the erosion process occurring through the meander, extensive suspended and river bed samples were retrieved as well as soil perforation over the banks to evaluate maximum scour depth. Using a digital elevation model the field data was incorporated in a 2DH flow model using the IBER freeware based on the finite volume method in a non-structured mesh environment. Model calibration was carried out comparing available historical data of a nearby hydrologic gauging station. Although the model was able to effectively predict overall flow processes in the region, its spatial characteristics and limitations did not allow for an accurate representation of erosion processes occurring over the specific endangered bank areas and dwellings. Field data indicates that, as may be expected, a helical flow is occurring through the meander across the channel width. The IBER model was unable to identify or represent such flow condition. It is unclear at this stage if this situation is related to channel bed characteristics, effects of turbulence or non-hydrostatic pressure condition across the channel. Furthermore, the rapidly changing channel cross section as a consequence of severe erosion has hindered the model's ability to provide decision makers with a valid up to date planning tool.

Key words: meandering river, erosion processes, secondary helicoidal flow, vortex flow.

1 INTRODUCTION

The Magdalena River, the most important in Colombia, has been of great relevance in the national development of the country stretching over 1550 km with an annual mean discharge of 7200 m3/s, the river displays high water level fluctuation and is characterized by a series of meanders midway upriver. Approximately 80% of Colombia's population lives within the limits of the river basin and a large part of the country's economy has been developed around it. (Sociedad de Acueducto Alcantarillado y Aseo de Barranquilla S.A E.S.P 2004)

La Dorada Township is located amongst the central range of the Andes Mountains and has historically been characterized by its social, economic and geographical-spatial relevance connecting the Colombian capital with the country's southern districts. The foundation of the city dates back to the building of the railway in the early 1897 and the implementation of river transport using barges. The city from its origin has been a main commercial hub for the whole central region. (Angulo 1995)

The hydrodynamic characteristics and the severe erosion process currently occurring along the urban perimeter of the city along the banks of the river is related to the development and progression of a series of meanders. The channel is characterized by a high sinuosity and the presence of two alternate curves of the channel way forming a semi regular basis upstream of the city. Furthermore, the construction of an artificial reservoir and run-of-the-river hydroelectric power stations upstream may be significatively affecting the morphology of the river.

This paper reports on the analysis, completed field work and modeling process undertaken in an attempt to correctly predict and explain the flow characteristics of the river and erosion processes in this region. The study also endeavoured through the modelling tool to assess the seriousness of the situation, propose mitigation measures and generally provide decision makers with a forecasting tool.

2 CONTEXT OF THE STUDY

The Colombian Andes are located in the intertropical convergence zone (ITCZ) and display a typical bimodal weather pattern, causing intense precipitation over two rain seasons usually in April and November of each year (Narváez and León 2001).

Over the past decades, La Dorada has suffered a large number of floods and a persistent shift of the river bank westwards, as well as severe erosion in a neighborhood called El Conejo (Aristizabal 2013).

In this area, limited urban planning and uncontrolled zoning has allowed for the construction of impoverished dwellings along the riverbank. The latter have been exposed to repeated flooding, and the constant danger of collapse caused by bank erosion (Amaya et al. 2013). Also, environmental concerns with the presence of debris and rubbish over the area has exacerbated the situation. Figure 1 shows the situation along the bank in this area in June of 2017.

Figure 1. West bank of the Magdalena River, La Dorada. (La patria 2014).

Using photogrammetric analysis over the past 50 years through Google Earth and aerial photos of the area (1969, 2007, 2009, 2010, 2013, 2015), it was estimated that the center of curvature of the meander is moving in a south western direction at a rate of approximately 5 meters annually. Figure 2 presents the general layout of the channel with respect to the urban built area.

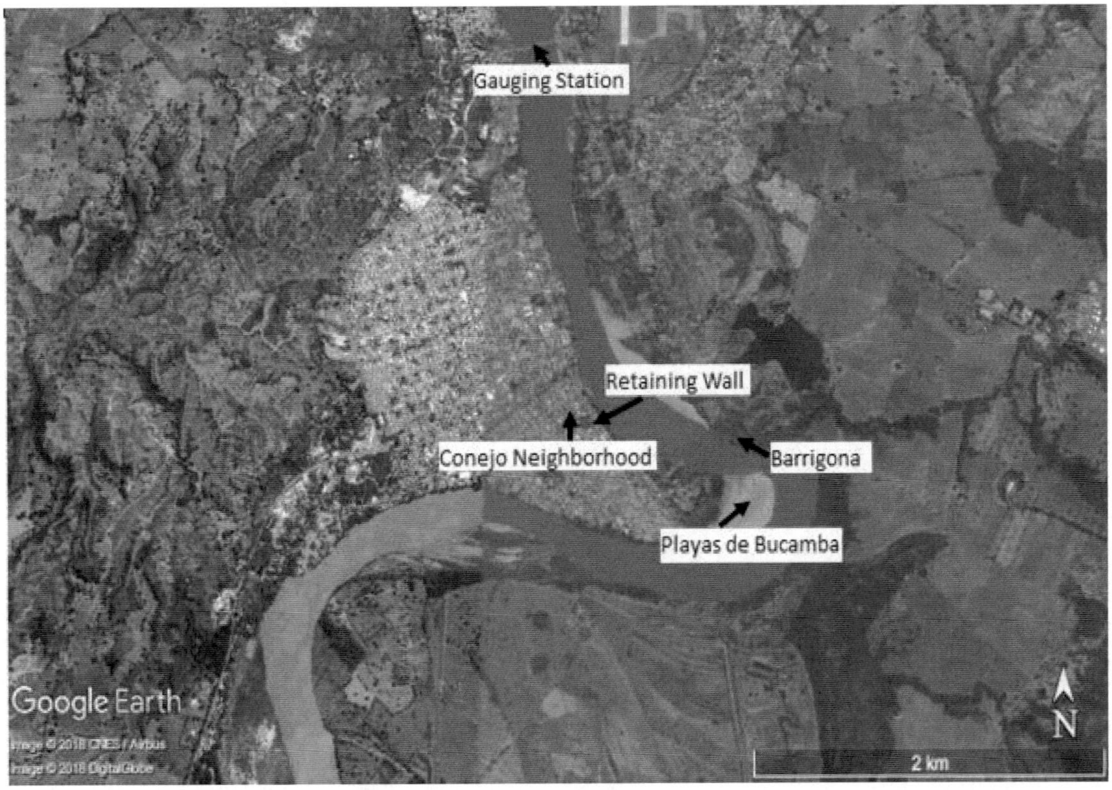

Figure 2. Meander of the Magdalena River along La Dorada Township (Google Earth)

The presence of the meander generates downriver from the bend a helicoidal flow across the channel width. The flow dynamic through a meander may be explained considering the balancing forces acting along the channel way which results in flow recirculation from the outer bank upwards and towards the inward part of the channel curve hence the observed particular flow pattern (Blanckaert 2010). The presence of higher shear forces along the outer bank results in a progressive erosion process over the outer bend and deposition on the opposite bank (Blanckaert and de Vriend 2003).

In order to characterize the Magdalena River in the sector of La Dorada, two field campaigns were completed in January and March of 2017, allowing for the recollection of topographical and bathymetric data using a Topcon GR5 DGPS and a River Surveyor ADCP. The field campaigns also included the collection of water quality, river bed and soil samples along the banks to determine the concentration of suspended sediment transport, characterize the channel roughness, and establish the in-situ soil granulometry.

Furthermore, a hydrological analysis was completed to establish the historical flow rates and water levels based on a nearby gauging station located in Puerto Salgar.

Finally, a hydrodynamic flow model of the channel along the "El Conejo" neighborhood of La Dorada using the IBER 2DH freeware was prepared in order to better understand the erosion process occurring in this sector.

3 MODEL MATHEMATICAL FORMULATION

The analysis and study of open channel flow dynamics for river applications is well established. The overall spatial characteristics of rivers, i.e. its length, to depth, to width ratio generally allows one to correctly disregard processes occurring in the vertical or transverse dimensions, thus imposing hydrostatic pressure conditions and considering solely a 1D flow model along the river length (Chow 1959). When channel width is also considered, for example when analysing a short stretch of a river over a larger channel width, a 2DH is preferable, allowing for the analysis of transverse flow processes while still averaging flow conditions over the channel depth (Wesseling 2001).

Through a calibration process an accurate flow model may generally be developed allowing for channel study and extrapolation of various scenarios.

The IBER software flow model is based on the usual Saint Venant two-dimensional equations taking into consideration the effects of turbulence and surface friction based on the usual conservation of mass (Eq. 1) and conservation of momentum (Eq. 2 and 3) equations in the usual integral form and a (x,y,z) coordinate system (Corestein, Bladé, and Niñerola 2014).

Mass Conservation:

$$[1]\quad \frac{\partial h}{\partial t}+\frac{\partial hU_x}{\partial x}+\frac{\partial hU_y}{\partial y}=0$$

Where h corresponds to water depth and U_x, U_y are the horizontal velocities averaged over depth.

Conservation of Momentum:

Equations 2 and 3 present the conservation momentum equation.

$$[2]\quad \frac{\partial}{\partial t}(hU_x)+\frac{\partial}{\partial x}\left(hU_x^2+g\frac{h^2}{2}\right)+\frac{\partial}{\partial y}(hU_xU_y)=-gh\frac{\partial Z_b}{\partial x}+\frac{\tau_{s,x}}{\rho}-\frac{\tau_{b,x}}{\rho}+\frac{\partial}{\partial x}\left(v_th\frac{\partial Ux}{\partial x}\right)+\left(v_th\frac{\partial U_y}{\partial y}\right)$$

$$[3]\quad \frac{\partial}{\partial t}(hU_y)+\frac{\partial}{\partial x}(hU_xU_y)+\frac{\partial}{\partial y}\left(hU_y^2+g\frac{h^2}{2}\right)=-gh\frac{\partial Z_b}{\partial y}+\frac{\tau_{s,y}}{\rho}-\frac{\tau_{b,y}}{\rho}+\frac{\partial}{\partial y}\left(v_th\frac{\partial U_y}{\partial x}\right)+\left(v_th\frac{\partial U_y}{\partial y}\right)$$

Where g the acceleration of gravity, Z_b the channel bed, τ_s free surface friction due to the wind using the Van Dorn's equation, τ_b bottom friction evaluated through Manning's equation and v_t turbulent viscosity calculated using a specific turbulence model for the shallow water equation. (Cea et al. 2016; Beffa 2008; Chow, Maidment, and Mays 1994)

3.1 Modelling Process

3.2 Model

Based on the finite volume method, the computational domain was divided into a finite number of nonoverlapping subdomains, in this case triangles and quadrilaterals and related nodes. The finite volume scheme is derived by applying integral balance equations to every cell that transforms the set of partial differential equations into a system of linear algebraic equations. Temporal discretization was completed using Euler's explicit scheme. (Blazek 2001; Moukalled, Mangani, and Darwish 2016; López, Alavez, and Hernández 2009).

A steady flow analysis was performed using the free software IBER 2.3.2. The domain mesh of 19283 elements and 9920 nodes was created using ArcGis software 10.2.2 based on the digital elevation model (DEM) obtained from the two field campaigns and was then imported as an ASCII file to the IBER environment with a tolerance value of 0.1.

The hydrodynamic boundary conditions were selected at the inlet and outlet, given the unique affluent in the study.

Although flow turbulence may be of limited impact as indicated by Iber's Manual (FLUMEN 2010) in determining channel mean flow velocity, a mixture length turbulence model was nonetheless implemented.

3.3 Bathymetric and topographic survey

The bathymetric and topographic surveys of the transversal and longitudinal sections of the river were carried out over an 800m stretch, 200 meters upstream of “El Cerro La Barrigona” to approximately 200 meters downstream of the retaining wall located in “El Conejo” neighborhood using an ADCP M9 River Surveyor Sontek and a Topcon GR5 DGPS. The field data mapping survey was then analyzed to produce the digital elevation model (DEM) implemented in IBER.

3.4 Hydrological frequency analysis

The hydrological study of the Magdalena watershed was not undertaken as it would involve the physiographic analysis of an extensive area. Given the magnitude of the river basin, the hydrological study was based on measurements taken from the Puerto Salgar gauging station operated by the *Instituto de estudios ambientales e hidrometeorológicos* (IDEAM 2017) #23037010 from 1956 up until 2014.

The hydrological analysis determined the multi-annual, annual and monthly flow rates and stages of the Magdalena over various return periods. The IBER model calibration was completed based on the observed historical data series.

3.5 Sediment transport

Particle removal and transport is related to the assessment of the limit tractive force characteristics over the river's bed and the channel embankments. Granular transport is function of grain size, the specific weight and concentration of suspended sediments present in the flow. The characteristic grain size of bed and suspended sediments was determined through the two field investigation. The IBER model allowed for the evaluation of corresponding tractive forces along the channel bank.

3.6 Channel Roughness

Channel roughness was assessed based on Manning's roughness coefficient as determined by recollected data during the two field campaigns. Table 1 shows the established roughness Manning coefficient values and Figure 3 the general Manning's values distribution over the banks and the main channel.

Table 1. Manning's values.

Location	Manning *n*	Color
Main Channel	0.014	Yellow
Right Bank	0.034	Green
Left Bank	0.027	Blue

Figure 3. Distribution of Manning's values.

3.7 Simulated Events

Two specific simulations were specifically examined. In the first case the model simulated the flow conditions as were observed during the first field campaign of January 30th, 2017, observing specifically flow conditions in front of the retaining wall of the "El Conejo" neighborhood. The second simulation examined the historical 2012 maximum discharge over the same area. Table 2 details both events specifically.

Table 2. River levels and flow rates

Event	Water Level El Conejo retaining wall (m)	Flow rate (m^3/s)
1. January 30th, 2017	5.91	1306.22
2. Registered maximum Discharge 2012	8.30	2025.00

4 RESULTS

Simulation results are presented in Figure 4.

Simulation 1 **Simulation 2**

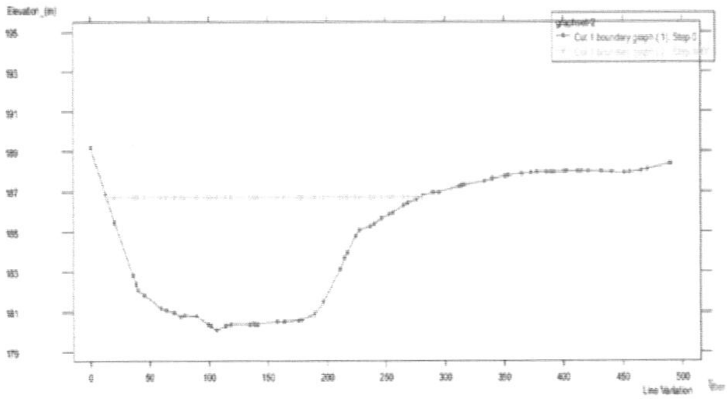

Fig 4a. Channel cross section Q = 1306.2 m^3/s.

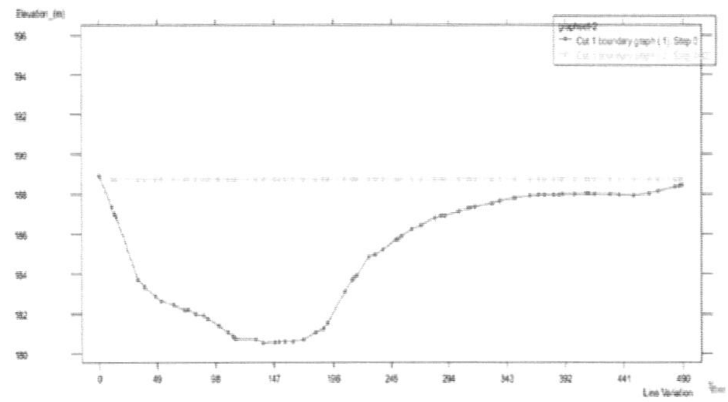

Fig 4b. Channel cross section Q = 2025.00 m^3/s.

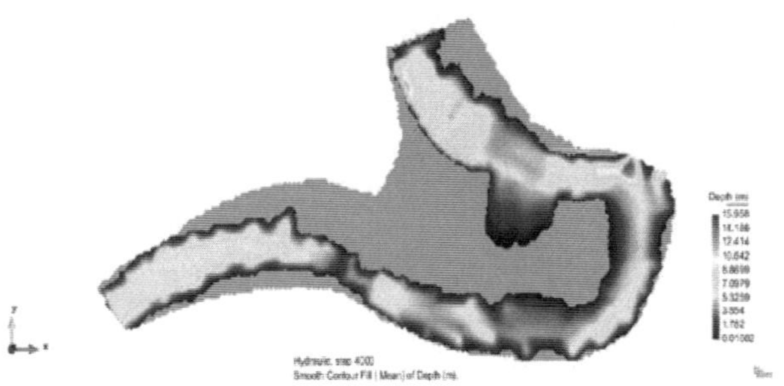

Fig 4c. Depth flooded area Q =1306.2 m^3/s.

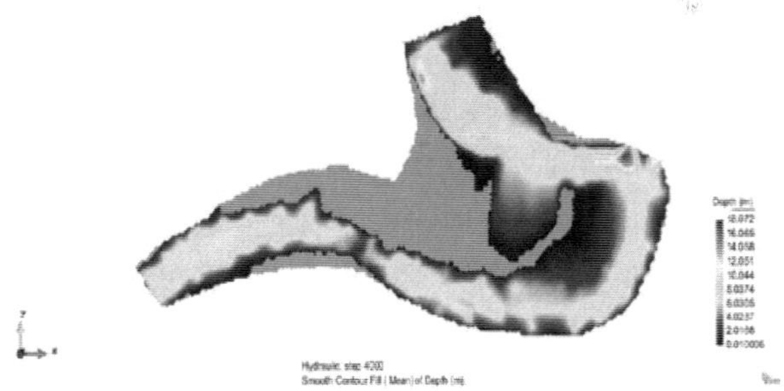

Fig 4d. Depth flooded area Q = 2025.00 m^3/s.

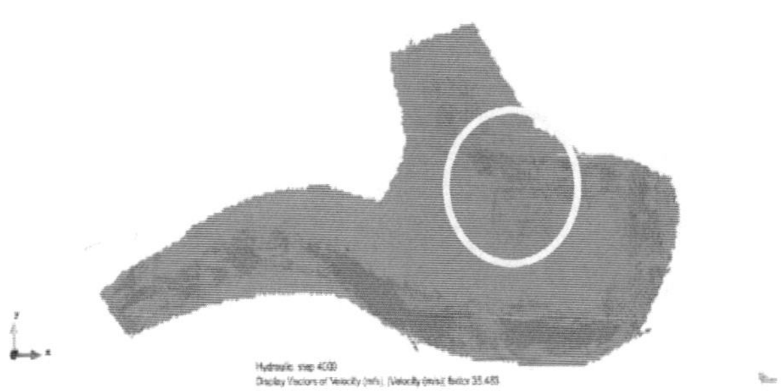

Fig 4e. Velocity vector field Q = 1306.2 m^3/s.

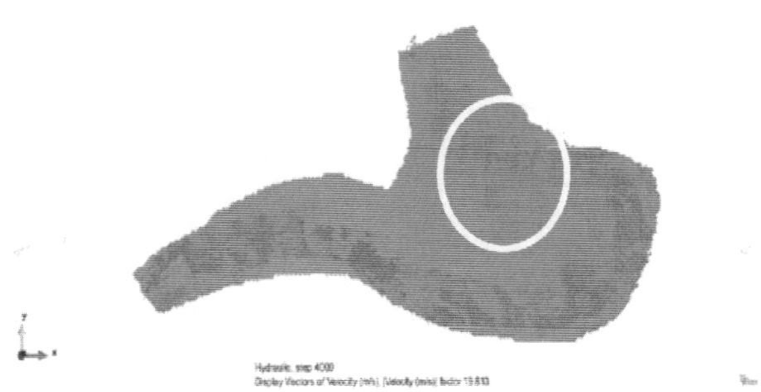

Fig 4f. Velocity vector field Q = 2025 m^3/s.

Figure 4. Simulation results.

Figures 4a and 4b compares the evolution the channel cross section for both flow conditions. In addition, figures 4c and 4d denote the river's floodplain for the two discharge conditions and greater depth can be observed towards the three external curves of the meander. Finally, figure 4e and 4f contrast the two-dimensional velocity vector field where flow acceleration is observed from "El Cerro La Barrigona" towards "El Conejo" neighborhood where the retaining wall is located.

Table 3 highlights the predicted and simulated water depth as considered in front of the retaining wall.

Table 3. Predicted vs Observed water depth

Simulation	Predicted level (m)	Observed level (m)	Agreement
1	6.00	5.91	+2%
2	8.10	8.30	-3%

5 DISCUSSION

The performed simulation correctly predicted the general velocity vector field orientation across the meander, predicting elevated embankment shear stress areas in the neighborhood of "El Cerro de la Barrigona" and "El Conejo" in front of the retaining wall. Predicted water depth was also consistent with observed field data, the model delivering results within a 5% agreement for the simulated conditions.

On the other hand, the model's two-dimensional depth averaged scheme could not explain or detail the erosion process occurring in this area or across the channel width including deposition towards the inner bend. During the two field campaigns the presence of a helicoidal secondary flow was observed acting initially downwards from the outside bend in the region of the retaining wall of El Conejo, then across the channel width and finally upwards. This was believed to be prompting the severe erosion of the bank in this area. This flow conditions were not correctly represented by the model, where the velocity vector field after impacting the outer bank was redirected downstream which did not correspond with field observation. Figure 5 shows the flow unsteadiness observed in front of the "El Conejo" neighborhood retaining wall in July 2017. It is believed that downwards vertical velocities are generating a significant vortex in this area.

Figure 5. Ripples observed in front of the retaining wall, "El Conejo" neighborhood (Vargas July 2017).

As was correctly identified by Blanckaert (Blanckaert and Schnauder 2009) in the case of 1D models, this study also infers that the current 2DH IBER model would only be able to capture the velocity field distribution through channel bends if it can account for non-linear effect between the horizontal and vertical distribution of the flow.

Although it may be argued that the current model is lacking in near bank hydrodynamics to correctly assess flow patterns near the bank, in this case, it is unclear if the structure of the turbulence is dominant or rather if channel bed configuration is the overriding factor in explaining the erosion process.

The model development was also hampered by the difficult site conditions with in some cases limited access to the river bank, limiting bathymetric data and requiring additional interpolation of the DEM. Furthermore, the constant evolution of the river stage and discharge, and the rapid progression of the erosion from January to March hindered field data recollection and required constant DEM update.

This study also highlights the difficulty in calibrating a dynamic channel and the limitations of 2DH open channel flow model. As has been shown although the model may be deemed "calibrated", correctly relating discharge to channel stages, it may be nonetheless inadequate in explaining specific processes, including important erosion processes that may be occurring in the area.

Finally, it may be concluded that the 2DH IBER model, as it is unable to predict the observed helicoidal flow occurring in this region, is of limited value in providing decision makers with a valid up to date planning and management tool.

6 CONCLUSION

The present paper reports on some initial findings concerning the characterization and study of the meander located along the urban perimeter of La Dorada. The current analysis is based on observations supported by two field campaigns leading to a 2DH hydrodynamic model of the river stretch.

Although the IBER model is correctly predicting the general flow field downriver, in this particular case, it is unable to predict the downwelling vertical velocities impinging on the channel bed and causing significant unsteadiness of the water surface in this area. It is believed that the natural river configuration is the main factor in explaining the important erosion processes occurring in this area, although it is unclear at this stage if this situation is related to channel bed characteristics, the effects of turbulence or non-hydrostatic pressure condition across the channel.

This study may highlight how the influence of bed topography may be a dominant factor in sediment transport dynamics. This assessment would be contrary to the generally accepted idea that river flow can be interpreted as solely a one or two-dimensional horizontal problem considering the river's depth, to length, to width ratio. If determined to be true, this situation considerably deteriorates the predictive capacity of 1D or 2DH river model, suggesting that the major weakness of the model is related to the concealed hydrodynamic processes occurring over the channel depth.

It is suggested that a full three-dimensional simulation be implemented in order to evaluate the cross section pressure conditions and improve model prediction of the observed flow conditions, hence better explaining the sediment erosion processes currently occurring.

References

Amaya, Oscar Dario, Maria Carolina Hernández, Maria Carolina Silva, and Juliana Hurtado. 2013. *Río Magdalena: Informe Social, Económico Y Ambiental.* Instituto de Estudios del Ministerio Publico.

Angulo, M.G. 1995. *Ciudades de Caldas. La Dorada Cien Años Su Historia Y Proyección.* La Dorada: Publicaciones Alcaldía Municipal de La Dorada.

Aristizabal, Víctor Mauricio. 2013. Modelos hidrológicos e hidráulicos de zonificación de la amenaza por inundación en el municipio de La Dorada Caldas Corpocaldas 1–134.

Beffa, Cornel. 2008. "2D-Shallow Water Equations Basics - Solutions - Applications," 1–11.

Blanckaert, Koen. 2010. "Topographic Steering, Flow Recirculation, Velocity Redistribution, and Bed Topography in Sharp Meander Bends." *Water Resources Research* 46 (9): 1–23. https://doi.org/10.1029/2009WR008303.

Blanckaert, Koen, and I. Schnauder. 2009. "Meandering: Field Experiments, Laboratory Experiments and Numerical Modeling." *River, Costal and Estuarine Morphodynamics.* http://infoscience.epfl.ch/record/146621/files/2009-695-Blanckaert_et_al-Meandering_field_experiments_laboratory_experiments_and_numerical.pdf.

Blanckaert, Koen, and Huib J de Vriend. 2003. "Nonlinear Modeling of Mean Flow Redistribution in Curved Open Channels." *Water Resources Research* 39 (12): 1–14. https://doi.org/10.1029/2003WR002068.
Blazek, J. 2001. *Computational Fluid dynamics:Principles of Grid Generation. Computational Fluid Dynamics: Principles and Applications.* https://doi.org/http://dx.doi.org/10.1016/B978-008044506-9/50013-8.
Cea, L., M. Bermudez, J. Puertas, E. Blade, G. Corestein, E. Escolano, A. Conde, B. Bockelmann-Evans, and R. Ahmadian. 2016. "IberWQ: New Simulation Tool for 2D Water Quality Modelling in Rivers and Shallow Estuaries." *Journal of Hydroinformatics* 18 (5): 816–30. https://doi.org/10.2166/hydro.2016.235.
Chow, VT. 1959. *Open-Channel Hydraulics.* New Yersey: The Blackburn Press.
Chow, VT, DR Maidment, and LW Mays. 1994. "Hidrología Aplicada." *Hidrologia Aplicada.* http://bases.bireme.br/cgi-bin/wxislind.exe/iah/online/?IsisScript=iah/iah.xis&src=google&base=REPIDISCA&lang=p&nextAction=lnk&exprSearch=158911&indexSearch=ID%5Cnhttp://www.sidalc.net/cgi-bin/wxis.exe/?IsisScript=BINAI.xis&method=post&formato=2&cantidad=.
Corestein, G, E Bladé, and D. Niñerola. 2014. "Modelling Bedload Transport for Mixed Flows in Presence of a Non-Erodible Bed Layer." *River Flow 2014* M (1999): 1611–18. https://doi.org/10.1201/b17133-214.
FLUMEN, CEDEX. 2010. "Manual Básico de Usuario" 3: 22.
IDEAM. 2017. "Suministro de Información Hidrometeorológica." 2017. www.ideam.gov.co.
López, Jorge, Justino Alavez, and Juan L Hernández. 2009. "Solución Numérica Del Modelo de Saint-Venant Vía Volúmenes Finitos" 2 (Diciembre): 34–53.
Moukalled, F, L Mangani, and M. Darwish. 2016. *The Finite Volume Method in Computational Fluid Dynamics. Fluid Mechanics and Its Applications.* Vol. 113. https://doi.org/10.1007/978-3-319-16874-6.
Narváez, Germán, and Gloria León. 2001. "Caracterización Y Zonificación Climática De La Región Andina." *Meteorología Colombiana*, no. 4: 121–26.
La Patria. 2014. "Rio Magdalena." 2014. http://www.lapatria.com/node/85861.
Sociedad de Acueducto Alcantarillado y Aseo de Barranquilla S.A E.S.P. 2004. "Nuestro Río Magdalena." 2004. http://guardianes.aaa.com.co/magdalena.htm.
Wesseling, Pieter. 2001. *Principles of Computational Fluid Dynamics.* Springer.

Building Tomorrow's Society
Bâtir la Société de Demain

Fredericton, Canada
June 13 to 16, 2018 / *13 au 16 juin, 2018*

PERFORMANCE AND DAMAGE ASSESSMENT OF GFRP-RC WALLS

Mohamed, Nayera[1], Farghaly, Ahmed[2], Saatcioglu, Murat[3,4] and Benmokrane, Brahim[2]
[1] Assiut University, Egypt
[2] University of Sherbrooke, Canada
[3] University of Ottawa, Canada
[5] Murat.Saatcioglu@uottawa.ca

Abstract: The performance-based assessment and design become more wide-spread in structural engineering practice. The relationship between deformation limits and performance levels (immediate occupancy, life safety, and collapse prevention) are reasonably well established for steel-reinforced concrete (RC) walls. However, performance of RC walls reinforced with fiber-reinforced-polymer (FRP) bars can be substantially different. Four full-scale shear wall specimens were tested; two specimens were reinforced with glass-FRP (GFRP) bars, while the other two specimens were reinforced with steel bars. Analysis of the experimental observation defined the three performance levels and compared to the specified limits provided in the ASCE/SEI 41-06 in the drift ratio format. It was found that, the GFRP-RC shear walls developed significant deformability, exceeding the ASCE/SEI 41-06 deformation performance limits due to the elastic nature of the GFRP bars. This suggested proposing new deformation limits for the GFRP-RC walls associated with the damage description of each of the three performance levels.

1 INTRODUCTION

Reinforced concrete (RC) shear walls provide strength, stiffness and energy dissipation capacity to earthquake resistant buildings, forming a preferred lateral force resisting system in buildings (Fintel 1991). The walls provide drift control, thereby reducing both structural and non-structural damage to buildings during strong earthquakes. They can also be designed to provide sufficient strength and ductility, providing resistance to earthquake forces while dissipating seismic-induced energy. The preferred mode of behavior in structural shear walls is flexural mode. Therefore, the current seismic-resistant design philosophy is based on promoting flexural yielding and preventing brittle shear failure (ACI 318-14, CSA A23.3-14). Significant research has been conducted on flexure-dominant steel-RC structural walls. However, research on shear walls reinforced internally with fiber-reinforced-polymer (FRP) bars for new buildings is scarce in the literature. Limited research was conducted by Mohamed et al. (2014a, 2014b, 2015), and provided valuable test data on large scale FRP-RC shear walls under reversed cyclic loading.

As performance-based building assessment and design become more wide-spread in structural engineering practice, the required analytical tools gain importance to quantify shear wall performance levels. In ASCE/SEI 41 (2006) four performance levels were defined as operational, immediate occupancy, life safety and collapse prevention, in order of severity in damage. The operational performance level is governed by non-structural damage, but the latter three are specified as "structural performance levels". They are quantified in terms of drift ratios for overall structural performance limits. Accordingly, immediate occupancy limit for concrete shear walls is specified as 0.5% drift ratio. This level is associated with the damage in structural members is limited to cracking of concrete in the effective elastic range of deformations prior to the yielding of reinforcement. Life safety performance level is specified as 1.0% drift ratio for the walls with Major cracking and some spalling of concrete is expected with reinforcing steel entering into the post-yield range, but the structure remains stable with sufficient strength and deformability to ensure life safety. Collapse prevention corresponds to a maximum story drift ratio of 2.0% for RC shear walls, and marks the ultimate limit prior to collapse with possibility of demolishing the building after the earthquake. At this level of deformation, up to 20% of strength decay is usually permissible because of the inherent redundancy in monolithically built multi-story RC buildings (Park 1989).

While the relationship between deformation limits and performance levels are reasonably well established for steel-RC walls, performance of FRP-RC walls can be substantially different with experimental data and research lacking for this type of application. FRP reinforcement has lower modulus of elasticity, resulting in softer response upon concrete cracking. Furthermore, FRP re-bars show higher rupturing strength in tension, lower strength in compression and brittle material behavior with linear stress-strain relationships. These aspects of FRP behavior may raise questions on deformability of FRP-RC shear walls and their suitability as seismic resistant elements. However, as later illustrated in the paper, FRP-RC shear walls develop significant deformability, meeting and overstepping the ASCE/SEI 41 (2006) drift performance levels described above.

Although the deformation limits specified in ASCE/SEI 41 are given either in the form of drift ratio or plastic rotations (ASCE/SEI 41-06 and ASCE/SEI 41-13, respectively) for concrete structural walls, this paper focuses mainly on calibrating the performance and damage levels to assess the GFRP-RC shear walls based on drift ratio. The three performance levels; immediate occupancy, life safety, and collapse prevention, were identified according to the specified definition of each of the damage level in ASCE/SEI 41-06. New drift ratios for each performance level was proposed accordingly. The results provide much needed guidance to structural engineers for use of FRP reinforcement in earthquake resistant shear walls.

2 EXPERIMENTAL PROGRAM

The presented experimental program consists of four full-scale shear walls, representing one-story prototype mid-rise shear walls, tested under quasi-static reversed cyclic loading and a constant axial concentric load of 7% of the wall axial capacity. The walls were cycled twice at each displacement level with increments of 0.06% up to 0.3% lateral drift, followed by increments of 0.15% up to 1.5%, and then increments of 0.3% to failure. A series of linear variable differential transducers (LVDTs) and strain gauges were used to measure critical-response quantities as shown in **Figure 1**. Reinforcement and geometric details of the test specimens in addition to the instrumentations are presented in **Table 1** and **Figure 1**. The details of test setup, loading procedure, and instrumentation can be found in Mohamed et al. (2014a).

The four walls consist of specimens ST15, ST15-R, G15, and C15. The experimental results of specimens ST15 and G15 were reported earlier by Mohamed et al. (2014a), while specimens ST15-R and C15 are newly tested and added to the current performance assessment investigation.

Table 1: Wall specimens

Specimen	h_w/l_w	Web RFT		Boundary RFT		Reinforcement ratio				f_c'
		Vl bars	Hz bars	Vl bars	Spiral stirrups	ρ_v%	ρ_h%	ρ_l %	ρ_t %	
ST15*	2.33	Steel 8 mm	Steel 8 mm	4 Steel 8 mm	Steel 8 mm	0.23	0.63	0.5	0.63	39.2
ST15-R		Steel #3	Steel #4	8 Steel #3	Steel #3	0.58	1.25	1.42	1.6	39.5
G15*		GFRP #3	GFRP #4	8 GFRP #3	GFRP #3	0.58	1.58	1.43	0.89	39.9
C15					CFRP #3	0.58	1.58	1.43	0.89	40.1

* reported in Mohamed et al. (2014a).
h_w/l_w wall aspect ratio; h_w wall height; l_w wall length; ρ_v vertical web-reinforcement ratio; ρ_h horizontal web-reinforcement ratio; ρ_l boundary longitudinal-reinforcement ratio; ρ_t boundary-tie reinforcement ratio; $N/A_g f_c'$ axial-load ratio using the measured axial load and the measured compressive strength of concrete; f_c' concrete compressive strength (MPa).

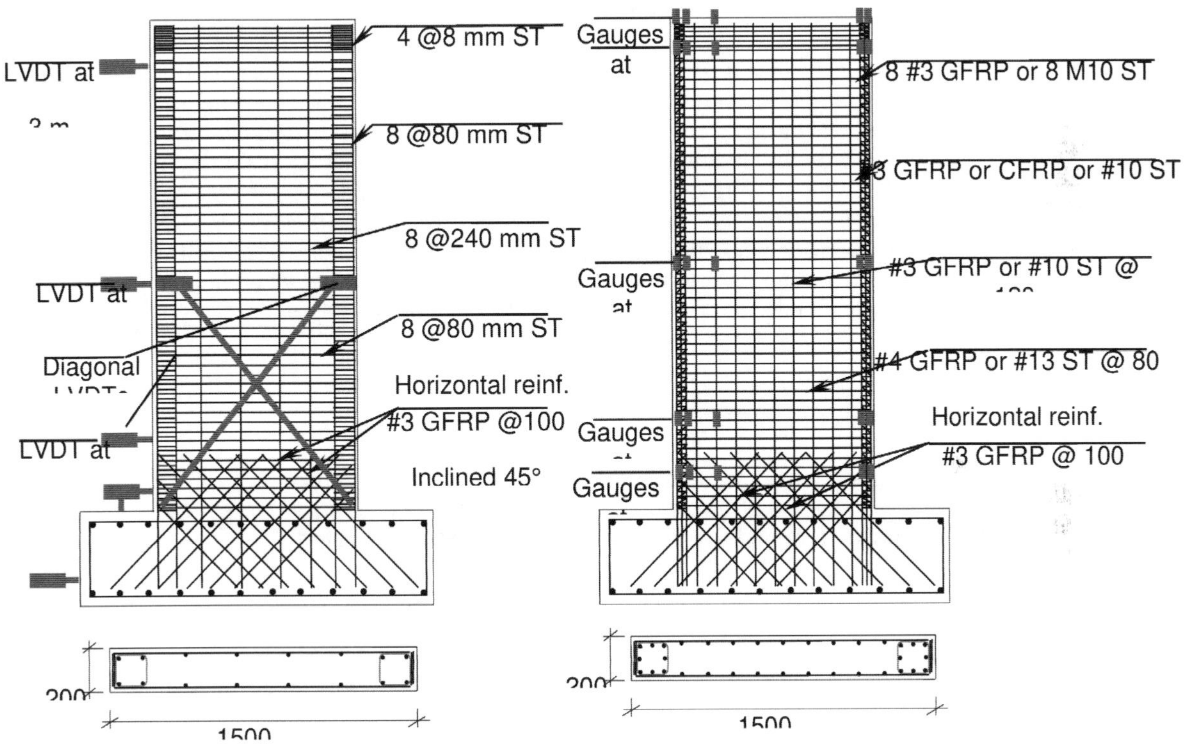

(a) ST15

(b) ST15-R, G15, or C15

Figure 1: Reinforcement details and concrete dimensions of the tested shear-wall specimens

Table 2: Material properties of reinforcement

Bar	Straight Bars					Bent Bars			
	8 mm steel	#3 steel	#4 steel	#3 GFRP	#4 GFRP	#3 GFRP Straight portion	#3GFRP Bent portion	#3 CFRP Straight portion	#3 CFRP Bent portion
d_b (mm)	8	9.5	12.7	9.5	12.7	9.5		9.5	
A_f (mm²)	50.3	71	129	71.3	126.7	71.3		71.3	
E_f (GPa)		200		66.9	69.6	52	---	120	----
f_{fu} (MPa)		$f_y = 400$		1412	1392	962	500	1596	1000
ε_{fu} (%)		$\varepsilon_y = 0.2$		2.11	2.00	1.85	---	1.33	----

d_b bar nominal diameter, A_f nominal cross-sectional area, E_f modulus of elasticity, f_{fu} guaranteed tensile strength, ε_{fu} ultimate strain, f_y steel yielding strength, ε_y steel yielding strain.

The two new specimens (ST15-R and C15) along with specimens ST15 and G15 represent mid-rise walls with an aspect ratio (height-to-length ratio h_w/l_w) of 2.33. They were 3500 mm high, 1500 mm long, and 200 mm thick. This implies that the walls were subjected to shear, developing sizeable shear distortions, though they were designed to prevent premature shear failure. Two of the walls were reinforced with steel bars and the other two with FRP bars. ST15 was reinforced with deformed 8- mm diameter steel bars to provide the minimum vertical web and boundary element reinforcement ratios specified in ACI 318 (2014) and CSA A23.3 (2014). G15 and C15 were designed to have GFRP bars with the same reinforcement axial stiffness ($A_f \times E_f$) as ST15, except for the boundary element transverse reinforcement. G15 had GFRP spiral stirrups in the wall boundary elements, whereas C15 had equal percentage of CFRP spiral stirrups. ST15-R had steel reinforcement ratio as the reinforcement ratio of G15. The walls' reinforcement ratio is presented in **Table 1**. **Table 2** provides mechanical properties of all the reinforcement used in the experimental program.

3 TEST RESULTS AND DISCUSSION

The observed damage is presented and discussed to assess the performance of FRP-RC shear walls concerning the three main structural performance levels: immediate occupancy, life safety, and collapse prevention. The limitations of the performance levels were identified based on the observation and evaluation of damage pattern for each tested wall based on the prescribed definition of damage state for each performance level according to ASCE/SEI 41 (2006).

Similar hysteretic behavior was attained by the newly tested specimens ST15-R and C15 in comparison to their counterpart specimens ST15 and G15, respectively (**Figure 2**). The specimen ST15-R achieved higher ultimate load and drift capacities than ST15 due to its higher reinforcement ratio. The CFRP spiral stirrups confining the boundary element of the specimen C15, instead of the GFRP spiral stirrup that had been used in specimen G15, had no significant effect on the general behavior of C15 except for slight increase in the ultimate load capacity. Failure was due to concrete crushing after buckling of the longitudinal steel bars in ST15-R and was associated with longitudinal and transverse rupture of the GFRP bars and CFRP spiral stirrups, respectively, in C15.

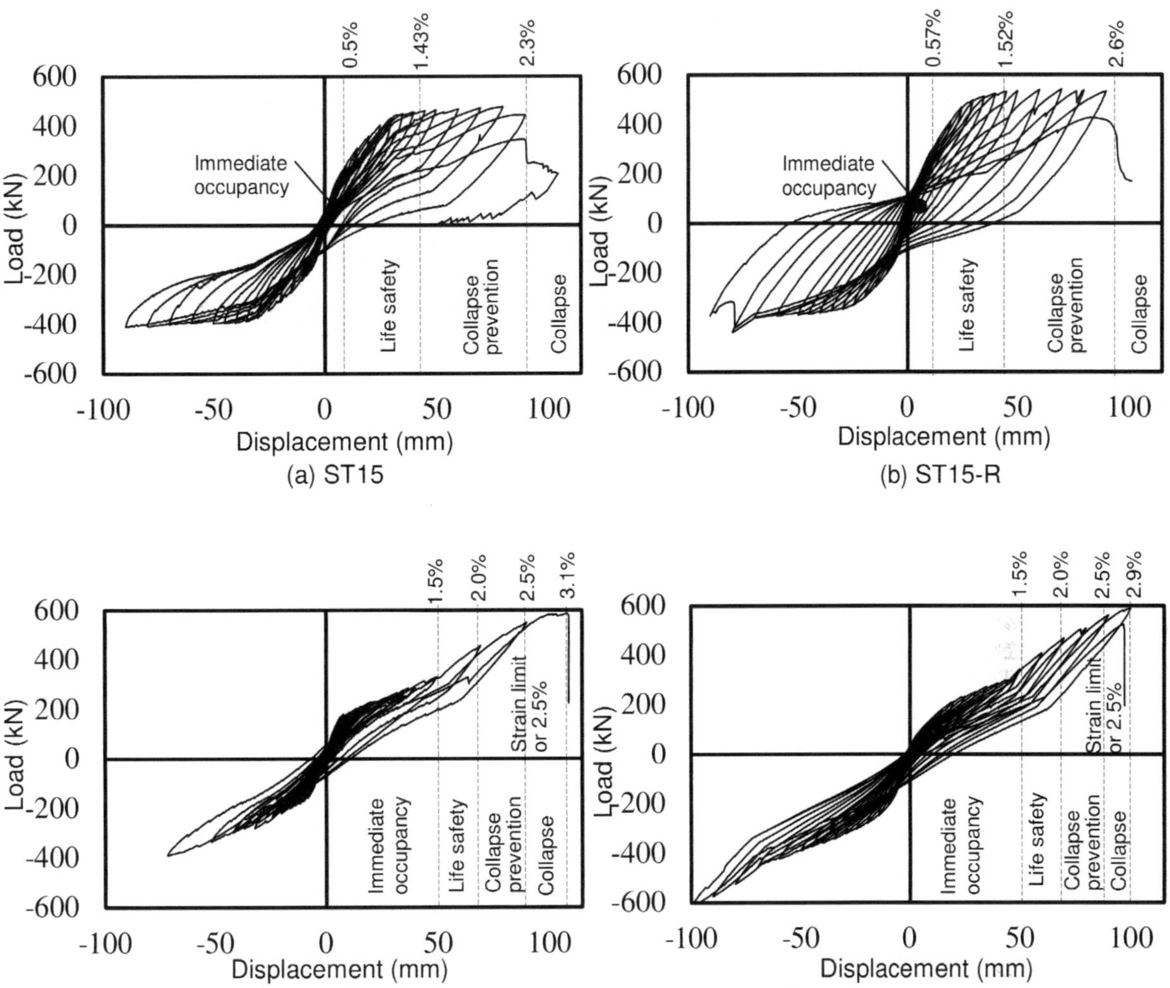

Figure 2: Hysteretic response

Figure 2 shows the hysteretic response of all tested walls. In the early loading stage (immediate occupancy level), the steel-RC wall achieved a higher load than the GFRP-RC walls, due to the softened response of the GFRP-RC walls, which is an advantage of using GFRP reinforcement. The yielding plateau was evidenced in the steel-RC shear walls by the yielding of the longitudinal steel reinforcement, creating the transition point to the life safety level at 0.5% and 0.57% drift ratio for specimens ST15 and ST15-R as shown in **Fig. 2a** and **b**, respectively. The two steel-RC walls showed a stable behavior through the life safety level.

The load capacity of the FRP-RC shear walls kept increasing with increasing lateral displacement due to the linear behavior of the FRP bars. Therefore, the concrete was the source of plasticity in the FRP-RC shear walls, and the start of the life safety level could be identified based on the determination of concrete plasticity. Experimental observations of specimens G15 and C15 showed the start of concrete plasticity at 1.5% drift ratio (**Fig. 2c** and **d**), which corresponds to a concrete compressive strain ε_c = 0.0035. The transfer to the life safety level was evidenced by no new crack propagation, and residual forces beginning to accumulate at zero displacement. Meanwhile, during the life safety level, the FRP-RC shear walls showed recoverable and self-centering behavior, in addition to the stable behavior. **Figure 2** confirms the significant difference in strength and stiffness degradation between the steel and GFRP reinforcement in the shear walls in the immediate occupancy level. The FRP-RC shear walls recorded a significantly less stiff response after initial cracking. This is excellent behavior in resisting earthquakes, as it would increase the displacement demand which is considered as an advantage of using GFRP bars. This is due to the fact that the lower structural stiffness results in a longer natural period of vibration and, consequently, lower seismic-force demand (Sharbatdar and Saatcioglu 2009).

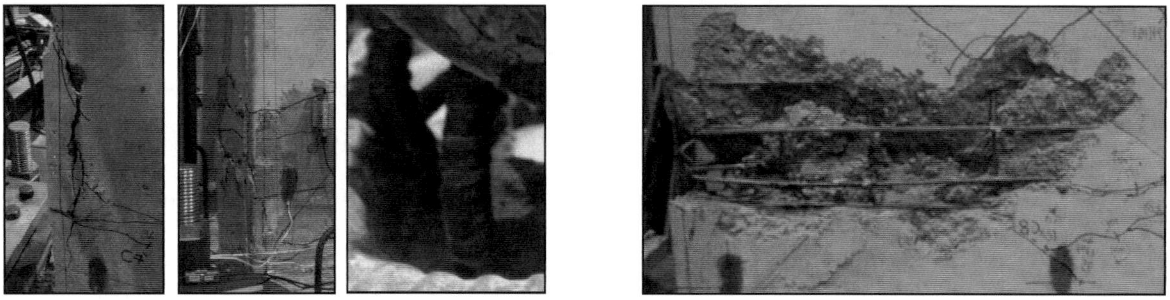

(a) (b) (c) (d)

Figure 3: Failure progression of specimen ST15: (a) vertical cover splitting; (b) spalling of concrete cover; (c) buckling of longitudinal bars; and (d) concrete crushing causing failure.

Figures 3 and **4** illustrates the typical progression of flexural-shear cracking for ST15, ST15-R, G15, and C15 shear walls. For all tested walls, flexural horizontal cracks were propagated in the early loading stages, followed by diagonal cracks. In both steel-RC shear walls, the flexural cracks were propagated and were followed by yielding of longitudinal reinforcement representing the end of the immediate occupancy level. Up to the end of life safety level identified at 1.43% and 1.52% drift ratio for ST15 and ST15-R, respectively, the crack width kept increasing on the tension side of the walls, while concrete-cover splitting was observed on the compression side at the boundary zone (Fig. 3a), causing significant permanent deformation. That formed the plastic hinge in the steel-RC shear walls and the transition to the collapse prevention level. Concrete-cover spalling and buckling of longitudinal bars in the boundary region under compression were observed (Figs. 3b and c, respectively) at around 2.0% drift ratio for the both walls (ST15 and ST15-R), at which point the steel-RC shear walls reached their ultimate capacity. In addition to the typical flexural failure, ST15 experienced major horizontal crack through the length of the wall which can be attributed to less vertical and horizontal web reinforcement ratio compared to ST15-R. Such a major horizontal crack led to an extensive concrete spalling in the compressed zone of ST15 through the plastic hinge region (**Fig. 3d**). Finally, specimens ST15 and ST15-R had 80% of the ultimate capacity at 2.3% and 2.6% drift ratio, respectively, defining the end of the collapse prevention level. In this level, the steel-RC shear walls experienced reduction in load capacity due to softening of tensioned steel bars and buckling of compressed steel bars (**Fig. 2a** and **b**).

In the GFRP-RC shear walls, the cracks were able to close and realign after each cycle up to 1.5% drift ratio, i.e., the transition from the immediate occupancy level to the life safety level. At the end of the life safety level, the concrete cover close to the wall-base in the compressed boundary zone started to split vertically (Fig. 4a) and formed a plastic-hinge region, followed by the onset of the concrete-cover spalling at 2% drift (Fig. 4b). Although GFRP bars have a linear stress–strain behavior, Mohamed et al. (2014b) observed a plastic hinge in GFRP-RC shear walls. With increasing load, the failure of specimens G15, and C15 occurred at 3.1%, and 2.9% drift ratio, respectively. The concrete was crushed in the boundary zone under compression, associated with the fracture of longitudinal GFRP bars and rupture of GFRP and CFRP spiral stirrups (Figs. 4c, d, and e, respectively). Specimens G15, and C15 had their load-carrying capacity reduced by around 30% of their maximum capacity, proceeding to failure.

(a) (b) (c) (d) (e)

Figure 9: Failure progression of GFRP specimens: (a) vertical cover splitting; (b) spalling of concrete cover; (c) concrete crushing causing failure; (d) fracture of longitudinal bars; and (e) rupture of GFRP tie

The FRP bars in the FRP-RC shear walls did not experience tension softening behavior due to their elastic nature. In addition, concrete softening under compression is ignored in comparison to the high strain that could be achieved by the FRP bars. Therefore, the FRP-RC shear walls attained high deformation capacity within the elastic strain capacity of the FRP bars, which needs to be limited to set a limit for the collapse prevention level. Accordingly, as shown in **Fig. 2 c** and **d**, the collapse prevention level in the GFRP-RC shear walls should be defined in terms of material strain limits or by setting a drift limit. The drift limit could be 2.5% to restrict non-structural damage (Mohamed et al. 2014b). Meanwhile, the strain limits could be set based on the experimental observation. The maximum recorded compressive strains were 0.0086 and 0.011 for the specimens G15 and C15, respectively, which is in the range of 0.4 to 0.5 of the ultimate tensile strain of the FRP bar. Generally, the suggested material strain limits under tension and compression could be related to the ultimate tensile strain in FRP bars; the strain could be limited in tension as $\varepsilon_{frp\,t} = 0.9\ \varepsilon_{frp\,max}$ and, in compression, $\varepsilon_{frp\,c} = 0.4\ \varepsilon_{frp\,max}$.

4 ASSESSMENT OF STRUCTURAL PERFORMANCE LEVELS

The deformation limits specified in ASCE/SEI 41 are given either in the form of drift ratio or plastic rotations, depending on the type of structural element or component under consideration. According to ASCE/SEI 41 (2006), the drift ratio for the three structural performance levels of the steel-RC shear walls are 0.5% for immediate occupancy, 1% for life safety and 2% for collapse prevention levels. **Figure 5** and **Table 3** show comparison between the ASCE/SEI 41 (2006) performance drift limits and the drift ratio experimental observation at different drift ratios corresponding to performance limits identified based on the experimental observations and associated with the damages specified in ASCE/SEI 41 (2006) and ASCE/SEI 41 (2013) for the tested steel- and GFRP-RC walls.

For the steel-RC shear walls, the yielding of steel reinforcement was recorded at 0.5% and 0.57% drift ratio for specimens ST15 and ST15-R, respectively, which is within the limitation of 0.5% drift ratio for immediate occupancy according to ASCE/SEI 41 (2006). For the life safety performance level, the major flexural cracking and concrete-cover splitting in addition to the start of gaining permanent drift were observed, while the walls were still stable. These evidences were observed at 1.43%, 1.52% drift ratio for ST15 and ST15-R, respectively, which exceeded the 1% drift ratio limitation. The specimens ST15 and ST15-R had their 80% of the ultimate capacity at 2.3% and 2.6% drift ratio, respectively, identifying the collapse prevention performance level, which also exceeded the limitation of 2% (**Fig. 5a**). The conformity between the specified performance limits in ASCE/SEI 41 (2006) and the experimental observation of ST15 and ST15-R for each performance level supports the analysis of such tested walls.

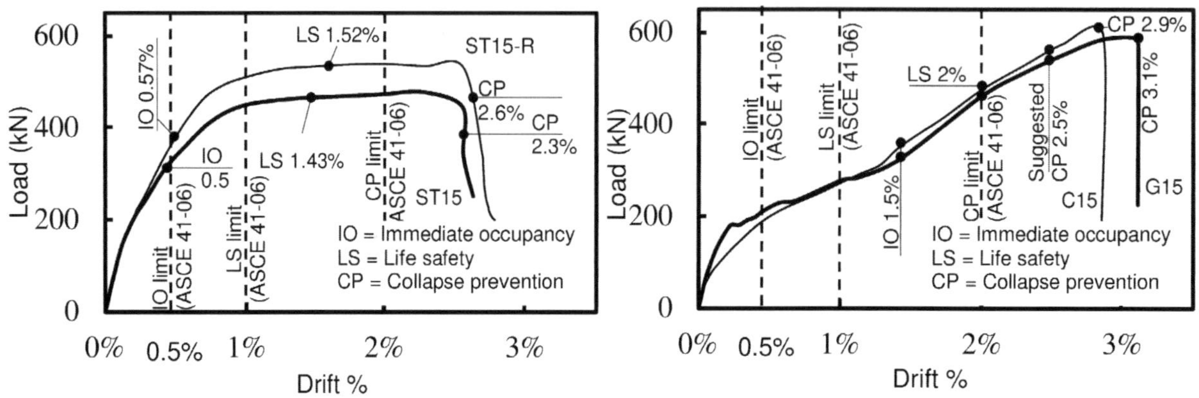

a – steel-reinforced concrete shear walls

b – GFRP-reinforced concrete shear walls

Figure 5: Assessment of structural performance levels

Table 3 – Structural performance levels based on drift ratio (ASCE/SEI 41-06)

Performance levels	Level associate damage	Drift limit according to ASCE/SEI 41-06	Drift ratio according to experimental observation			
			ST15	ST15-R	G15	C15
IO	Minor structural damage	0.5%	0.5%	0.57%	1.5%	
LS	Substantial structural damage due to inelastic material behavior. Major cracking and some spalling of concrete, but the structure remains stable with sufficient strength and deformability	1.0%	1.43%	1.52%	2%	
CP	Major flexural and shear cracks. Sliding at joints. Extensive crushing and buckling of reinforcement. Severe boundary element damage. Mark the ultimate limit prior to collapse	2.0%	2.3%	2.6%	2.5%*	
Failure	Failure of test specimens		2.3%	2.6%	3.1%	2.9%

* Collapse prevention was set to 2.5% for the GFRP-reinforced shear to restrict nonstructural damage (Mohamed et al. 2014b).

For the GFRP-RC shear walls (**Fig. 5b** and **c**), the immediate occupancy performance level was identified experimentally at 1.5% drift ratio with negligible permanent deformation which clearly exceeds the immediate occupancy (0.5% drift ratio) and life safety (1% drift ratio) performance levels according to ASCE/SEI 41 (2006). This is mainly attributed to the elastic nature of the GFRP bars and the concrete is the source of plasticity (Mohamed et al. 2014b). The GFRP-RC shear walls kept stable and had the ability of self-centering till a drift ratio of 2% identifying the life safety performance level which exceeds the 1% drift ratio of life safety and even reached the 2% drift ratio limit for collapse prevention level according to ASCE/SEI 41 (2006). The damage signs associated with the collapse prevention performance level; like major flexural and shear cracks and concrete cover spalling at boundary elements, were observed at 2% drift ratio, which is specified as the drift limit for the collapse prevention performance level according to ASCE/SEI 41 (2006).

5 CONCLUSIONS

Through the presented experimental program, which consists of four tested full-scale shear walls representing one-story prototype mid-rise shear wall, the performance-based assessment of GFRP-RC shear walls was investigated. The assessment was based on comparing the drift ratio limits at immediate occupancy, life safety, and collapse prevention specified in ASCE/SEI 41-06 and the drift ratio obtained based on the experimental observation corresponding to the damage level associated with each of the three performance levels. The findings are promising with respect to applications for GFRP bars as follows:

- All tested steel-RC and GFRP-RC shear walls clearly comply with the ASCE/SEI 41-06 drift ratio limitations when comparing the prescribed damage level associated with each performance level with the identified damage level based on the experimental observations. However, GFRP-RC walls overstepped these limitations.
- During immediate occupancy level, the softer response of the GFRP-RC shear walls comparing to the steel-RC shear walls is considered as an advantage of using GFRP bars as it would increase the displacement demand and attracting lower seismic force demand.
- Due to the elastic nature of the GFRP bars and to prevent the brittle failure, setting the collapse prevention at 2.5% is recommended to restrict non-structural damage.
- The GFRP bar strain could be limited at the collapse prevention level in tension and compression to 90% and 40% of the ultimate tensile strain of GFRP bar.
- Limiting the GFRP-RC walls' drift ratio according to ASCE/SEI 41-06 at the immediate occupancy and life safety levels is not representative as the walls still experience self-centring behavior with negligible permanent deformation with low damage levels.

Acknowledgements

The authors wish to acknowledge the financial support of the Natural Sciences and Engineering Research Council of Canada (NSERC), the Canada Research Chair in Advanced Composite Materials for Civil Engineering, and the Fonds québécois de la recherche – Nature et Technologies - (FQRNT) of Quebec.

References

ACI Committee 318, "Building Code Requirements for Structural Concrete and Commentary (ACI 318-14)," American Concrete Institute, Farmington Hills, MI, 2014, 503 pp.

ASCE/SEI 41, "Seismic rehabilitation of existing buildings (ASCE/SEI 41-06)." American Society of Civil Engineering (Structural Engineering Institute), Reston, VA, 2006, 427 pp.

ASCE/SEI 41, "Seismic Evaluation and Retrofit of Existing Building (ASCE/SEI 41-13)." American Society of Civil Engineering (Structural Engineering Institute), Reston, VA, 2013, 555 pp.

Canadian Standards Association A23.3, "Design of concrete structures standard (CSA A23.3-14)," Mississauga, ON, Canada, 2014, 240 pp.

Fintel, M., 1991. "Shear walls—An answer for seismic resistance?" Concrete International, 13(7), 48–53.

Mohamed, N., Farghaly, A. S., Benmokrane, B., and Neale, K.W., 2014a. "Experimental investigation of concrete shear walls reinforced with glass fiber–reinforced bars under lateral cyclic loading." Journal of Composites for Construction, ASCE, 18(2), A4014001.

Mohamed, N., Farghaly, A. S., Benmokrane, B., and Neale, K.W., 2014b. "Drift Capacity Design of Shear Walls Reinforced with Glass Fiber-Reinforced Polymer Bars." *ACI Structural Journal*, 111(6), 1397-1406.

Mohamed, N., Farghaly, A. S., Benmokrane, B., 2015. "Aspects of Deformability of Concrete Shear Walls Reinforced with Glass Fiber–Reinforced Bars." Journal of Composites for Construction, ASCE, 19(5), 06014001.

Park, R., 1989. "Evaluation of ductility of structures and structural assemblages from laboratory testing." Bull. N. Z. Nat. Soc. Earthquake Eng., 2(3), 155-166.

Sharbatdar, M. K., and Saatcioglu, M., 2009. "Seismic Design of FRP Reinforced Concrete Structures." *Asian Journal of Applied Sciences*, 2(3), 211-222.

Building Tomorrow's Society
Bâtir la Société de Demain

Fredericton, Canada
June 13 to 16, 2018 / *13 au 16 juin, 2018*

REDUCTION OF BRIDGE PIER SCOUR THROUGH THE USE OF A NOVEL COLLAR DESIGN

Valela, Christopher[1,2], Nistor, Ioan[1] and Rennie, Colin[1]
[1] University of Ottawa, Canada
[2] cvale034@uottawa.ca

Abstract: Bridge pier scour is the cause of many bridge failures as the pier foundations can become undermined, therefore jeopardizing the structural integrity of the bridge. Due to the potential harm associated with pier scour, numerous prevention methods have been designed and practised which either alter the riverbed around the pier or the shape of the pier itself. However, there is not one method that is easy to install, minimally intrusive, low cost, and maintenance free. The purpose of this study is to present a novel three-dimensional collar as a new scour counter measure alternative. The collar works by containing the horseshoe vortex that forms on the front and sides of the pier and then directs the vortex in the downstream direction away from the critical region around the pier. The erosive vortex behaviour is then unable to touch the riverbed, thus avoiding scour. To guide the collar design process, OpenFOAM (CFD) software was utilized in an iterative manner to provide the flow field behaviour of successive designs. A benefit of this novel collar is that it can be constructed offsite in multiples pieces, transported to the pier location, and simply installed on the pier, making it an easy and minimally intrusive option, while requiring little to no maintenance in the future. The goal of this novel scour counter measure is to provide a new viable solution to the industry which can overcome many of the existing issues.

1 INTRODUCTION

When constructing bridges over rivers or bodies of water, it is ideal to avoid the use of piers positioned within the water; however, this is often unavoidable due to the distances that need to be spanned. In such cases, piers are anchored into the riverbed and protrude out of the water to support the bridge. However, when anchoring into the riverbed, the piers become susceptible to local scour. Scour, which is the reduction in riverbed elevation and in this case, around the base of bridge piers, has proven to be detrimental to the structural integrity of bridges (Chiew 1992). Rapidly or over time, sediment becomes removed within a close proximity to bridge piers, which can lead to the undermining of the pier foundation, ultimately jeopardizing the safety of the bridge (Deng and Cai 2010).

Local scouring occurs because of the presence of a structure interrupting the passing flow. In this case, the structure is a bridge pier which induces flow separation throughout the entire depth as piers always extend out above the water surface. As a result, a portion of the flow is split and redirects itself around the pier causing a local acceleration on either side of the pier. As the faster moving water passes the sides of the pier, alternating vortices, revolving about a vertical axis, are formed and shed off either sides of the pier travelling downstream (Stevens et al. 1991). Due to the orientation of the vortices, sediment along the sides and rear of the pier can be drawn up the low pressure center of the vortices and then deposited a short distance downstream of the pier (Ettema et al. 2006). The remainder of the flow that is obstructed by the pier is drawn downwards to the riverbed due to a pressure gradient. Upon reaching the riverbed, the flow circulates outwards forming a vortex that propagates around the upstream face of the pier and down the sides eventually dissipating downstream into the passing flow (Dey and Raikar 2007). This vortex is known as the horseshoe vortex and is the largest contributing factor in pier scour as sediment is dislodged upstream and along sides of the pier and is transported downstream (Stevens et al. 1991). The combination of both vortex behaviours, but mainly the horseshoe vortex, form the common scour holes present around many bridge piers which can be seen in Figure 1.

Figure 1: Scour hole downstream of a pier at Agassiz Bridge, Fraser River, BC.

Bridge pier scour has been a popular topic of study for many years due to the potential to harm the public and property. As a result, various different approaches have been taken to address scouring, which can be summarized into four categories: hydraulic, structural, biotechnical, and monitoring counter measures (FHWA 2009). The first, hydraulic counter measures, are intended to address the passing flow or the riverbed surrounding the pier. This can be done by altering the flow near the pier, utilizing sacrificial piles or vanes for example, which helps to deflect the flow and create a region of lower velocity directly upstream of the pier. Another hydraulic approach, being the most common, is to apply armouring around the base of the piers with for example riprap, gabions, or grout filled containers. This method does not alter the flow but rather protects the vulnerable riverbed areas from the erosive forces of the vortices formed around the piers (FHWA 2009). The second counter measure, structural, pertains to addressing the structural aspects of the pier and bridge itself. This can be accomplished by strengthening the portion of the pier that is below the riverbed through extending or underpinning the existing footing. Another approach is to alter the portion of the pier that is above the riverbed by streamlining the shape of the pier, adding structures to the pier such as collars, or cutting holes into the pier, all to reduce the vortex generating factors (FHWA 2009). The third counter measure is biotechnical as it incorporates vegetation such as root wads to perform tasks similar to those of hydraulic or structural counter measures already discussed (FHWA 2009). The last counter measure is monitoring, which does not attempt to directly reduce the scour but instead can give a very informative indication as to the level of erosion that has occurred (FHWA 2009). The four counter measures can be used individually or together, depending on the severity of the situation.

As seen above, there are various types of scour mitigation methods available; however, there are very few that are easy to install, cost effective, minimally intrusive, and low maintenance. One of the most widely used methods is riprap, which is simply larger rocks installed around the base of the pier that protects the sediment from being washed away (Lauchlan and Melville 2001). The problem with this method is that if the riprap is not sized correctly or extreme flooding occurs then some of the rocks can become ejected and transported downstream, reducing the protection around the pier. In addition, upon becoming aware of the dislodged riprap, reinstallation is required to provide the original level of protection necessitating future inspections and maintenance (Lauchlan and Melville 2001). Therefore, there is not one ideal method that can satisfy all of the criteria while guaranteeing protection under extreme flow conditions, which as a result is what motivated the research presented in this paper. There are three main objectives of this paper and they are: 1) to introduce a new, state-of-the-art scour counter measure for cylindrical piers which takes the shape of a collar, 2) to explain the workings of the new collar and its ability to overcome the problems associated with existing scour counter measures, and 3) to perform a numerical modelling flow comparison between a pier with and without the new collar.

2 METHODOLOGY

2.1 Numerical Model

The design of the new pier scour counter measure was guided intensively by the use of a numerical model. The modelling software chosen was the open-source computational fluid dynamics (CFD) software, OpenFOAM v1706, which was used to illustrate the flow behaviour in a channel in the presence of a bridge pier. The solver used within OpenFOAM was *pimpleFoam* operating in *piso* mode which is a transient solver applicable for turbulent, incompressible flows (Hosur et al. 2014). In order to replicate the experimental flow conditions best, the k-omega-SST Delayed Detached Eddy Simulation (DDES) turbulence model was utilized. Detached Eddy Simulation (DES) models switch between Reynolds-Averaged Navier-Stokes (RANS) and Large Eddy Simulation (LES) turbulence models based on the grid size and turbulent length scale for the purpose of reducing the computational cost that is normally associated with LES models. DDES turbulence models improve on the DES models by addressing the issue where the turbulence models switch by shielding the RANS model from the DES model (Gritskevich et al. 2011).

The model domain consisted of a rectangular flume 10m in length, 0.85m in width, and 0.15m in depth, filled entirely with water. A cylindrical pier, 0.09m in diameter by 0.15m tall, was generated used *snappyHexMesh* and situated 1.5m from the outlet. An 8.5m approach was used because, based on a sensitivity analysis, it was found that the flow becomes fully developed after approximately 8m. A flume width of 0.85m was used because when the walls are situated that far apart, they do not interfere with the wake generated by the pier.

In order to achieve the most accurate results while maintaining a reasonable computational cost, two refinement regions were used. First, a level one rectangular refinement, yielding a cell size of 0.004m, was used at the downstream end of the flume stretching 2m upstream by 0.5m across by 0.15m high, therefore encompassing the pier and downstream wake. Then, a level two cylindrical refinement, yielding a cell size of 0.002m, was used around the pier, 0.2m in diameter by 0.06m high, to ensure a high resolution result in the focus area at the base of the pier. In total, this complete mesh yielded a cell count of approximately 5.5 million cells and an average y+ value of 25 at the base of the pier. This mesh was chosen because a sensitivity analysis was performed consisting of six different meshes each possessing an equally finer mesh and the velocities at specified locations, for this chosen level of refinement, no longer changed in the focus region and remained constant for finer meshes.

The inlet velocity condition used was *turbulentInlet,* as it consists of a constant inflow which was set to 0.335m/s but also includes a turbulent fluctuation which was set to ten percent of the inflow velocity. The pier and the bottom of the flume were given a *noSlip* condition, while the walls were given a *slip* condition to ensure the walls had as little impact on the passing flow as possible. The top of the flume, which was filled to the surface with water, utilized a *symmetryPlane* to reduce the influence of the top boundary on the flow behaviour especially around the pier. Wall functions were used for the pier and bottom of the flume specifically employing the nutkWallFunction and the nutkRoughWallFunction, respectively. The rough wall function was used only for the bottom of the flume because it allows for the bed roughness to be specified, which in this case was 0.001m. To ensure the most efficient computational time, *adjustTimeStep* was turned on such that it adjusted the time step to maintain a maximum courant number of 0.8. In order to calibrate and further validate the results obtained from the numerical model, preliminary experimental tests were performed in a laboratory setting.

2.2 Experimental Setup

The experimental tests were conducted within a straight rectangular flume possessing a length of 30m, a width of 1.5m, and a depth of 0.5m, located in the University of Ottawa's Civil Engineering Hydraulic Laboratory. The focus area was a 4m long section located 3m from the downstream end of the flume. This section possessed a rigid concrete floor surfaced with sieved 0.001m sand affixed to the concrete surface in a uniform layer. The purpose of using a fixed floor was to replicate the conditions within the numerical model since OpenFOAM is not equipped with sediment modelling capabilities. In the center of the 4m long focus area of the flume, a cylindrical acrylic pier with a diameter of 0.09m was installed to the floor vertically. The tests were run under constant flow conditions where the depth was set to 0.15m and the velocity was maintained at 0.36m/s at a height of 0.07m above the bottom. To measure the velocity, an acoustic Doppler velocimeter (ADV) was used. The values used for both experimental and numerical tests were based off of a 1:30 Froude scale.

2.3 Model Validation

In order to validate the numerical model, velocity profiles of the flow were taken in the laboratory flume using the ADV at three locations: 0.72m upstream of the pier where the flow is fully developed yet undisturbed by the pier, 0.14m upstream of the pier at the beginning of the stagnation region, and 0.75m downstream of the pier in the turbulent wake. The velocity profiles consist of a series of point measurements taken 0.05m apart in the vertical direction. A comparison between the velocity profiles obtained numerically and experimentally at the three locations can be seen in Figure 2. At all three locations the numerical model matched the experimental results respectively well indicating a high level of trust in the numerical results. The downstream location possesses error bars, representing the standard deviation, that are substantially larger than the other two locations due to the turbulence induced by the bridge pier.

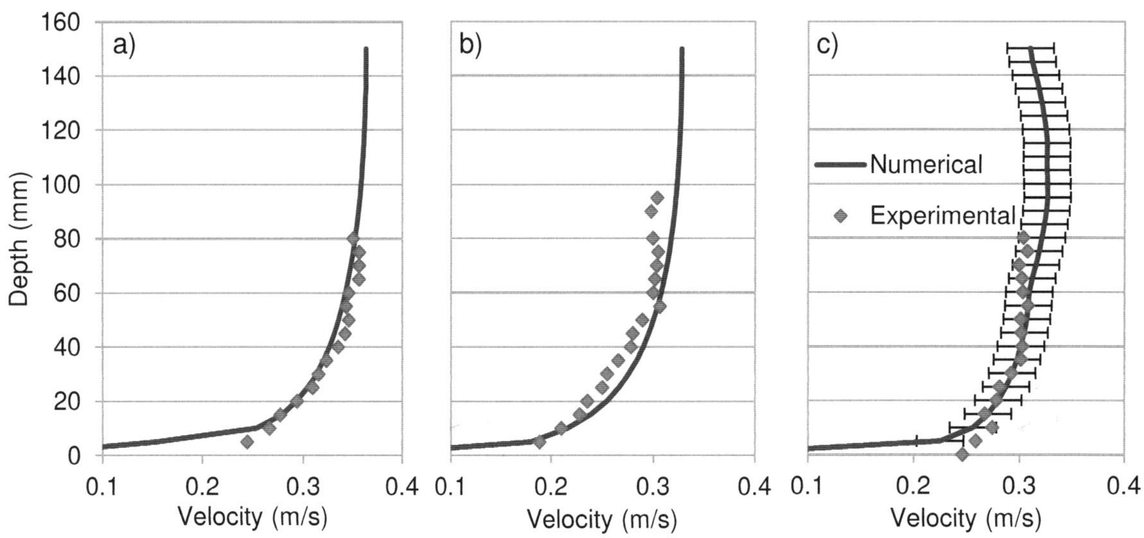

Figure 2: Velocity profile comparison. (a) 0.72m upstream of pier; (b) 0.14m upstream of pier; and (c) 0.75m downstream of pier.

3 SCOUR COUNTER MEASURE DESIGN

It is known that the largest contributor to pier scour is the horseshoe vortex and then to a lesser degree, the wake vortices (Stevens et al. 1991). Therefore, if the goal is to reduce scour then the horseshoe vortex must be addressed. The approach taken with this new counter measure was to harness the horseshoe vortex rather than trying to interrupt or impede its very powerful forces. In doing so, the objective was to disallow the horseshoe vortex from contacting the riverbed by containing it within a structure. This structure was formed into the shape of a circular collar which mounts around the pier and rests on the riverbed. The only similarity between this design and existing pier scour prevention collars is that both revolve around the pier, otherwise the differences are significant. Specifically, as the flow approaches the upstream face of the pier and is drawn downwards to the riverbed, this new collar design allows the flow of water to enter into a rounded cavity within the collar where it can circulate forming the common horseshoe vortex and propagate around the upstream face of the pier as it naturally would. This occurs within the collar cavity until it passes the sides of the pier, then the circulating flow is directed out of the rear of the collar and the energy is dissipated downstream. This can be seen in Figure 3 as the new collar design, more formally known as collar prototype number one, is presented. The main concern is to protect the immediate vicinity around the base of the pier especially on the upstream side and transfer the turbulent flow downstream away from the pier where it cannot harm the stability of the pier foundation.

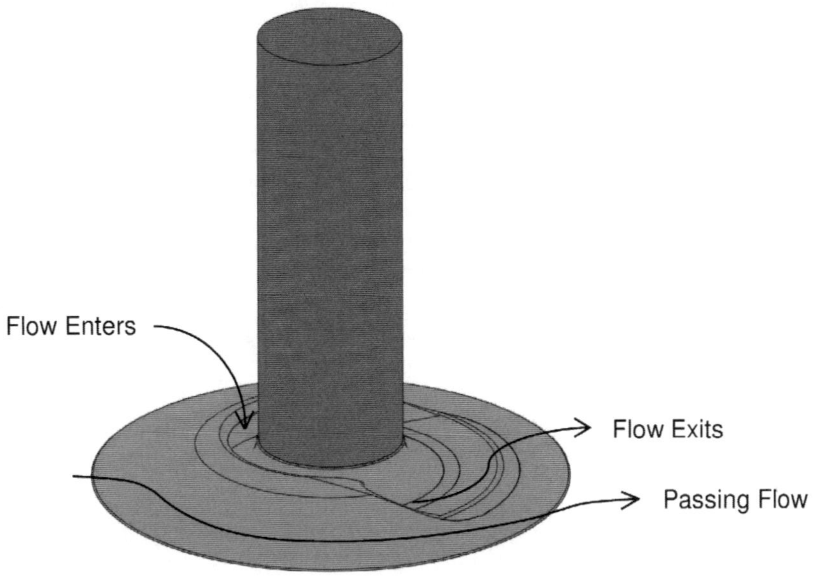

Figure 3: New counter measure design (collar prototype number one).

The remainder of the flow that does not enter into the collar cavity passes over the outer surface. The outer surface was designed in such a way that it slopes down to the riverbed, then curves outwards reaching a thin edge in an asymptotic fashion. This thin flat edge was incorporated to allow the passing flow to step smoothly onto and off of the collar generating as little turbulence as possible, therefore reducing the chances of scour occurring beneath the collar. In addition, the outer diameter of the collar was designed to be 3.3 times the pier diameter because the optimal size for reducing scour is approximately 3 times the pier diameter but in order to accommodate the inner cavity while maintaining the correct outer shape, a slightly larger diameter was required (Kumar et al. 1999). Therefore, any residual turbulence generated by the presence of the pier or raised collar region will not be able to contact the riverbed in the critical region around the pier because it is shielded by the large diameter collar. Lastly, the gradual change in slope was then utilized to gently transition the flow that does not enter the collar cavity to bypass over the collar and around the pier with the goal of minimizing any additional vortices that may form as a result of the collar's presence.

4 RESULTS

The numerical model was utilized throughout the design process as a tool to guide the changes made necessary to achieve a successful design. Specifically, stream tracers were used, as one of the main goals was to entirely contain the horseshoe vortex while using the smallest cavity possible for the purpose of reducing the influence on the passing flow. This took a series of trial and error iterations of analysing the stream tracers to obtain the correct diameter of the inner cavity and the correct outer slope. The shape was deemed successful and the design iterations were seized when the entire horseshoe vortex occurred inside the collar cavity and the redirected flow that did not enter the cavity remained on top of the collar's edge for the entire upstream half of the collar. Initial designs possessed a slope that was too steep which caused the flow that did not enter the cavity to be separated forming a secondary vortex immediately upstream of the collar on top of the riverbed. As a result, this would induce greater amounts of scour counteracting the purpose of the collar.

In Figure 4, a time-averaged comparison is presented showing the flow behaviour around a standard cylindrical pier and the same pier with collar prototype number one. Figure 4(a), which shows the pier without the collar, illustrates the standard flow behaviour where the flow separates in the presence of the pier causing a majority to be redirected around the sides of the pier but also a large portion to be drawn down to the riverbed. In the stagnation region immediately upstream of the pier, almost all of the streamlines approaching the pier are drawn downwards to some degree in the z-direction; however, only the streamlines in approximately the bottom one-third of the water depth are drawn all the way down to the riverbed. This portion of the flow that does reach the riverbed, redirects itself into a circular path, propagating around the upstream face of the pier forming the famous horseshoe vortex. Meanwhile, the remainder of the flow that does not become drawn all the way down to the riverbed redirects itself around the sides of pier but increases in velocity while doing so. The combination of the horseshoe vortex and the increased velocity around the sides of the pier, especially at the base, is what causes the sediment in that region to be dislodged and transported downstream.

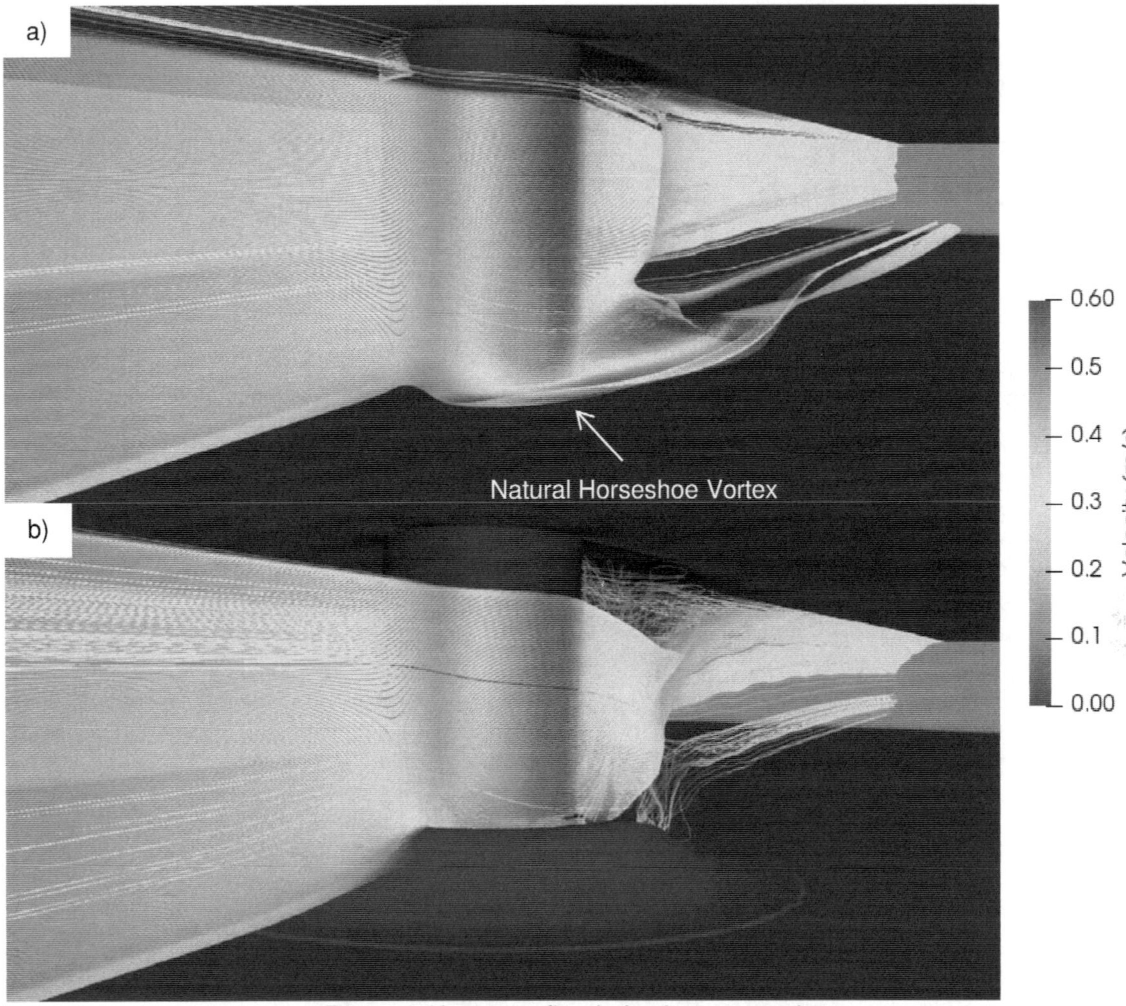

Figure 4: Upstream flow behaviour comparison.

Figure 4(b) shows the flow behaviour as a result of adding the new collar. The streamlines approach the pier and separate in the same manner as when the collar is not present but any of the streamlines that are drawn down to the riverbed enter directly into the collar cavity, circulate within, and then exit out the rear without being interrupted. Due to the collar presence, both scour causing mechanisms, seen when the collar is not installed, are addressed. Firstly, the entire horseshoe vortex no longer contacts the riverbed because it occurs inside of the collar cavity. While secondly, the high velocity flow along the sides of the pier also can no longer induce higher shear stress on the riverbed as the collar protects the entire critical region around the pier. In addition, since the critical region is protected, wake vortices that are naturally generated and shed off of the sides and rear of the pier are not able to scour around the base as it is protected by the collar.

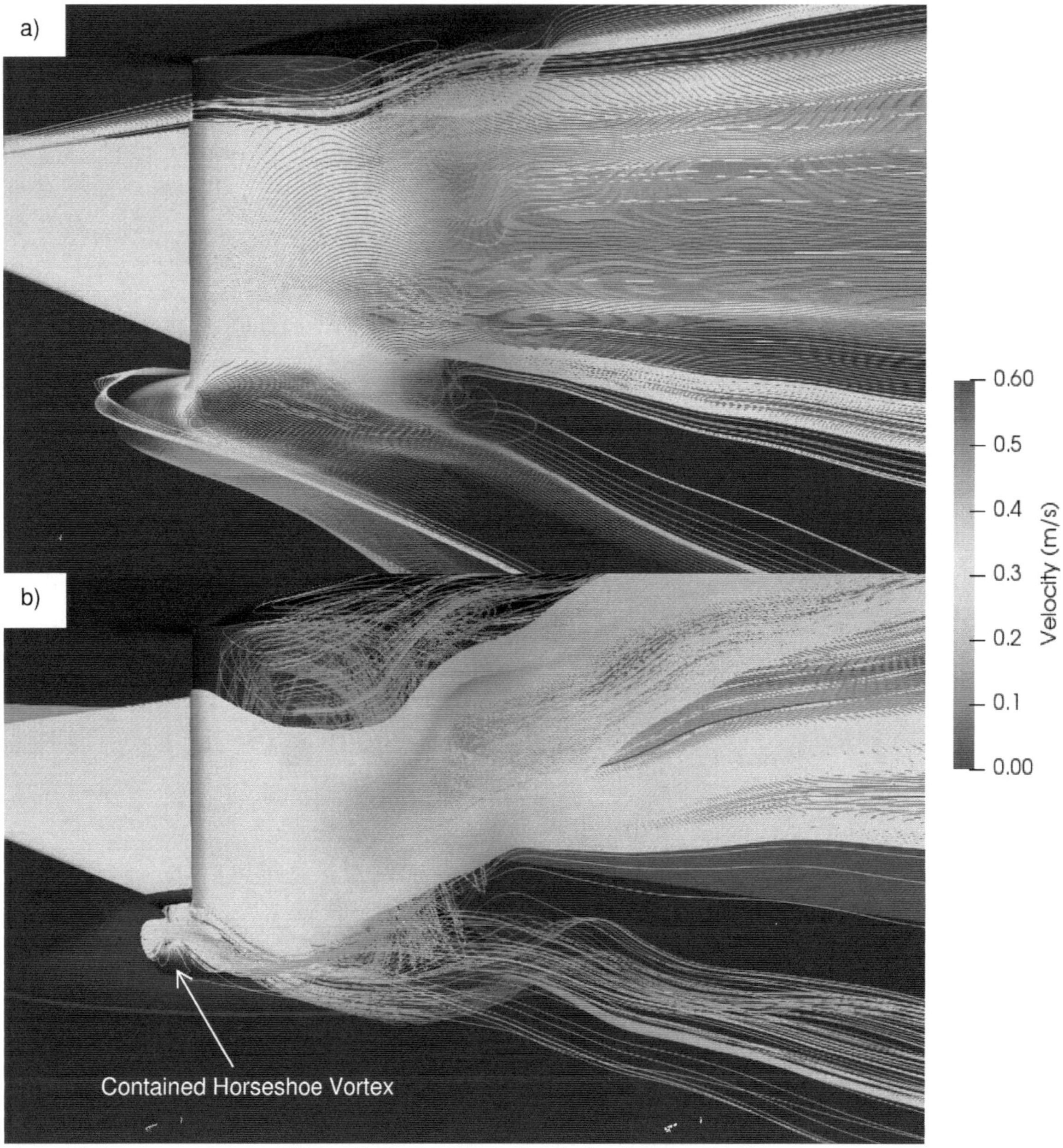

Figure 5: Downstream flow behaviour comparison.

Figure 5 is another comparison at a different angle, looking upstream, of the time-averaged flow behaviour around a pier with and without the collar. When the collar is not present, as seen in Figure 5(a), the horseshoe vortex extends further out from the pier and naturally contacts a large area of riverbed. However, when the collar is installed, the vortex is contained in a more compact manner close to the pier, allowing the remaining flow that is passing by to smoothly traverse over the collar and around the sides of the pier. The benefit to such collar over other counter measures is that the location and size of the horseshoe vortex is stationary and known despite varying flow conditions because it is confined to the predetermined shape of the collar. Figure 5(b) confirms that the horseshoe vortex does actually occur inside the collar cavity because the stream tracers demonstrate the circulating motion within. Once the vortex passes the sides of the pier, it exits out the rear of the collar in an upward direction off of a spoiler. This helps lessen the contact between the exiting vortex and the riverbed, even though there is little harm in small amounts of scour occurring downstream of the pier outside of the critical region.

5 DISCUSSION

Collar prototype number one overcomes the installation issue faced with many other counter measures because it can be constructed offsite in a specific facility. In doing so, the quality of the final product and the accuracy of the shape will be significantly better because there is access to more equipment in a weather controlled environment. To further simplify the installation, the collar will be constructed in pieces at the offsite location and transported to the pier site in smaller manageable sections. Upon arriving to the pier site, these sections will be bolted together around the pier and rested directly on the riverbed. It is crucial that when the collar is installed there is as small of a gap between the collar and pier as possible because the downward flow upstream of the pier will otherwise pass through the gap and erode underneath. This method of construction and installation is minimally intrusive as the pier does not need to be altered and no excavation of the existing riverbed is required, which ultimately results in less of an impact on the surrounding environment.

Another positive aspect of collar prototype number one is the lack of maintenance required. Once the collar is installed on the pier, for the remainder of the collar's life there is no foreseeable maintenance that is needed since increased flow will not harm the collar. If a large quantity of sediment is transported from upstream and enters into the collar cavity, due to the high velocity that occurs within the horseshoe vortex, the sediment will be washed out and sent downstream. Any stones or debris that are too large to be flushed out will be blocked from entering through the use of a screen affixed to the top of the cavity opening.

To further confirm the versatility and robustness of the design, experimental tests are underway in the laboratory. Numerous other existing scour counter measures, in addition to collar prototype number one, are being examined for comparison purposes under varying flow conditions.

The research presented in this paper possesses some limitations which are mainly attributed to the fact that the collar design is still in its infancy stage. The first limitation is that the collar was made for cylindrical piers only. The intention was to validate the design concept on cylindrical piers due to simplicity, then following validation adapt it for different shaped piers such as square, rectangular, and elongated. Once the shape has been adapted, a solution will then be available for bridge piers of every shape. The second limitation is that the collar was designed for unidirectional flow only, while it is understood that some locations where piers are present experience tides or flows that change direction. The goal is to eventually adjust the design in such a way to accommodate flood and ebb tides but the current focus is on perfecting unidirectional flow. The third limitation is that the physical construction and material of the actual full-size collar has yet to be determined as a thorough investigation into these areas still needs to be performed. The final limitation is that the collar was tested under only one flow velocity and that is due to the high computational cost associated with numerical modelling. However, upon achieving satisfactory results from the experimental tests for this collar design, additional velocities will be investigated to fully evaluate the collar's capabilities.

6 CONCLUSION

Bridge pier scour has been an issue for many years and a significant amount of time and money has been invested into researching the topic. Many counter measures have been designed; however, there is not one solution that is ideal. The introduction of a new state-of-the-art three-dimensional collar looks to overcome issues existing in the previous scour counter measures by being easy to install, minimally intrusive, maintenance free, and low cost. The newly proposed design works by containing the horseshoe vortex that occurs on the upstream face and sides of the pier within a collar structure that prevents it from contacting the riverbed. As a result, the erosive forces associated with the horseshoe vortex occur within the inner collar cavity and scour is avoided. Therefore, this novel design has the potential to change the way bridge pier scour is currently being addressed and will significantly improve the safety of bridges.

Acknowledgements

The authors gratefully acknowledge J.L. Lara, M. Maza, and G. Barajas from IH Cantabria, in Santander, Spain, for their assistance with numerical modelling. C. Valela also acknowledges A. Brévot and Q. Giraud from the University of Grenoble, France, for their research assistance. Lastly, the authors acknowledge the support from the Ontario Graduate Scholarship (OGS) and the Natural Sciences and Engineering Research Council of Canada (NSERC).

NSERC
CRSNG

References

Chiew, Y. 1992. Scour Protection at Bridge Piers. Journal of Hydraulic Engineering, 118(9): 1260-1269.

Deng, L. and Cai, C. S. 2010. Bridge Scour: Prediction, Modelling, Monitoring, and Countermeasures – Review. Journal of Hydraulic Engineering, 15(2): 125-134.

Dey, S. and Raikar, R.V. 2007. Characteristics of Horseshoe Vortex in Developing Scour Holes at Piers. Journal of Hydraulic Engineering. 133(4): 399-413.

Ettema, R., Nakato, T., and Muste, M. 2006. An Illustrated Guide for Monitoring and Protecting Bridge Waterways Against Scour. IIHR Technical Report No. 449, Iowa Highway Research Board, Project TR-515.

FHWA. 2009. Bridge Scour and Stream Instability Countermeasures: Experience, Selection, and Design Guidance – Third Edition, Volume 1. Hydraulic Engineering Circular No. 23, Publication No. FHWA-NHI-09-111.

Gritskevich, M.S., Garbaruk, A.V., Schütze, J., and Menter, F.R. 2012. Development of DDES and IDDES Formulations for the k-ω Shear Stress Transport Model. Flow, Turbulence, and Combustion, 88(3): 431-449.

Hosur, S.M., Ramesha, D.K., and Basu, S. 2014. Transient Simulation of Flow Past Smooth Circular Cylinder at Very High Reynolds Number Using OpenFOAM. Applied Mechanics and Materials, 592-594: 1972-1977.

Kumar, V., Raju, K.G.R., and Vittal, N. 1999. Reduction of Local Scour Around Bridge Piers Using Slots and Collars. Journal of Hydraulic Engineering, 125(12): 1302-1305.

Lauchlan, C. and Melville, B.W. 2001. Riprap Protection at Bridge Piers. Journal of Hydraulic Engineering, 127(5): 412-418.

Stevens, M.A., Gasser, M.M., and Saad, M.B.A.M. 1991. Wake Vortex Scour at Bridge Piers. Journal of Hydraulic Engineering. 117(7): 891-904.

Building Tomorrow's Society
Bâtir la Société de Demain

Fredericton, Canada
June 13 to 16, 2018 / *13 au 16 juin, 2018*

RESIDUAL DRIFT BASED SEISMIC DESIGN OF SHAPE MEMORY ALLOY REINFORCED CONCRETE BRIDGE PIER

Billah, AHM Muntasir[1], and Alam, Shahria[2,3]
[1] Parsons, Canada
[2] The University of British Columbia, Canada
[3] shahria.alam@ubc.ca

Abstract: Experience from past earthquakes and laboratory experiments suggest that structures designed according to current seismic standards are susceptible to residual deformation even during a design level earthquake. With the advancement in performance based seismic design, residual deformation has received very little attention in design codes and guidelines. Under seismic excitation, residual deformation is a direct outcome of nonlinear structural response. Researchers have developed several recentering structural systems to reduce the residual deformation during a seismic event. This paper proposes a residual drift-based design methodology for Shape Memory Alloy reinforced concrete (SMA-RC) bridge piers, which consists of defining the performance objectives, developing performance based damage states and formulating a performance based design guideline considering residual drift. The procedure anticipates the allowable residual drift based on target performance level, calculates the maximum allowable drift, and ensures that those deformation demands remain below the allowable residual and maximum drift.

1 INTRODUCTION

Over the last decade, there is growing consensus among researchers to consider residual drift as a design parameter for seismic design of structures. Observation from recent earthquakes and research results have evidenced that residual drift sustained by a structure after an earthquake plays a significant role in defining the seismic performance of a structure and need to be considered in the seismic design (Christopoulos et al. 2008, Erochko et al. 2011). Specifically for highway bridges, significant residual deformations can result in the total loss of a bridge if static incipient collapse is reached. Furthermore, they can also result in restriction of traffic loads and speed if not collapsed. These aspects are not reflected in current seismic design procedures. Researchers have developed different recentering structural systems for example, unbonded post-tensioned RC bridge columns (Kwan and Billington 2003), shape memory alloy (SMA) reinforced concrete structures (Billah and Alam 2015), and bracing systems with SMA (Haque and Alam 2017) to control or completely eliminate residual deformations. The objective of this study is to develop a performance-based seismic design guideline for shape memory alloy reinforced concrete (SMA-RC) bridge piers considering residual drift as the key performance indicator.

2 RESIDUAL DISPLACEMENT-BASED DAMAGE STATES

Performance-based seismic design largely relies on the identification and selection of proper limit/damage states. Often damage states are defined in terms of drift or displacement. Damages are usually defined as discrete observable damage states (e.g., rebar yielding, concrete spalling, longitudinal bar buckling, bar fracture) (Marsh and Stringer 2013). Although residual drift dictates the post-earthquake functionality of highway bridges, no other design guidelines except the Japanese code for highway bridge design (JRA 2006) provide any residual drift limit of bridge piers.

Five different SMAs are used in this study to develop the residual drift-based damage states for SMA-RC bridge piers. The bridge pier is assumed to be located in Vancouver, BC and was seismically designed following Canadian Highway Bridge Design Code (CHBDC 2014). Figure 1 shows the cross section and elevation of the bridge pier. The diameter of all the columns was fixed to be 1.83 m; the columns were reinforced with 48 longitudinal reinforcement of different diameter bars for different SMAs and 16 mm-diameter steel spirals at 76 mm pitch. The height of the pier is 9.14m with an aspect ratio of 5 which ensured the flexure dominated behavior.

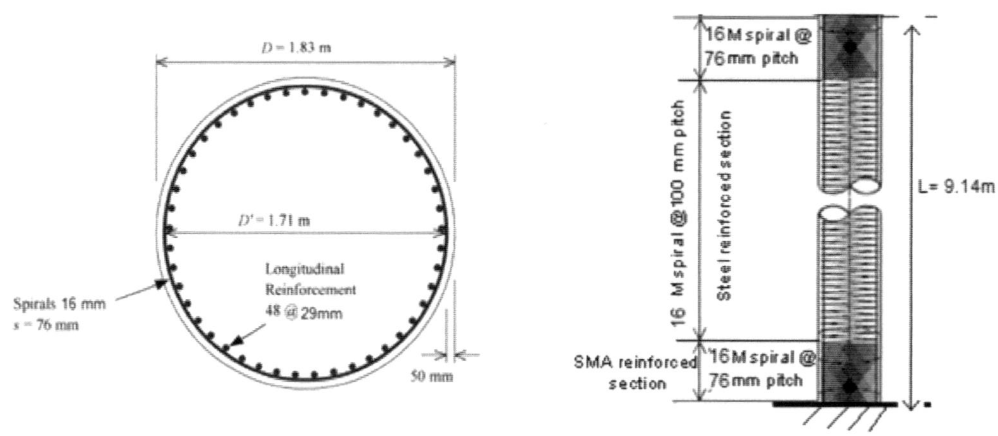

Figure 1: Cross section and elevation of SMA reinforced concrete bridge pier

In order to develop the residual drift based damage states (DS) for SMA-RC bridge pier a probabilistic approach has been adopted in this study. Based on the existing literature (O'Brien et al. 2007, Billah and Alam 2015), four different damage states have been identified and a range of limiting residual drifts were considered. It was assumed that a residual drift below 0.25% would meet the serviceability requirement (DS-1) while a residual drift larger than 1% would be characterized as a collapse damage state (DS-4). The intermediate damage states DS-2 and DS-3 are assumed to take place at a residual drift larger than 0.5% and 0.75%, respectively.

Incremental dynamic analysis (IDA) (Vamvatsikos and Cornell, 2002) was employed to determine the performance limit states of different bridge piers using an ensemble of ten selected ground motions. The incremental dynamic analyses were carried out using the 10 selected ground motions obtained from the PEER NGA ground motion database (2011).

Once the damage sates have been identified, fragility curves for residual drifts were developed using the IDA results for three different seismic hazard levels. In this study, fragility functions were developed using equation 3 which take the form of lognormal cumulative distribution functions having a median value of θ and logarithmic standard deviation or dispersion of β.

[1] $$F(RD) = \phi\left(\frac{\ln(RD/\theta}{\beta}\right)$$

where, *F(RD)* represents the conditional probability that the bridge pier will be damaged to a given DS as a function of the residual drift (*RD*); *F* denotes the standard normal cumulative distribution function; and θ and β are the median value of the probability distribution and the logarithmic standard deviation corresponding to the DS, respectively.

Figure 2 shows the fragility curves for SMA-RC bridge piers for different damage states at three different hazard levels. Here, the fragility curves are plotted irrespective of the SMA types to generalize the associated damage states. Using these fragility curves, the residual drift based damage states for SMA-RC bridge pier have been developed. From the fragility curves corresponding to each damage state, the RD value with a 50% probability of occurrence indicates the limiting value for the corresponding damage state. The limiting RD values with a 50% probability of occurrence at different damage states and hazard levels are developed as outlined in Table 1. From Table 1 it can be observed that as the ground motion return period decreases (probability of occurrence increases) the limiting residual drift corresponding to different DS decreases. Observation from Table 6 indicates that, as the damage level increases (DS-1 to DS-4) the difference in limiting RD values at different hazard levels decreases. For instance, at DS-2, the limiting RD value corresponding to 2475 years return period is 11% and 22.5% higher than that of 975 and 475 years return period, respectively. However, this difference goes down to 6.5% and 13.1% for DS-4.

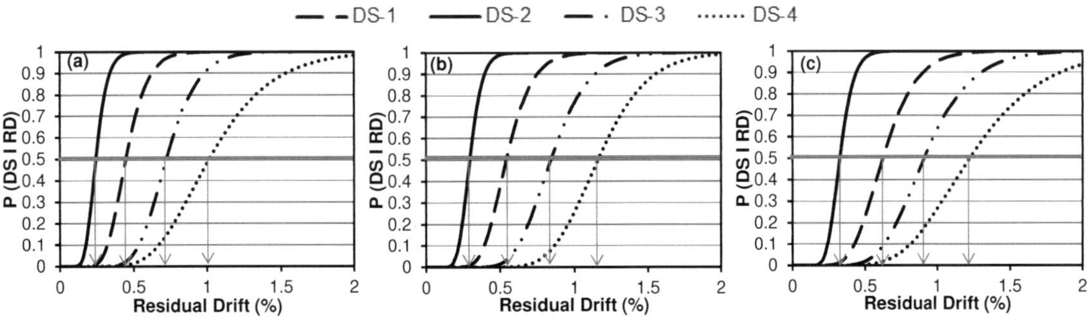

Figure 2: Fragility curves in terms of residual drift at (a) 10% in 50 years (b) 5% in 50 years and (c) 2% in 50 years probability of exceedance

Table-1: Residual drift damage states of SMA-RC bridge pier

Damage State	Functional Level	Description	Residual Drift, R_{Δ} (%) Probability of Exceedance		
			10% in 50	5% in 50	2 % in 50
Slight (DS=1)	Fully Operational	No structural realignment is necessary	0.24	0.28	0.33
Moderate (DS=2)	Operational	Minor structural repairing is necessary	0.48	0.55	0.62
Extensive (DS=3)	Life safety	Major structural realignment is required to restore safety margin for lateral stability	0.73	0.82	0.87
Collapse (DS=4)	Collapse	Residual drift is sufficiently large that the structure is in danger of collapse from earthquake aftershocks	1.04	1.16	1.22

3 RESIDUAL DISPLACEMENT-BASED DESIGN OF SMA REINFORCED BRIDGE PIERS

The proposed design method for SMA-RC bridge piers is developed following a displacement-based approach. Unlike other displacement-based approach, the required design base shear is calculated corresponding to a target residual drift and target performance level corresponding to a selected seismic hazard. The procedure adopted in this study follows the procedure developed by Kowalsky et al. (1995) and Priestley et al. (2007), but is specifically tailored to SMA-RC bridge piers using the damping-ductility relationship developed in this study. The design steps adopted in this study are outlined in a simple flowchart in Figure 3.

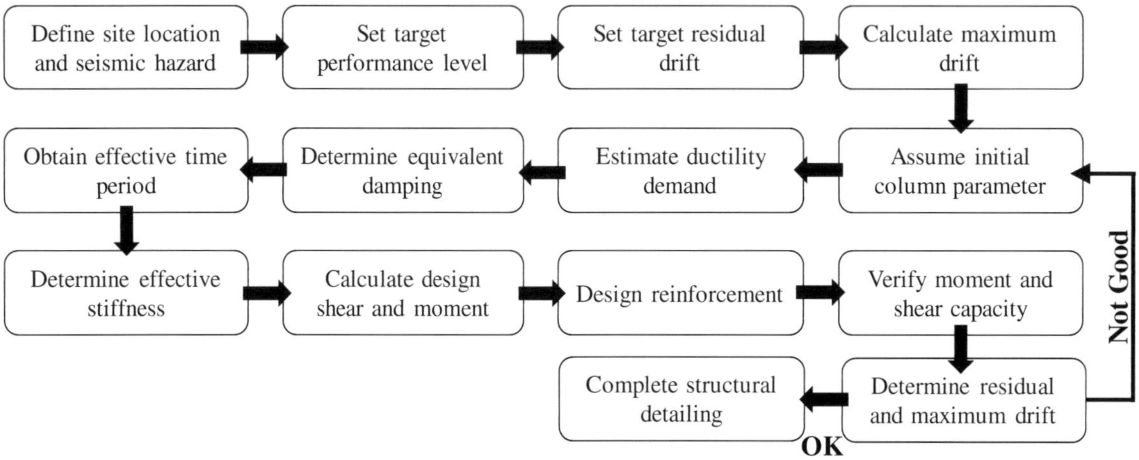

Figure 3: Flow diagram for residual drift based design of SMA-RC bridge pier

In this study, a damping-ductility relationship for SMA-RC bridge pier has been developed using a total of 100 ATC55/FEMA440 ground motions (Miranda and Garcia 2003) and following the method described by Dwairi et al. (2007). A set of new damping-ductility equations, in accordance with the previous expressions proposed by other researchers were developed in order to best approximate the damping-ductility relationship. Equation 2 represents the general form of the proposed equivalent viscous damping equation based on ductility for the SMA-RC bridge pier:

[2] $$\xi_{eq} = \xi_0 + \frac{a}{\pi}\left(1 - \frac{1}{\mu^b}\right)$$

In this equation, a and b are the two regression coefficients, and μ is the ductility demand. In order to obtain a generic damping-ductility relationship for SMA-RC bridge piers, all the examined bridge piers were considered together and the following expression was developed for the SMA-RC bridge pier:

[3] $$\xi_{eq} = 5 + \frac{32}{\pi}\left(1 - \frac{1}{\mu^{0.56}}\right)$$

The coefficient of determination or R^2 value obtained from this expression was higher than 85%.

3.1 Illustrative example

The following example is presented to demonstrate the performance-based design procedure for SMA-RC bridge piers.

The bridge pier is assumed to be located at Vancouver, BC in site soil class-C (stiff soil). The corresponding design spectrum is selected according to CHBDC-2014 which corresponds to 2% probability of exceedance in 50 years with a return period of 2475 years (Figure 4).

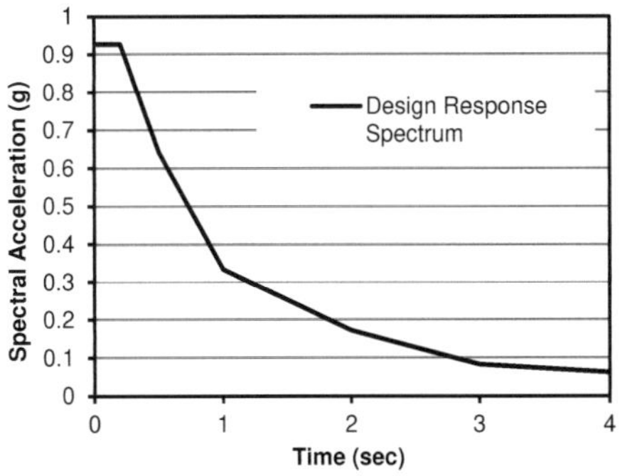

Figure 4: Design Acceleration Response Spectrum

The considered bridge is a lifeline bridge and according to CHBDC-2014 performance requirement, the bridge should be operational with limited service at the selected seismic hazard level. For the considered damage level, a target residual drift of 0.6% is selected to meet the performance objective. To restrict the residual drift within the target level, a Nitinol shape memory alloy with 6% superelastic strain (ε_s) is selected. In this design example, Nitinol SMA is considered since it is the most commonly used and commercially available SMA in the market. Moreover, a good number of research studies are also available on SMA-RC bridge piers using nitinol SMA.

Based on the target residual drift (RD) and superelastic strain (ε_s), the maximum drift (MD) is calculated using equation 3:

[4] $$RD = 0.5 \times \left(\frac{\varepsilon_s}{100} \times MD^2 \right) - \left(\frac{\varepsilon_s}{100} \times MD \right) + \frac{1}{\varepsilon_s}$$

or,

$$0.6 = 0.5 \times \left(\frac{6}{100} \times MD^2 \right) - \left(\frac{6}{100} \times MD \right) + \frac{1}{6}$$

Solving this quadratic equation we get, maximum drift, $MD = 4.92\%$

Maximum displacement, $\Delta_m = 0.0492 \times 5 = 0.246$ m

Initial column parameters:

Height of the pier = 5m

Lumped mass at the top of pier = 500,000 kg

Selected material properties of concrete, steel, and SMA are provided in Table-2.

Table-2: Material Properties

Material	Property	
Concrete	Compressive Strength (MPa)	42.4
	Elastic modulus (GPa)	23.1
Steel	Elastic modulus (GPa)	200
	Yield stress (MPa)	400
	Ultimate stress (MPa)	672
SMA	Modulus of Elasticity (GPa)	58.8
	Austenite-to-martensite starting stress (MPa)	401
	Austenite-to-martensite finishing stress (MPa)	510
	Martensite-to-austenite starting stress (MPa)	370
	Martensite-to-austenite finishing stress (MPa)	130

Superelastic strain (%)	6.0

For the considered hazard level, the yield drift is selected as 1.68% as developed by Billah and Alam (2016).

Yield displacement, Δy_T = 0.0168 × 5= 0.084 m

Equivalent viscous damping value corresponding to the design ductility is calculated using the following damping-ductility relationship developed in this study:

$$\xi_{eq} = 5 + \frac{32}{\pi}\left(1 - \frac{1}{\mu^{0.56}}\right) = 5 + \frac{32}{\pi}\left(1 - \frac{1}{2.93^{0.56}}\right) = 9.6\%$$

The spectral reduction factor (R_ξ) is calculated as:

$$R_\xi = \left(\frac{0.10}{0.05 + \xi}\right)^{0.5} = \left(\frac{0.10}{0.05 + 0.096}\right)^{0.5} = 0.83$$

Using the spectral reduction factor (R_ξ) of 0.83, the displacement spectrum corresponding to 9.6% damping is obtained. With this reduced displacement spectrum and the maximum displacement, Δ_m, the effective time period of the pier (T_{eff}) is calculated as 3.42 sec.

The effective stiffness (K_{eff}) based on the effective period (T_{eff}) is calculated as:

$$K_{eff} = \frac{4\pi^2 M}{T_{eff}^2} = \frac{4\pi^2 \times 500000}{3.42^2} = 1.68 MN/m$$

The design base shear is calculated as: $V_{base} = K_{eff}\Delta_m = 1.68 \times 10^6 \times 0.246 = 413.3kN$

The design moment is calculated as: $M_d = V_{base} \times L = 413.3 \times 5 = 2066.5kN - m$

Finally, for the design moment of 2066.5 kN-m, the column section is designed according to CHBDC 2014 [19] considering a column diameter of 1 m. For this design moment, a longitudinal steel ratio of 1.73% is required which is provided using 28-25M SMA rebar (24.9 mm diameter) in the plastic hinge region and 28-25M steel (diameter 25.2 mm) rebar in the remaining portion. The shear reinforcement was designed following CHBDC 2014 seismic design requirements which yielded 15M spirals at 50mm pitch providing a spiral reinforcement ratio of 1.49%. The shear capacity of the column is checked using modified compression field theory (Vecchio and Collins, 1986) which predicts the experimentally determined shear failure within 1% error. The shear resistance of the pier is found to be 2264 kN which is much higher than the applied shear force.
Performance evaluation of SMA-RC bridge pier

In order to validate the proposed design approach, the performance of the designed bridge pier is evaluated using NLTHA with ten earthquake records. The bridge pier was modelled in Seismostruct, a fiber based finite element software. The bridge piers were modelled through a 3D inelastic beam–column element (force based element). The objective of this evaluation is to compare the performance objectives (residual drifts and maximum drifts) with the predicted performance under the ensemble of 10 selected ground motions.

The results of the analyses in terms of maximum and residual drifts are presented in Figures 5a and 5b, respectively. These figures show the maximum and residual drift response obtained from each nonlinear time history analysis along with the target maximum and residual drift (horizontal dotted line) used in the design.

From these figures it is evident that the bridge pier sustained maximum and residual drifts within 15% of the target maximum and residual drift. It was observed from the analysis that among ten earthquake records, two marginally exceeded the target residual drift of 0.6% and maximum drift of 4.92%. The remaining eight are below the design level residual drift and targeted maximum drift. These minor discrepancies can be attributed to the linearization of the displacement spectrum adopted during the design and scaling of ground motions. However, the average response in terms of both residual and maximum drifts was very close to the targeted drift levels. Priestley et al. (2007) suggested that the differences in the target drift and obtained drift from NLTHA is acceptable if the mean of the peak drifts remains close to the design drift.

4 SUMMARY AND CONCLUSIONS

This study presented a new residual drift based design method for shape memory alloy reinforced concrete bridge piers. The approach outlined in this paper is a comprehensive approach for performance-based design of SMA-RC bridge piers. This study developed necessary design equations and graphs for PBSD of SMA-RC bridge piers. The proposed method provides the owner to select expected performance of the bridge pier and allows the designer/engineer to select multiple hazard and performance expectation combinations. Following the DDBD guidelines of Priestley et al. (2007) the authors developed their own design method and damping-ductility relationship for SMA-RC bridge piers. In contrast to the conventional DDBD approach, the proposed procedure anticipates a target residual drift based on the expected performance during design earthquake, calculates the maximum drift demand, and ensures that those drift demands (maximum and residual) remain below acceptable limits for the design level earthquakes. The performance of the bridge pier was validated using NLTHA, and the maximum and residual drifts at the design level earthquakes were found to satisfy the performance expectations.

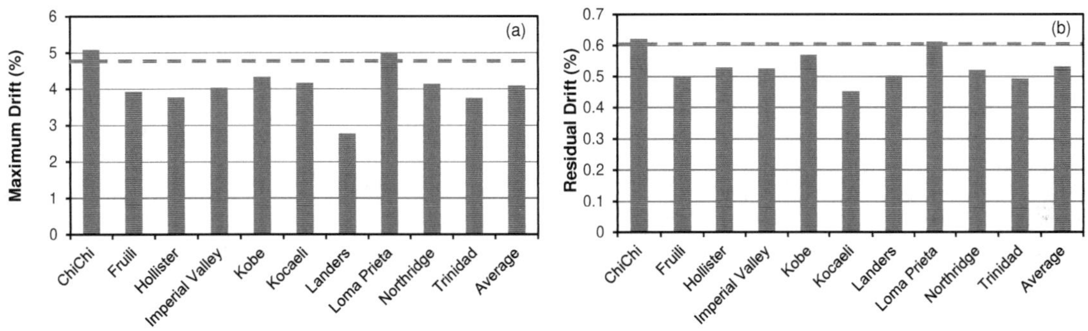

Figure 5: (a) Maximum and (b) residual drift value obtained from time history analysis of the designed pier (Red line showing the target maximum and residual drift)

Acknowledgment

The financial contributions of Natural Sciences and Engineering Research Council of Canada (NSERC) through Discovery Grant and Industrial Postgraduate Scholarship Program are gratefully acknowledged.

References

Billah, A.H.M.M., and Alam, M.S. (2015). "Seismic fragility assessment of concrete bridge piers reinforced with superelastic shape memory alloy." Earthquake Spectra, 31(3):1515-1541.

Billah AHMM and Alam MS. (2016). "Performance based seismic design of concrete bridge pier reinforced with Shape Memory Alloy- Part 1: Development of Performance- Based Damage States." ASCE J. Struct. Eng. 142 (12):1-11.

Canadian Highway Bridge design code, 2014. CAN/CSA-S6-14. Draft version for public review. National Research Council of Canada, Ottawa; ON.

Christopoulos, C., Pampanin, S. and Priestley, M.J.N. (2003). "Performance-based seismic response of frame structures including residual deformations – Part I: single-degree-of-freedom systems." J. Earthq. Eng., 7(1): 97–118.

Dwairi, H.M., Kowalsky, M. J. and Nau, J.M. Equivalent Damping in Support of Direct Displacement-Based Design. Journal of Earthquake Engineering, 2007, 11(4), 512-530.

Erochko, J., Christopoulos, C., Tremblay, R. and Choi, H. (2011). "Residual drift response of SMRFs and BRB frames in steel buildings designed according to ASCE 7-05." J. Struct. Eng., ASCE, 137(5): 589-599.

Haque, ABMR and Alam, MS. (2017). "Hysteretic Behaviour of a Piston Based Self-centering (PBSC) Bracing System Made of Superelastic SMA Bars – A Feasibility Study." Structures. 12: 102-114.

Japan Road Association, 2006. Specifications for Highway Bridges. Japan Road Association: Japan.

Kowalsky MJ, Priestley MJN, MacRae GA Displacement-based design of RC bridge columns in seismic regions. Earthq Eng Struct Dyn. 1995, 24(12), 1623–1643

Marsh, L. K. and Stringer, S.J. (2013). Performance-based seismic bridge design, A Synthesis of highway practice, NCHRP Synthesis-440, TRB, Washington, D.C.

Miranda E. and Ruiz-García J. Evaluation of approximate methods to estimate maximum inelastic displacement demands. Earthq Eng Struct Dyn. 2003; 31, 539–560.
O'Brien, M., Saiidi, M.S. and Zadeh, M.S. (2007). "A study of concrete bridge columns using innovative materials subjected to cyclic loading." CCEER, Department of Civil Engineering, University of Nevada, Reno, Nevada, Report No. CCEER-07-01.
Priestley, M. J. N., Calvi, G. M., and Kowalski, M. J. Displacement-based seismic design of structures. IUSS press, Pavia, 2007.
Vamvatsikos, D. and Cornell, C.A. (2002). "Incremental dynamic analysis." Earthq. Engr. Struct. Dyn. 31(3): 491-514.
Vecchio, F.J., and Collins, M.P. The modified compression-filed theory for reinforced concrete elements subjected to shear. ACI Structural Journal, 1986; 83(2): 219–231.

Building Tomorrow's Society
Bâtir la Société de Demain

Fredericton, Canada
June 13 to 16, 2018 / *13 au 16 juin, 2018*

RESILIENCE ASSESSMENT OF WATER NETWORKS AGAINST SEISMIC HAZARD

Assad, Ahmed[1,4], Nasiri, Fuzhan[2], and Zayed, Tarek[3]
[1] Ph.D. Student, Concordia University, Canada
[2] Assistant Professor, Concordia University, Canada
[3] Professor, Hong Kong Polytechnic University, Hong Kong
[4] ahmed.assad@mail.concordia.ca

Abstract: Water Supply Networks (WSNs) are critical infrastructure systems that provide water to the public. Maintaining functionality of WSNs is desirable especially after hazard events to facilitate firefighting and rescue activities. Aging of WSNs increases their vulnerability and the likelihood of function interruption. There is a growing need to investigate the resilience of infrastructure systems to assess their ability to withstand a disruption and to recover rapidly after service interruption. Several models have been developed to consider resilience during the design of WSNs, however there is a setback regarding incorporation of resilience in repair strategies. This study introduces a resilience-based approach that aims at both evaluating the resilience of WSNs as well as enhancing their recoverability after a hazard event. Firstly, a resilience metric is suggested based on the physical condition of the pipelines in the WSN. Secondly, causal loop diagrams are established to capture the interactions among the parameters related to asset condition, hazard intensity, performance loss, rehabilitation and restoring activities. Thirdly, stock and flow diagrams are generated, relevant mathematical and logical relations between the parameters are defined, and several checks are performed to verify the developed model. Finally, the model is applied on a portion of Montreal WSN to simulate an earthquake that causes around 40% performance loss. A budget of 750,000 dollar is found sufficient to recover the network resilience to 75% of its initial value. The practical value of this study is signified by the tool, which can be used by the stakeholders to select the best management and mitigation strategies that increases the resilience absorptive and restorative capacities of their WSNs.

1 INTRODUCTION

Following any hazard, water infrastructures play a dominant role in firefighting and rescue efforts. Hence, maintaining the functionality of such critical systems is of a paramount importance after any hazard event or disruption (Farahmandfar et al. 2016). While the traditional focus was on the physical protection of water systems, the emerging trend is highly raising the issue of resilience. The American Society of Civil Engineers defines the resilience of infrastructure as the ability to mitigate all-hazard risks and rapidly recover critical services with minimum harm to the public safety, health, economy and national security (Ayyub 2014). In this context, it is desirable for the water networks to be strong enough to withstand any disruption with a minimum impact on its performance and to recover quickly in case of service loss (Cimellaro et al. 2015). Consequently, there has been a rapidly increasing attention both, in practice and academia, to define resilience and derive quantifiable resilience metrics. The literature includes several frameworks that have been developed to include during the design and operation of civil infrastructures. Nevertheless, most of the previous work focused on a single hazard event instead of assessing the resilience of water assets to multi-hazard events. Moreover, there are very limited efforts that consider the various aspects and parameters of WNs resilience. Therefore, this paper introducers a comprehensive resilience assessment framework and demonstrates its use in evaluating resilience of water networks as well as selecting optimal recovery strategies.

1.1 Background

Reviewing the literature reveals two main approaches for measuring and assessing resilience of water networks: qualitative approaches and quantitative approaches. Most of the qualitative approaches comprise listing resilience attributes, assigning them numerical weights, and then aggregating these weighted metrics into a single composite index. The main drawback of this approach is ignoring the relations between the indicators. Fisher et al. (2010) introduced a resilience index to measure the resilience of critical infrastructures including water systems. The model required extensive data collection about 1,500 variables categorized under robustness, recovery, and resourcefulness. The variables were then weighted and summed to generate a single global resilience index that facilitates the comparison between different infrastructure systems (Fisher et al. 2010). Such approaches are usually subjective, and the results cannot be generalized on a large scale. Quantitative approaches of resilience assessing involve modeling the impact of a specific hazard on the network. Such approaches account for two main components of resilience: the hazard severity and the system response. Some studies employed resilience as a system property and aimed to investigate the dynamic relations between the components of the system. Jayaram and Srinivasan (2008) formulated a resilience assessment model that accounts for multiple sources within a network, a major limitation of Todini's model (Jayaram and Srinivasan 2008). Gay and Sinha (2013) integrated MATLAB and EPANET2 software packages to develop a stochastic simulation-based methodology to assess the resilience of water distribution system. Simulations were run to quantify the performance loss, recovery time and recovery cost following a hazard event (Gay and Sinha 2013). Another quantitative approach to evaluate system's resilience employs it as the opposite of the system's vulnerability to various disruptions. Principles of graph theory were utilized in many of these approaches to model the water network as nodes and links (Farahmandfar et al. 2016). Despite the previous efforts in trying to address the performance assessment of water networks, several draw backs can be realized. Most of the proposed models aimed at enhancing the resilience with the minimum cost. Such approach overlooks other chief objectives that should be considered such as time of service disruption. Moreover, very few models addressed the issue of repair and recovery actions considering the scarcity of non-monetary resources. Accordingly, this study is utilizing system dynamics simulation to introduce a scenario-based approach for assessing the resilience of water networks. This approach is intended to provide a deeper insight on the inter-relations between the different system parameters, hazard characteristics, and recovery strategies. To achieve the ultimate aim of this study the below objectives needs to be satisfied:

- Identify causal relationships between the parameters of water network resilience.
- Construct stock and flow diagrams based on the developed causal relationships.
- Simulate a hazard scenario and investigate the effect of different restoration strategies on the time of service disruptions.

2 DATA COLLECTION

The City of Montreal granted an access to their data base of water distribution network. The city of Montreal is responsible for managing a huge inventory of water pipeline some of them was installed as far back as 1895 and is still active till today. Figure 1 shows that over 92% of the City's network is either cast iron or ductile iron. The extracted data include data related to the characteristics of the pipeline, installation data, and breakage data. The breakage data include the number and time of each observed break in the network.

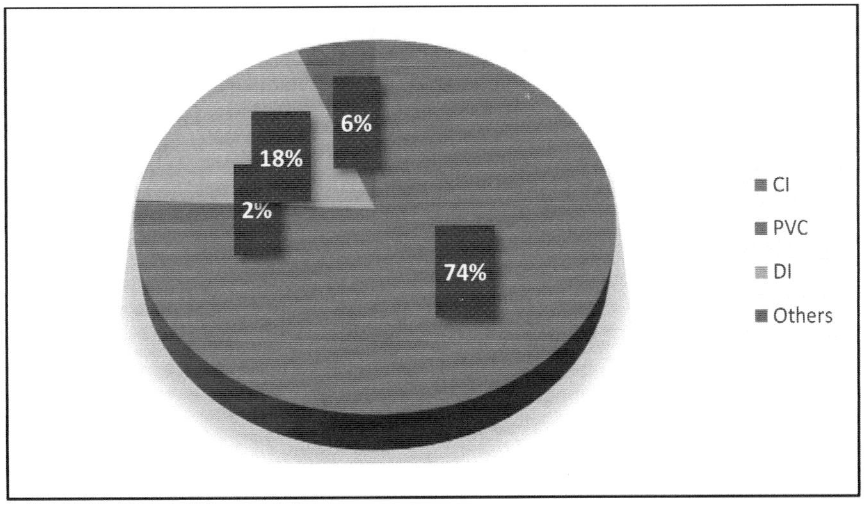

Figure 1 : Distribution of Pipe size

The breakage rate along with the pipe age were used as indications of the pipe's condition. In this study, three condition ratings are considered namely: Excellent Condition, Good Condition, and Poor Condition. Each of these conditions is mapped to a certain resilience state such that an excellent condition corresponds to withstanding resilient, good condition to resilient, and poor condition to non-resilient.

3 MODEL DEVELOPMENT

A literature review was done to investigate the different available models on resilience of WSNs. afterwards, the needed factors to develop a causal loop diagram were identified. Based on this diagram, stock and flow diagrams were generated and the associated logical and mathematical relations were defined. Figure 2 is the causal loop diagram that serves as the basic rational of this study. The main parameters in this diagram are hazard characteristics and its impact on resilience state as well as the factors controlling the restoring activities along with their impact on the resilience state. There is a reinforcing, positive, relationship between the hazard intensity and the performance loss it causes to the system. More over, the exhibited loss is less when the absorptive capacity of this asset is higher. The second loop in this diagram is capturing the ownership policies and the owner willingness to pay. As the asset is in a better resilience state, the owner will be less likely to spend on it. Hence a lower budget will be allocated for this particular asset, less bargaining power. However, with a continuous decrease in the possible repair activities the overall resilience estate will deteriorate. Figure 3 illustrates the stock and flow diagram which is generate based on the aforementioned causal loop diagram. In this diagram, there main stock variables were defined namely: withstanding resilient, resilient and non-resilient. Each stock represents the number of water mains in a certain resilience state. An asset would typically flow from one resilience state to a lower one causing a performance loss due to the hazard event occurrence. In this model, two attributers are used to quantify the hazard characteristics: the severity of the hazard and its duration. On the other hand, the restoring activities will cause an asset to flow on the opposite direction, from a lower to a higher resilience state. The parameters that control these resorting activities are related to the share of the budget allocated for each repair type as well as the resource availability ratio. The earlier one is usually set a head of time during the evaluating of losses and planning phase while the latter is a property that describes the entity's ability to address more than one repair task at once. This ability is generally a function of the entity's manpower, machinery, and available money. It should be noted that for further clarity, some intermediate variables are hidden. Figure 4 shows the remaining parameters of the model and it includes the life cycle costing, budget allocation and network resilience computations. The overall network resilience can be found using equation 1:

$$[1]\ \text{Network Resilience} = \frac{3\times\text{Withstanding Resilient}+2\times\text{Resilient}+1\times\text{Non-Resilient}}{\text{Withstanding Resilient}+\text{Resilient}+\text{Non-Resilient}}$$

Where: withstanding resilient is the number of assets that are in a withstanding resilience state, resilience is the number of assets that are in a resilient state, and non-resilient is the number of assets that are in non-resilient state.

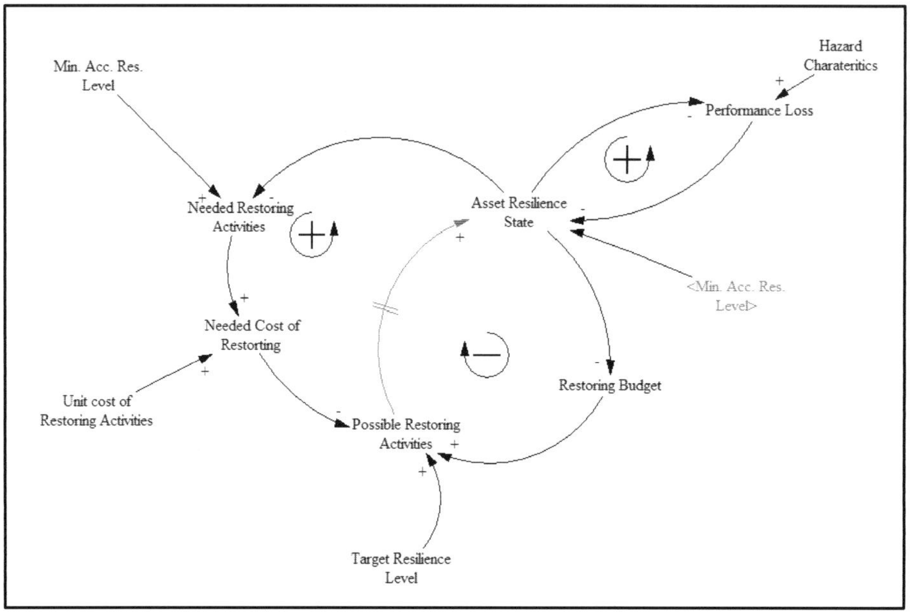

Figure 2 Causal loop diagram

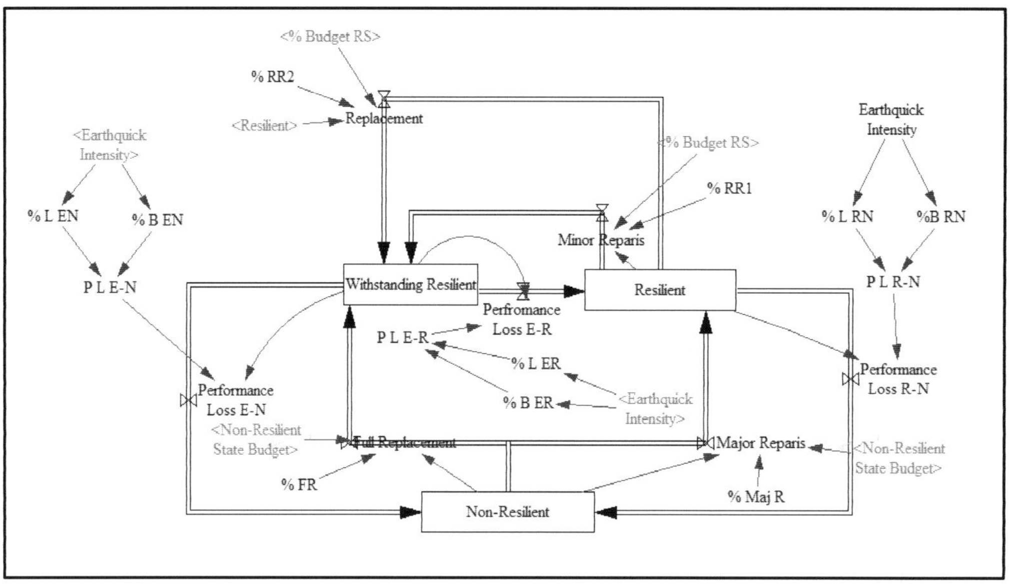

Figure 3 Stock and flow diagram (technical)

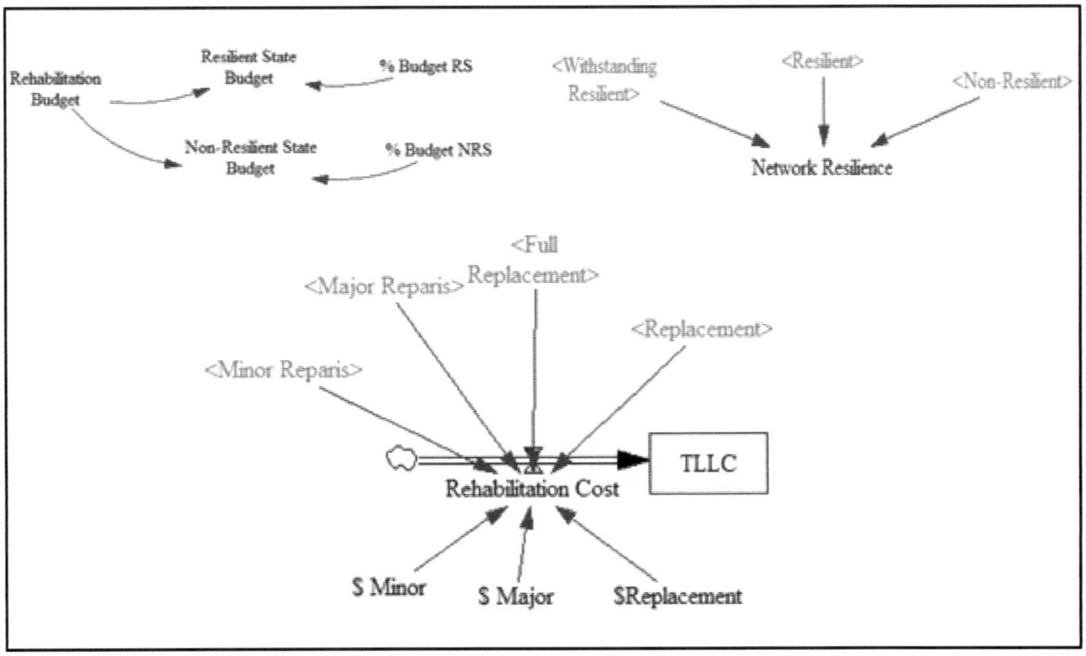

4

Figure 4 Stock and flow diagram (financial)

In Figure 4, the stock variables are calculated as the difference between the inflows and outflows at any time instant. For example, the withstanding resilient stock which represents the number of assets in a withstanding resilient state is computed using equation 2:

[2] $Withstanding\ Resilient = \int full\ replacment + minor\ repairs - perfromance\ loss_{E-N} - perfromance\ loss_{E-R}$

Where full replacement represents the number of pipe assets that are fully replaced with new ones, major repairs represent the number of assets that experience a major repair, performance $loss_{E-N}$ and performance $loss_{E-R}$ represents the number of assets that deteriorates from withstanding resilience to non-resilience or from withstanding resilience to resilience states respectively. The number of water mains that will experience a minor repair, major repair, or full replacement is determined based on the pre-set budget allocated for each restoring type. The number of assets that the performance loss that identifies the number of assets deteriorating from one state to another in correlated to the hazard intensity.

5 RESULTS AND DISCUSSIONS

A preliminary test of the model was performed on a virtual water distribution network that consists of 115 pipelines. The condition of the assets was forecasted and mapped to the corresponding resilience state. Twenty-five pipelines were classified in withstanding resilience state, seventy as were classified as resilient, and twenty were classified as in the non-resilient state. An earthquake is simulated as the hazard event of this model. The probability that the Liquification Potential Index exceeding a threshold of five, P(LPI>5) is the basis of estimating its intensity in this study (American lifeline Alliance, 2005). The assumed earthquake has P(LPI>5) between (45%-95%). Subsequently, performance loss indices were assumed based on the hazard intensity. These assumptions can be changed based on the operational and environmental condition of the network as well as its topology and location. Any variance in assigning the values of these parameters can be readily handled by sensitivity analysis. In this study, three repair (resorting) activities are included: minor, major and full replacement. It is assumed that a minor repair would transfer an asset from resilient to withstanding resilient stare, a major activity would transfer an asset from non-resilient to a resilient state, and a full replacement would transfer an asset from non-resilient to withstanding resilient state. Figure 5 depicts the variation of the three stock variables over time because of both the performance losses and the resorting actives. The network resilience dropped initially from 2.04 – 1.27 due to the earthquake. It was recovered to a value of 1.47 in a period of two months. It can be noticed from figure 5 an increase in the value of non-resilient state compared to a drop in the resilient and withstanding resilient states at the beginning of the simulation. That is due to the fact that the hazard event causing the witnessed performance loss started at that instant. On the other hand, the graph shows an increase in the values of the resilient and withstanding resilient states upon the end of the simulation period affected by the accumulated restoring activities that were achieved. Figure 6 shows that the allocated budget for the recovery in this scenario is 750,000, however the total encountered cost is just around 513,000. This is because of the hidden delays that are implemented in this model. Such delays are known as performance delays which reflect the real time spent between initiating the repair action and achieving it. various scenarios were then simulated to investigate the effect of changing the total budget on the overall network resilience. The effect of changing the budget on the overall network resilience at the end of the recovery process is shown in Figure 7. It was possible to increase the network resilience to a value of 1.60 at the end of the second month by increasing the resorting budget to 1.5million dollar.

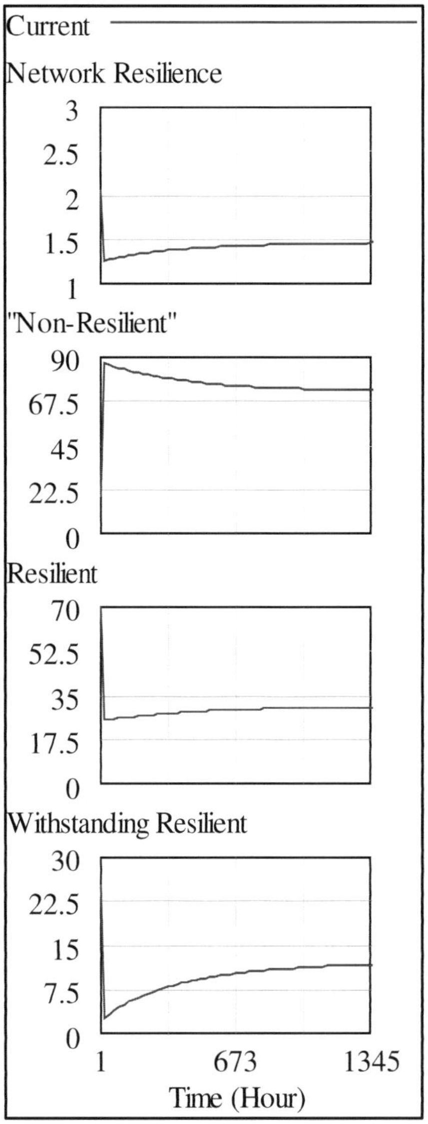

Figure 5 Variation of stock variables

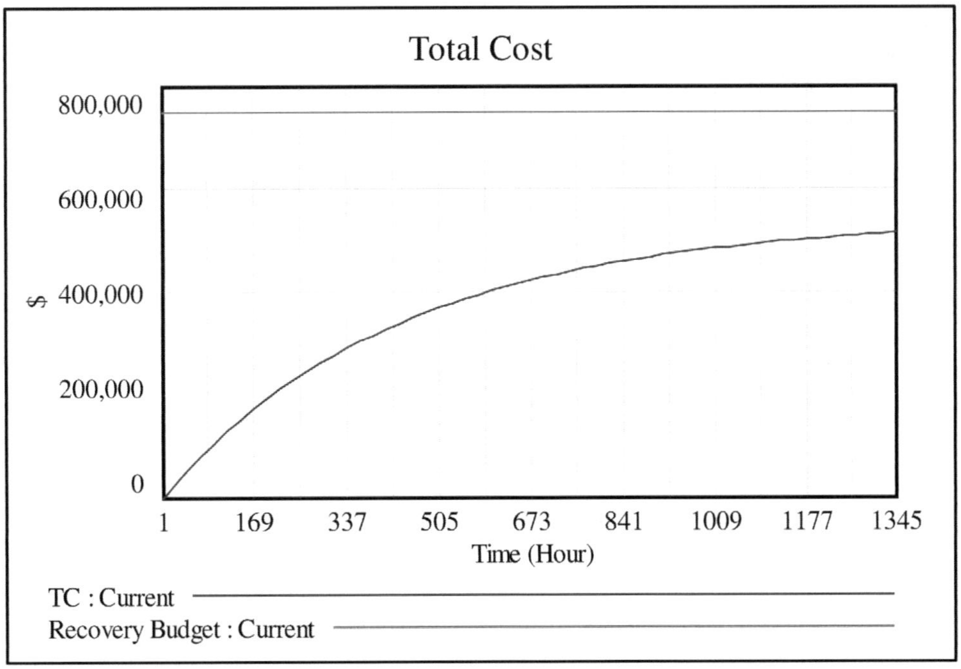

Figure 6 Total accumulated cost

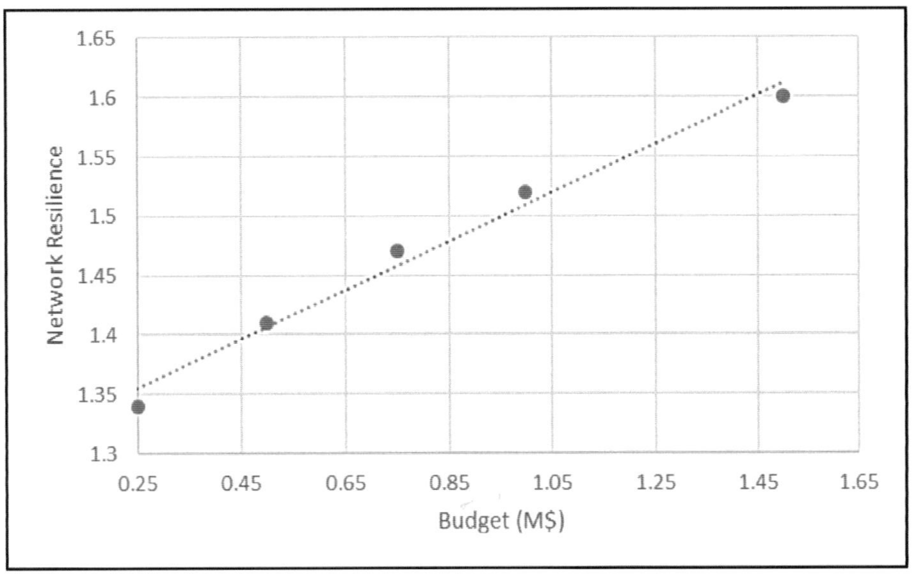

Figure 7 Effect of changing the budget on overall network condition

6 CONCLUSION

A novel resilience assessment model is presented in this study. The model utilizes the use of system dynamics simulation to establish a scenario-based resilience assessment for water network to multi-hazard events. This paper demonstrates the feasibility of using system dynamic as a way to capture all the different aspects of resilience including the system attributes, the hazard characteristics, and the resorting polices and procedures. Different scenarios might be investigated in future work based on the formulation of this model. Examining the effect of different budget allocation strategies, different hazard intensities, and different on the overall resilience process are just examples.

References

ALA (American Lifel inesAlliance). (2005). Seismicguidelines forwaterpipelines. ⟨http://www.americanlifelinesalliance.com/pdf/SeismicGuidelines_WaterPipelines_P1.pdf⟩ (Jan. 29, 2017).

Ayyub, B. M. (2014). Systems resilience for multihazard environments: Definition, metrics, and valuation for decision making. *Risk Analysis*, **34**(2), 340–355.

Cimellaro, G. P., Tinebra, A., Renschler, C., and Fragiadakis, M. (2015). New Resilience Index for Urban Water Distribution Networks. *Journal of Structural Engineering*, **142**(8), C4015014.

Farahmandfar, Z., Piratla, K. R., and Andrus, R. D. (2016). Resilience Evaluation of Water Supply Networks against Seismic Hazards. *Journal of Pipeline Systems Engineering and Practice*, **8**(1), 4016014.

Fisher RE et al. (2010) Constructing a Resilience Index for the Enhanced Critical Infrastructure Recovery Program. Lemont, IL: Argonne National Laboratory. ANL/DIS-10-9.

Gay, L. F., and Sinha, S. K. (2013). Resilience of civil infrastructure systems: Literature review for improved asset management. *International Journal of Critical Infrastructures*, **9**(4), 330–350.

Jayaram, N., and Srinivasan, K. (2008). Performance-based optimal design and rehabilitation of water distribution networks using life cycle costing. *Water Resour. Res.*, **44**(1), W01417.

Building Tomorrow's Society
Bâtir la Société de Demain

Fredericton, Canada
June 13 to 16, 2018 / *13 au 16 juin, 2018*

REVIEW OF EUROPEAN APPROACH FOR PERFORMANCE-BASED SEISMIC DESIGN OF STRUCTURES

Tsiotsias, K.[1,3], Pantazopoulou, S.J.[2]
[1] Graduate Research Assistant, Dept of Civil Engineering, York University, Canada
[2] Professor of Civil Engineering, the Lassonde Faculty of Engineering, York University, Canada
[3] kostasts@yorku.ca

Abstract: This paper aims to provide a coherent summary of the performance design methods adopted for seismic design by the European Code Framework (EC 8-I 2004), as background reference for the introduction of performance based design in the NBCC. Performance-based design begins with selection of performance objectives for the design earthquake, a decision made by the stakeholders with explicit acknowledgment regarding the degree of damage acceptable to occur at the design event. Performance objectives are mapped on the seismic demands, expressed in terms of displacements at the selected performance limit state. Depending on the selected structural system, global displacements are converted to local deformation demands, and acceptance criteria are used to establish realization of the performance objectives (member detailing to meet the deformation performance requirements). The manner in which this process affects detailing procedures in the European design practice is summarized with reference to reinforced concrete (RC) systems and members (beams, columns and walls).

1 INTRODUCTION

Seismic design in the 27 European member states follows a common standard known as "Eurocode-8 Design of structures for earthquake resistance", which comprises six different parts as follows: Part 1: General rules, seismic actions and rules for buildings (2004); Part 2: Bridges (2005); Part 3: Assessment and retrofitting of buildings (2005); Part 4: Silos, tanks and pipelines (2006); Part 5: Foundations, retaining structures and geotechnical aspects (2004); and Part 6: Towers, masts and chimneys (2005). The Eurocodes are published and maintained by CEN (i.e., European Committee for Standardization or Comité Européen de Normalization), but values for several parameters that are local or specific to the natural hazards of the member states are prescribed in respective National Annexes.

Motivation for the present paper is the intended revision of the NBCC so as to explicitly incorporate a performance based design framework (PBD). In fact, many of the essential elements of PBD (i.e. a distinction between low, moderate and high ductility demands) are already included in the NBCC – particularly so with respect to the design of structural walls; however design requirements are linked to demands and not to a uniform expectation of performance. In this paper the framewotk of the European performance based seismic design is summarized. The intent is not to make a comparison with other similar frameworks (e.g. ASCE/SEI 2007) or with the existing NBCC procedures, but rather to identify and present the essential components that are needed in any such design framework in order to accomplish the desing objective.

Both the design and the assessment framework in the Eurocode 8 (EC-8) are tailored to the so-called performance-based philosophy. The difference between performance-based design (PBD) and capacity design (CD) is that in the latter, a state of equilibrium is considered near failure without explicit acknowledgment of the magnitude of displacement (i.e. damage) that the structure would have to undergo before the intended mode of failure may be possible; whereas in PBD the magnitude of displacement and the extent of damage at the reference limit state is determined by the stakeholders or the code, and design is intended to secure, through pertinent detailing, that the structure meets this expectation without undue or premature failure.

The ultimate limit state therefore, for PBD, is referred to as a "Performance Point", (PP), which is the point on the resistance curve of a structure which defines the maximum displacement and associated force that will be attained during the design earthquake hazard (Figure. 1). The vertical axis represents the base shear and the horizontal axis represents the lateral displacement at the top of the building; with increasing displacement coordinate the degree of damage increases. If Δ_p defines the available plastic deformation capacity (which extends from yielding to failure, then the cost of retrofit is linked directly to the fraction of Δ_p that has been exhausted at the attainment of the PP. In design the PP is linked to specific performance objectives – that is, distinct performance limit states, identified as PL in Figure. 1, where the amount of damage a structure has sustained, and the degree of operationality in the primary and secondary members of the structural system may be mapped on the resistance curve; but can also be described in terms of visual indicators, as follows:

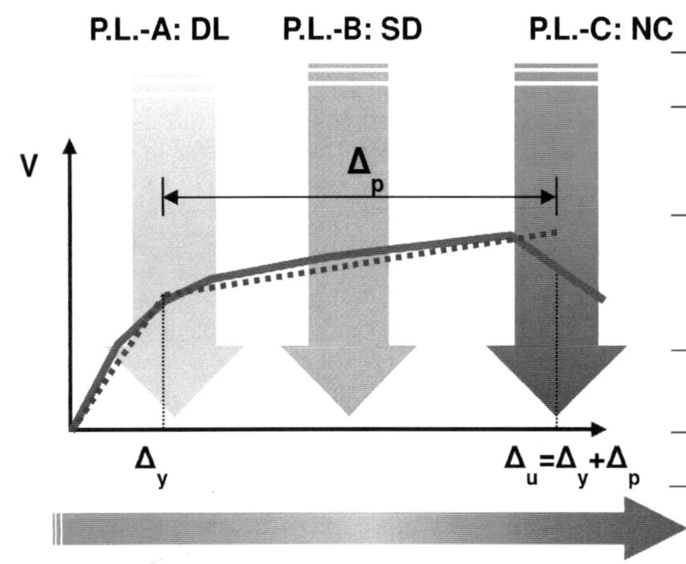

Limit state:	Max. Displ.	Cost of repair/replace
Damage Limitation: Operational: A1	Δ_y	0%-10%
Damage Limitation: Immediate Occupancy: A2	$\Delta_y+0.3\Delta_p$	5%-30%
Significant Damage – Life safe: B	$\Delta_y+0.6\Delta_p$	20%-50%
Near Collapse: C	$\Delta_y+\Delta_p$	>70%

Figure 1: The color code identifies the intensity of damage – failure is defined as the point in the post peak branch of the response curve where residual strength is less than 80% of maximum.

2 VISUAL INDICATORS OF ATTAINMENT OF A PERFORMANCE LIMIT:

A1. Damage Limitation (**DL-1**): The structure is operating normally during the earthquake: Fine cracks to structural elements; light cracking on masonry infills; no bar yielding; no disruption of utilities; function of building unaffected.

A2. Damage Limitation (**DL-2**): Immediate Occupancy, the structure can be used immediately after the earthquake: Light damage on the primary structural members (cracking up to 1-2mm); local delamination of cover without disintegration of concrete core; marginal yielding of reinforcement, concrete strains do not exceed the 0.0035 limit in compression; maximum lateral drift ratio <1% for frame systems, and <0.5% for wall systems; cracking in masonry infills up to 3mm wide; partial detachment of veneer and fracture of glass panes; Structure can be economically refurbished and used immediately after the earthquake. Permanent drifts are negligible.

B. Significant Damage (**SD**) : Substantial damage on structural system but with sufficient safety margin until total or partial collapse, and adequate stiffness to vertical and horizontal loads; maximum drift ratio for frame systems is <2% and <1% for wall systems; permanent drift <1% and 0.5% respectively; flexural and shear cracks up to 2-3mm; local failures of cover and local concrete disintegration (in compression); buckling of longitudinal reinforcement (no fractures) at distinct locations; local failures at the corners of openings in walls; disintegration of coupling beams; possible cracking of horizontal diaphragms; heavy damage to masonry infills; extensive detachment of stuccos and veneers; damage to nonstructural or decorative elements. The structure can sustain aftershocks of moderate intensity, and repair will be costly.

C. Near Collapse (**NC**): Structural system is critically damaged; vertical elements marginally maintaining strength to overbearing loads; maximum sustained drift is <4% for frame systems and <2% for wall systems; extensive and through cracks several millimeters wide and possible rupture of reinforcement at the cracks of the structural system; extensive loss of cover and complete disintegration of concrete in compression in critical locations (beam and column ends, captive columns, and wall ends); buckling of reinforcement; complete disintegration of external joints; through cracks in horizontal diaphragms; settlement of foundations and tilting of structure; collapse of masonry infills and detachment from the frame structure; complete loss of cover and veneers; obstructions of doors, corridors and strairwells due to rubble; with the exception of occasional accidents, inhabitants should be able to evacuate the premises before possible aftershocks. The structure is near collapse and would probably not survive another earthquake, even of moderate intensity.

3 SCENARIOS FOR DESIGN AND ASSESSMENT

The performance limit states described above, when considered as possible scenarios by the stakeholders of a prospective design, are necessarily mapped to the neighborhood of the three performance limits (PL) depicted in Figure 1, reflecting the tolerance to the intensity of the expected damage at the respective PL. Thus a possibility of scenarios are envisioned, each mapped to a different hazard level, i.e. design earthquake which is prescribed by the magnitude of peak ground acceleration (pga) in the site of interest, as the value that only stands a 10% probability of being exceeded in 50 years (return period of 475 years). A more frequent event corresponding to 2/3 of the acceleration intensity is also considered in cases with higher than usual importance (not residential construction); an event with a very rare possibility of exceedence (2% in 50 years, corresponding to acceleration levels that may be 30% higher than the design pga) may also be considered. In PBD, the design scenarios associate different intensities of the PL's to the three levels of hazard definition listed in Table 1: A1, A2, A3 are progressively more lenient in terms of damage definition associated with what is shown as PL-A in Figure 1 (so performance limit A is a range of values of lateral displacement that lead to different damage intensities, all however still lying near the end of the elastic range). Similarly the PL-B1, B2, B3 are states within the realm of repairability and life-safety, but of increasing damage intensity; and a commensurate definition for PL-C is possible (how close to collapse)– the code enables the stakeholder to select choices from the rubric below and determine their own design scenario.

Table 1: Mapping of Performance objectives (PL-A, PL-B and PL-C) with design earthquake intensity

	Performance objectives			
Probability of exceedance of design seismic action within the service life of 50 years.	50%: (Return Period: 225 years)	A_1	B_1	C_1
	10%: (Return Period: 475 years)	A_2	B_2	C_2
	2%: (Return Period: 2475 years)	A_3	B_3	C_3

Note however, that PBD does not abandon the principles of capacity design – rather, these still control the design of those members or response mechanisms where the inherent mechanism of behavior is naturally brittle or inaccessible (e.g. joints in beam-column connections); these members are designed to withstand by a margin the full development of post-yielding behavior of ductile members with which they interact: for example, a column is overdesigned in shear to enable development of flexural strength at the column ends (i.e., the design shear value, $V_{c,S}$, is calculated from the flexural strength of the column critical section, $M_{c,R}$, divided by the shear span L_s of the member). However the margin of overstrength is related to the performance objective (i.e. the safety factor α used in the expression: $V_{c,S} = \alpha \cdot M_{c,R}/L_s$, is a number ≥1, but it may take different values depending on the selected performance objective.)

In practical terms, and in order to bind the designer with his/her commitment to a specific performance objective, detailing for the design earthquake is linked to the required detailing in the structure. So, detailing rules are associated to the choice made in the design according to three classes of required ductility: (a) *Ductility Class Low (DCL)*, which corresponds to attainment of a displacement ductility $\mu_\Delta=1.5$ at the PL; (b) *Ductility Class Medium (DCM)*, which corresponds to attainment of a global displacement ductility of $\mu_\Delta=3.5$, and (c) *Ductility Class High (DCH)*, which corresponds to the attainment of a $\mu_\Delta=5$ during the design earthquake. Note that these values are ductility demands, and therefore a higher ductility class means that the structure is to endure a higher level of damage in the design event.

Ductility demand is related to the behavior factor, q (=R in Canadian nomenclature) through the so-called T-q-μ_Δ relationships: For structures with a predominant period T in the short period range of the design spectrum (see Figure A-I.1 in Appendix I, for $T<T_C$) the relationship adopted by EC-8 is: $\mu_\Delta = 1+(q-1)\cdot T_C/T$; otherwise it is taken that $\mu_\Delta=q$. The recommended value of q may be modified from the above reference values (referred to as basic values q_o) in consideration of the available experience regarding the general behavior of a structural system class: for example, for RC structures where the ratio $\alpha_u/\alpha_1 \leq 1.5$, and α_1 is lateral force at first yielding in the structure, and α_u is the maximum strength in the resistance curve of the structure. To determine these values as well as the design value of the nonlinear displacement associated with the performance point, analysis of structural performance under lateral loads is required.

Table 2: Recommended values of behavior factor, q :

Structural Type	DCM (q_o=3.5)	DCH (q_o=5)
Frame system, dual system, coupled wall	$3.0 \cdot \alpha_u/\alpha_1$	$4.5 \cdot \alpha_u/\alpha_1$
Uncoupled wall system	3.0	$4.0 \cdot \alpha_u/\alpha_1$
Torsionally flexible system	2.0	3.0
Inverted pendulum system	1.5	2.0

Different variants of analysis methods are possible, depending on the characteristics of the structure: Simple linear elastic analysis is allowable in the case of structures with uniform distribution of stiffness and mass both in plan and in height. Much more complex methods of analysis are required when designing structures with significant irregularities either in plan or heightwise, as depicted in Table 3. In concrete buildings, (but also in the case of other structural systems such as masonry and composite steel construction) cracked member stiffnesses are used in the computer models to conduct the elastic static or modal analyses (actually, secant member stiffness to yielding of reinforcement is the recommended stiffness value).

If linear elastic analysis is used (modal analysis or the lateral force analysis, see Table 3), then the displacement ductility demand, μΔ and the corresponding nonlinear displacement Δin, may be obtained from the elastic displacement Δel calculated for the design earthquake using: Δin = q·Δel. In cases where nonlinear analysis (e.g. pushover analysis) is conducted, then the inelastic displacement Δin, the corresponding behavior factor q, and the displacement ductility μΔ associated with the PP are explicitly identified on the response curve as described in Appendix II of the present study.

Table 3: Permitted type of analysis for seismic loads depending on regularity of the Structure (criteria are prescribed in the code, regarding definition of regularity in plan and in height relating to abrupt changes in stiffness, mass, and plan arrangement from floor to floor).

Regularity		Allowed Simplification		Behavior factor, q
Plan	Elevation	Model	Analysis Method (L.E.=Linear Elastic)	Used to reduce the elastic seismic forces ($F_y = F_{el}/q$)
Yes	Yes	Planar	L.E. - Lateral force	Reference value (e.g. q=3.5)
Yes	No	Planar	L.E. – Modal	Reduced value†
		Planar	Non-linear Pushover	Output of analysis: See Appendix II
		Planar	Non-linear Time History	Output of analysis
No	Yes	Spatial	L.E. - Lateral force	Reference value (e.g. q=3.5)
No	No	Spatial	L.E. – Modal	Reduced value†
		Spatial	Non-linear Pushover	Output of analysis: See Appendix II
		Spatial	Non-linear Time History	Output of analysis

† *The behavior factor is taken equal to 0.8 of the reference value if the regularity criteria are not satisfied.*

4 TRANSFORMING GLOBAL DISPLACEMENT DUCTILITY DEMAND TO LOCAL DEFORMATION DEMANDS

The performance point prescribes on the resistance curve the global displacement demand, Δ_{in}, and the displacement ductility demand μ_Δ; it is related to the displacement response from the spectra, corresponding to the fundamental period, T_1 (note that elastic spectral relative displacement demand, S_d, may be obtained from the Spectral total acceleration demand, S_e, from: $S_d = S_e \cdot T_1^2/4\pi^2$). However to quantify damage, this variable needs to be mapped locally at the critical member locations where inelasticity is expected to occur, on account of the fact that damage is a result of exceedence of material strain capacity. In this regard, it is necessary to establish the behavior of the structural system: What is the mechanism of plastic hinge formation where strain localization is expected, and what is the relationship between global displacement or displacement-ductility with local curvature and/or curvature ductility.

To conduct this transformation an implicit plastic mechanism is assumed as depicted in Figure 2, i.e., the location of the intended plastic hinges are identified, depending in the structural system selected in design. For example in the case of a system designed for frame-action (Figure 2a) where plastic hinge formation is allowed only at the ends of the beams and at the base of the lower storey columns, the rotation developing at the plastic hinges is equal to the building lateral drift ratio, and therefore, for members with predominant flexural response it may be said that $\mu_\Delta=\mu_\theta$ at beam ends and the first sorey column base. In the soft storey system shown in Figure 2b the local rotation demand at the column ends is N times the average drift ratio of the building, where N is the number of storeys in the structure. In the case of the wall-frame system the global lateral drift ratio $\theta=\Delta/H$, equals the rotation demand θ in the plastic hinge allowed to form at the base of the wall. So the detailing demands in the plastic hinge region are linked directly to the global drift demands in this type of structural system, however the rotation demand in the ends of the coupling beams is, $\theta' = \theta(1+\ell_w/\ell)$, where ℓ_w the wall section length and ℓ the span of the coupling beam (the rotational demands in the coupling beams are always excessive thereby rendering these elements extremely vulnerable).

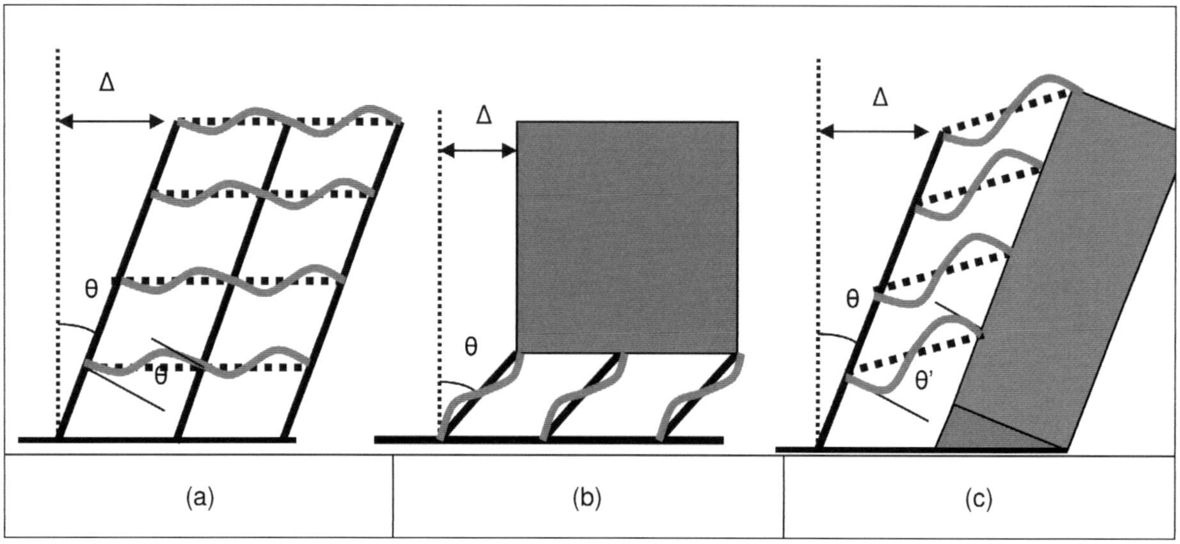

Figure 2: Effect of structural system selection on the relationship between global and local (at plastic hinges) drift ratios.

Rotation demand, θ, is transformed to curvature ductility demand which is directly related to material strains, and therefore is a more immediate indicator of material damage. Considering that for prismatic, flexural members the yield and ultimate end rotation of the deforming member (from its chord) is related to the yield and ultimate curvatures ϕ_y and ϕ_u and the length of plastic hinge l_{pl} and the member length H_m:

[1] $$\theta_y = \varphi_y \cdot H_m/6; \quad \theta_u = \theta_y + \varphi_{pl} \cdot l_{pl} = \theta_y + (\varphi_u - \varphi_y) \cdot l_{pl}$$

it can be shown that the curvature ductility is related to the displacement ductility and the corresponding behavior factor as follows:

[2a] $$\mu_\varphi = 2 \cdot q - 1 \quad \text{if } T_1 \geq T_C$$

[2b] $$\mu_\varphi = 1 + 2 \cdot (q-1) \cdot (T_C/T_1) \quad \text{if } T_1 < T_C$$

(For example if T_1=0.5s, for q=3.5 it follows that the curvature ductility demand is, μ_ϕ=6, i.e., a bar tension strain of ε_s = 0.002x6 = 0.012, and a commensurate compressive strain in concrete (if the depth of compression zone is 20% of the section depth, the strain in concrete may be as high as ε_c=0.012x(0.2/(1-0.2)) = 0.003). On account of this relationship between curvature (i.e. strain) ductility, and drift ductility demand the EC8-I specifies a range of detailing requirements that are directly linked to deformation demand as follows:

(a) For Beams:

The maximum tension longitudinal reinforcement ratio, ρ_{max}, is controlled by the maximum tensile strain in the reinforcement, $\varepsilon_{sd} = \mu_\phi \cdot \varepsilon_{yd}$ where the subscrip d signifies design values (i.e. these are the nominal values divided by a material factor =1.5 for concrete and 0.85 for steel):

[3] $$\rho_{max} = \rho' + \frac{0.0018}{\mu_\varphi \cdot \varepsilon_{sy,d}} \cdot \frac{f_{cd}}{f_{yd}}, \quad \text{where} \quad \rho' \geq \rho/2 \text{ for DCM \& DCH (but 4\% for DCL)}$$

where ρ' is the ratio of compressive reinforcement . To ensure commensure energy dissipation capacity, the maximum stirrup spacing limits in plastic hinge region is adjusted according to the ductility class:

[4a}] $$s_{max} = \min\{h_w/4, 8D_{bl}, 24D_{bw}, 225\text{mm}\} \quad \text{for DCM}$$

[4b] $s_{max} = \min\{h_w/4, 6D_{bl}, 24D_{bw}, 175mm\}$ for DCH

For shear, the design value of shear strength is obtained from capacity design requirements, considering the flexural strength scaled by the overstrength factor α=1.2 and 1.0 for DCH and DCM respectively. An additional requirement which applies only to DCH is to assess the ratio of minimum to maximum earthquake shear force demand considering the reversed cyclic nature of the earthquake; the ratio, $\zeta = V_{Emin} / V_{Emax}$ controls the fraction of shear reinforcement to be provided in the form of diagonal cross-inclined reinforcing bar pairs (forming X combinations to be more effective means of shear reinforcement than regular stirrups, Figure 3).

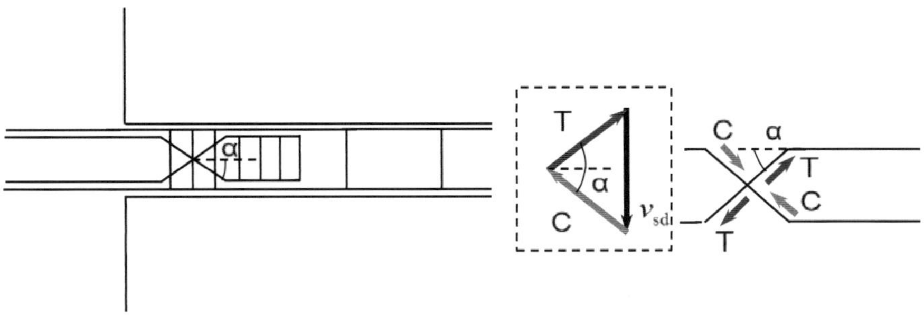

Figure 3: Bi-diagonal shear reinforcement taking 50% of the beam shear in case of ζ<-0.5.

(b) For Columns:

To promote ductile behavior the axial load ratio of the earthquake combination, $v_d = N_{ed}/(A_c \cdot f_{cd})$ is limited to 0.55 and 0.65 for DCH and DCM respectively, whereas the required confinement is obtained in terms of mechanical volumetric ratio of transverse reinforcement is related to the strain ductility demand of longitudinal column tension reinforcement and the design axial load ratio through:

[5] $$k_{eff} \cdot \omega_{wd} \geq 30 \cdot v_d \cdot \mu_\varphi \cdot \varepsilon_{sy,d} \cdot \frac{b_c}{b_o} - 0.035 \geq \begin{matrix} 0.08 \text{ for DCM} \\ 0.12 \text{ for DCH} \end{matrix}$$

where, b_c and b_o are the core and outside dimensions of the column, respectively. This condition is rather demanding and may only be satisfied if the axial load ratio is taken conservatively to be much lower than the values defined in the preceding. For example, considering a 400x400mm square column, perimeter stirrups 8mm diameter at 100 mm spacing, the confinement effectiveness is 48%, and therefore $k_{eff} \cdot \omega_{wd} = 0.48 \text{x} 0.18 = 0.0912$, which does not satisfy the DHC requirement but only the DCM; whereas for the value of μ_ϕ=6 which was considered in the preceding example, and for b_o=400mm and b_c=350mm, for 400MPa steel it follows that, $k_{eff} \cdot \omega_{wd} = 0.0912 \geq 30 \cdot v_d \cdot 6 \cdot (0.002/1.15) \cdot (400/350) - 0.035 = 0.357 \cdot v_d - 0.035 \rightarrow v_d \leq 0.355$; this example illustrates the tradeoff imposed through the limits of Eq. 5 between maximum allowed axial load ratio and confinement required in the plastic hinges in columns.

Shear design is conducted using capacity-design principles at two levels: at beam-column joints in frame structures the sum of column flexural strengths of the column sections adjacent to the joint is required to exceed by 30% the sum of the flexural strengths of beam cross sections converging to the joint both for DCH and DCM (strong column weak girder design concept). For columns, the design seismic shear force is calculated from the column flexural strength of the column end sections using a 1.3 and 1.1 overstrength factor for DCH and DCM.

(c) For Ductile Walls:

Requirements for walls are very similar to those of the Canadian Code (A23.3, 2014). Therefore in the plastic hinge region which is allowed to form at the base of the wall (extending over a plastic hinge distance along the wall height measured from the base, $\ell_p \geq \max\{\ell_w, H_w/6\}$) confinement in the ends is required in those parts of the cross section where normal compressive strains in the concrete exceed the delamination limit of 0.0035; confinement in these sections follows Eq. 6 – i.e., it is expressed in terms of the demanded curvature ductility at the critical section of the wall estimated from Eq. [1,2], both for DCH and DCM.

$$[6]\quad k_{eff}\cdot\omega_{wd}\geq 30\cdot\mu_{\varphi}\cdot(v_d+\omega_v)\cdot\varepsilon_{sy,d}\cdot b_w/b_o-0.035\geq\begin{matrix}0.08 \text{ for DCM}\\ 0.12 \text{ for DCH}\end{matrix}$$

where ω_v is the vertical reinforcement mechanical ratio of the web (mechanical ratio = geometric ratio multiplied by the ratio of material design strengths, f_{yd}/f_{cd}); b_o the web thickness and b_w is the thickness of the confined boundary elements.

Design shear force of the wall is obtained from the analysis fothe design seismic action, but amplified up by factor ε which is taken equal to 1.5 for DCM, whereas for DCH, ε is related to the ratio of flexural strength to the flexural demand if the wall is squat ($H_w/\ell_w\leq 2$) whereas Eq. [7] is used to define the amplification factor considering higher mode effects in slender walls ($H_w/\ell_w>2$):

$$[7]\quad \varepsilon=q\cdot\sqrt{\left(\frac{\gamma_{Rd}}{q}\cdot\frac{M_{Rd}}{M_{Ed}}\right)^2+0{,}1\left(\frac{S_e(T_C)}{S_e(T_1)}\right)^2}\leq q$$

where $S_e(T_C)$ is the spectral acceleration at the end of the constant acceleration range, and $S_e(T_1)$ is the response acceleration for the wall structure (T_1 the fundamental period of response).

5 CONCLUSION

The European seismic design practice links explicitly the displacement demand during the design earthquake (Performance Point) with the detailing rules of the critical regions of all types of structural members. In this manner the extent of acceptable damage (performance objective) during the design seismic hazard may be chosen by the stakeholders to vary depending on the importance and type of the structural system and the intensity of the design ground motion (i.e., buildings in regions of vastly different seismicity may be designed with comparable performance expectations, for events of equal probability of occurrence.)

The same concepts transcend both new design as well as assessment of existing structures in EC8; in all cases, the design inequaility is expressed by requiring that the displacement demand at the PP exceeds the displacement capacity of the structure. Therefore, the design inequality is enforced in all deformation-controlled members in the form: $\Delta_{dem}\leq\Delta_{cap}$ (where Δ_{dem} is the seismic displacement demand, and Δ_{cap} is the displacement capacity of the structural system at the performance objective targeted for). Success of the approach rides in the accuracy and marging of safety of the available expressions that link deformation capacity with reinforcement detailing and member geometry. On the other hand, force-controlled members are designed with capacity design principles. The severity of enforcement of these principles depends on the ductility-class targeted for, through the building design. The higher the ductility demand, the higher the overstrength of the force-controlled members to gueard against brittle failure.

A displacement-based approach that follows performance-based design is also used in Canadian practice, but only for the design of structural walls; all other structural systems are designed based on capacity design principles: although high, medium and low ductility classes are recognized, there is no link between detailing of reinforcement and anticipated strain or curvature ductility demands in the critical regions of beams and columns. Thus, the performance level is not prescribed as a target that would be served by the design decisions. Rather, in NBCC the design is driven by the estimated demands. This presentation was motivated by the emerging consensus in Canada to move towards a generalized Performance-Based seismic design framework for all types of structural systems in reinforced concrete and is intended to serve as a quick benchmark reference of similar established practices.

Acknowledgements

The work presented here was sponsored partially by NSERC DG – 2016- 05324.

References

CEN (2005). European Standard EN 1998-1: 2005 Eurocode 8: Design of structures for earthquake resistance. Part 1: General rules, Seismic action and rules for buildings. European Committee for Standardization, Brussels, Belgium.

Fardis, M. N. (2009). Seismic Design, Assessment and Retrofitting of Concrete Buildings (Vol. 8). https://doi.org/10.1007/978-1-4020-9842-0

Fardis, M. N., Carvalho, E., Fajfar, P., & Pecker, A. (2015). Seismic design of concrete buildings to Eurocode 8.

Fardis MN, Panagiotakos TB (1997c) Displacement-based design of RC buildings: Proposed approach and application. In: Fajfar P and Krawinkler H (eds) Seismic Design Methodologies for the Next Generation of Codes, Balkema, Rotterdam, pp. 195–206.

Chopra, A. (2007). Dynamics of structures. Theory and applications to earthquake engineering. Prentice Hall, Englewood Cliffs, NJ.

NRCC (2015). National Building Code of Canada, Associate Committee on the National Building Code, National Research Council of Canada, Ottawa, ON.

Gupta B. and Kunnath S. K. (2000) Adaptive Spectra-Based Pushover Procedure for Seismic Evaluation of Structures. Earthquake Spectra: May 2000, Vol. **16**, No. 2, pp. 367-392.

Vidic, T., P. Fajfar, and M. Fischinger. (1994). Consistent inelastic design spectra: Strength and displacement. Earthquake Engineering and Structural Dynamics **23**:502–521.

Fajfar, P. (1999). Capacity spectrum method based on inelastic demand spectra. Earthquake Engineering and Structural Dynamics **28**(9):979–993.

CSA. 2014. CSA Standard A23.3, Code for the design of concretestructures for buildings. Canadian Standards Association, Mississauga, ON.

Appendix I : Definition of the general form of the Seismic hazard – design spectrum

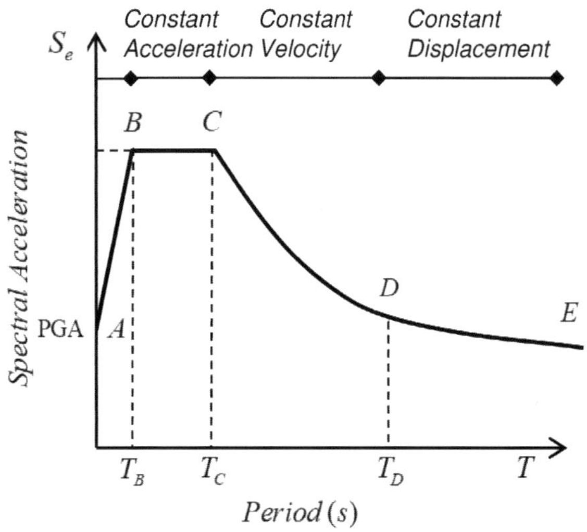

$T \leq T_B$:	$S_e = a_g \cdot S \cdot \left[1 + \frac{T}{T_B} \cdot (2.5 \cdot \eta - 1)\right]$ $S_d = a_g \cdot S \cdot \left[\frac{2}{3} + \frac{T}{T_B} \cdot (\frac{2.5}{q} - \frac{2}{3})\right]$
$T_B \leq T \leq T_C$:	$S_e = 2.5 \cdot a_g \cdot S \cdot \eta$ $S_d = \frac{2.5}{q} \cdot a_g \cdot S$
$T \geq T_C$:	$S_e = 2.5 \cdot a_g \cdot S \cdot \eta \cdot \left[\frac{T_C \cdot T_D}{T^2}\right]$ $S_d = \frac{2.5}{q} \cdot a_g \cdot S \cdot \left[\frac{T_C \cdot T_D}{T^2}\right] \geq \beta \cdot a_g$

Figure A-I.1: Elastic Response Spectra. Parameter S accounts for soil amplification (S=1 for dense soil, S=1.2 for more compliant soils; η accounts for damping according with: $\eta=[10/(5+\xi)]^{1/2}\geq 0.6$, where ξ is the damping ratio (e.g. =5 for 5% damping). The values of TB and TC and TD are specified in the National Annexes, but the recommended values are 0.15s, 0.4·S in s, and 2s respectively.

Appendix II: Finding the Performance Point from the Building Resistance Curve and the Design Spectrum.

The building resistance curve is calculated using non-linear static analysis under the combination of constant gravity loads and monotonically increasing horizontal loads. The resistance curve is also known as "pushover curve" since the building, during the analysis is pushed laterally till attainment of a state of failure. There is significant debate with regards the proper pattern of lateral load distribution that ought to be used in this analysis. At least two patterns – one uniform and one following the shape of the fundamental mode of lateral response has been advised in EC8-I Section 4.3.3.4. Lateral loads are applied at the loacations of lumped mass in the structure (e.g. at the center of floor diaphragms) and are gradually increased from zero to the maximum tolerable base shear that can be supported by the building. The resistance curve is a useful result as it may be used to determine the actual oversrength of the building after the onset of yielding, as well as the plastic mechanisms and the global to local deformation demand transformation at the Performance Point.

However in order to combine the pushover curve (capacity) with the response spectrum it is first required to convert the properties to those of an equivalent single degree of freedom (SDOF) system. To this end the fundamental mode shape of lateral vibration, Φ, is used (Φ_i is the shape coordinate at the i-th floor; m_i is the corresponding floor mass, and F_i the floor shear; F_b is the base shear and d_n is the top lateral displacement of the point in the n-th floor of the building, where the shape coordinate Φ_n equals 1.) Transformation of the properties and response indices between the SDOF and the actual building is effected by the transformation factor Γ, as depicted in the insert table of Fig. A-II.1. Terms marked with a superscript (*) refer to the equivalent SDOF that represents the structure (structural response parameters are referred to by the terms with no *).

To find the performance point the procedure followed involves (a) Equivalent linearization of the resistance curve (i.e. definition of a distrinct point of yielding and post-yielding branch as illustrated in

Figure A-II.1a. (b) Ploting of the design spectrum in ADRS format (i.e. the vertical axis is the spectral acceleration $S_e(T)$, whereas the horizontal axis is the relative spectral displacement d^* ($d^* = S_d(T) = S_e(T) \cdot T^2/4\pi^2$).

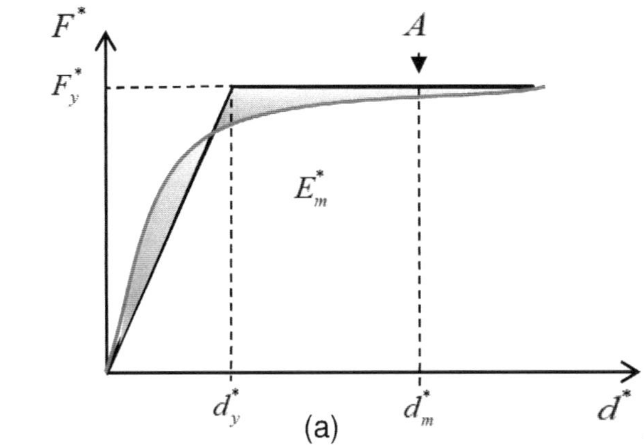

Equivalent SDOF mass	$m^* = \sum m_i \cdot \Phi_i = \sum \bar{F}_i$
Transformation factor	$\Gamma = m^* / \sum m_i \cdot \Phi_i^2 = \sum \bar{F}_i / \sum \left(\bar{F}_i^2 / m_i \right)$
Equivalent SDOF: F, d	$F^* = F_b / \Gamma$, $d^* = d_n / \Gamma$
Equivalent SDOF yield displacement and period	$d_y^* = 2\left(d_m^* - E_m^* / F_y^* \right)$ $T^* = 2 \cdot \pi \cdot \sqrt{m^* \cdot d_y^* / F_y^*}$

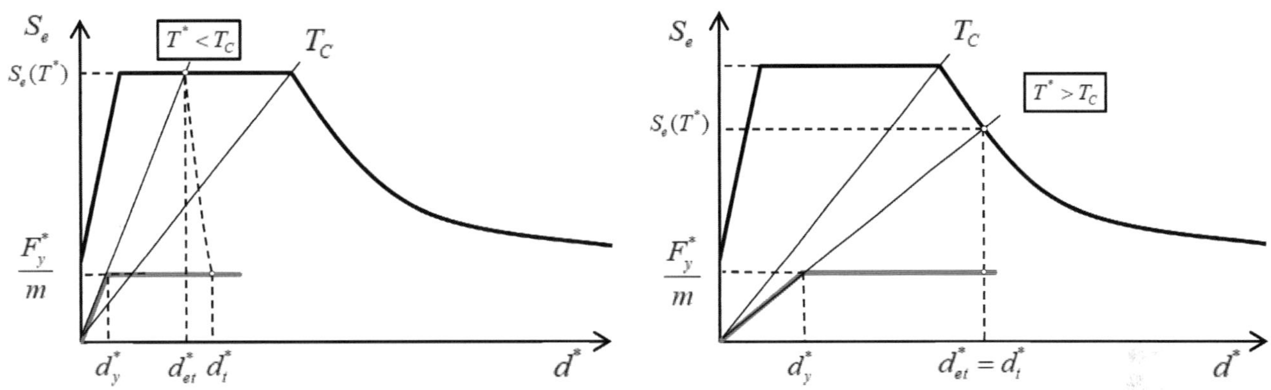

(b) (c)

Figure A-II.1: (a) Equivalent linearization of the pushover curve.

(b), (c): Determination of the performance point on the resistance curve (red line) depending on the period of the equivalent SDOF,T*, relative to the period at the end of the constant acceleration range, T_C. Note the two cases defined in Figure A-II.1b,c: (b) For $F_y^*/m > S_e(T^*)$ it shows that $d_t^* = d_{et}^*$ whereas for $F_y^*/m < S_e(T^*)$ it will be $d_t^* = d_{et}^*/q_u \cdot (1+(q_u-1) \cdot T_c/T^*) \geq d_{et}^*$, where $q_u = S_e(T^*) \cdot m^*/ F_y^*$.

(c) In the case of Figure A-II.1c, $d_t^* = d_{et}^*$.

AUTHOR INDEX